Engineered Nanomaterials: Modeling, Methodologies and Applications

Engineered Nanomaterials: Modeling, Methodologies and Applications

Edited by **Mindy Adams**

WILLFORD PRESS

New York

Published by Willford Press,
118-35 Queens Blvd., Suite 400,
Forest Hills, NY 11375, USA
www.willfordpress.com

Engineered Nanomaterials: Modeling, Methodologies and Applications
Edited by Mindy Adams

International Standard Book Number: 978-1-68285-128-9 (Hardback)

Contents

Preface

I am honored to present to you this unique book which encompasses the most up-to-date data in the field. I was extremely pleased to get this opportunity of editing the work of experts from across the globe. I have also written papers in this field and researched the various aspects revolving around the progress of the discipline. I have tried to unify my knowledge along with that of stalwarts from every corner of the world, to produce a text which not only benefits the readers but also facilitates the growth of the field.

Nanomaterials have been a major part of the fundamental research being conducted around the world in the field of materials science. These materials have unique properties which can be a leverage for prospective applications. This book presents latest case studies of internationally renowned experts revolving around nanoparticles, synthesis of hybrid nanomaterials, modelling of nanomaterials, etc. It will serve as a resource guide for students, academicians and professionals in the fields of materials science and nanotechnology.

Finally, I would like to thank all the contributing authors for their valuable time and contributions. This book would not have been possible without their efforts. I would also like to thank my friends and family for their constant support.

Editor

Magnetic Properties of Magnetic Nanoparticles for Efficient Hyperthermia

Ihab M. Obaidat [1,*], Bashar Issa [1] and Yousef Haik [2,3]

[1] Department of Physics, United Arab Emirates University, Al-Ain 15551, United Arab Emirates;
E-Mail: b.issa@uaeu.ac.ae

[2] Department of Mechanical Engineering, United Arab Emirates University, Al-Ain 15555,
United Arab Emirates; E-Mail: yhaik@uaeu.ac.ae

[3] Center for Research Excellence in Nanobiosciences, University of North Carolina at Greensboro,
Greensboro, NC 27412, USA

* Author to whom correspondence should be addressed; E-Mail: iobaidat@uaeu.ac.ae

Academic Editor: Thomas Nann

Abstract: Localized magnetic hyperthermia using magnetic nanoparticles (MNPs) under the application of small magnetic fields is a promising tool for treating small or deep-seated tumors. For this method to be applicable, the amount of MNPs used should be minimized. Hence, it is essential to enhance the power dissipation or heating efficiency of MNPs. Several factors influence the heating efficiency of MNPs, such as the amplitude and frequency of the applied magnetic field and the structural and magnetic properties of MNPs. We discuss some of the physics principles for effective heating of MNPs focusing on the role of surface anisotropy, interface exchange anisotropy and dipolar interactions. Basic magnetic properties of MNPs such as their superparamagnetic behavior, are briefly reviewed. The influence of temperature on anisotropy and magnetization of MNPs is discussed. Recent development in self-regulated hyperthermia is briefly discussed. Some physical and practical limitations of using MNPs in magnetic hyperthermia are also briefly discussed.

Keywords: magnetic nanoparticles (MNPs); hyperthermia; power dissipation; curie temperature; anisotropy

1. Introduction

1.1. Localized Magnetic Hyperthermia

Magnetic hyperthermia is the field of treating cancer by supplying heat to tumor cells using magnetic nanoparticles (MNPs) and an alternating magnetic field. This method could be promising to treat small or deep-seated tumors. Magnetic hyperthermia using MNPs is a multidiscipline research field which requires the involvement of physics, chemistry, material science and medical science. This technique, which started in 1957 [1], where maghemite nanoparticles (γ-Fe$_2$O$_3$) were used, is based on the observation that tumor cells can be destroyed by heating the cells for a duration of time to temperature between 43 and 46 °C while healthy cells are less affected [2,3]. The heating process is enabled by the application of an alternating magnetic field of suitable amplitude and frequency. One of the major issues that is being investigated in magnetic hyperthermia is the reduction of the amount of MNPs that can be used in living organs [2,3]. In order to achieve this goal, the power dissipation or heating efficiency of MNPs should be enhanced. Several factors influence the heating efficiency, such as the amplitude and frequency of the external alternating magnetic field, magnetic anisotropy, magnetization, particle-particle interactions, as well as the size and size distribution of the MNPs. There are several excellent reviews that discuss magnetic hyperthermia using MNPs [4–6].

There are also several excellent reviews on the physics of heating efficiency using magnetic nanoparticles in magnetic hyperthermia [7–9]. In this short review we focus on the physical and magnetic properties of MNPs that are related to heating efficiency in magnetic hyperthermia. We only discuss selective recent reports that display interesting results that could influence magnetic properties for magnetic hyperthermia.

1.2. Main Properties of MNPs

Other than their intense applications in data storage devices [10,11], MNPs have several other technological applications in biomedicine [12,13] such as magnetic resonance imaging, drug delivery and magnetic hyperthermia.

Magnetic properties of nanoparticles (NPs) are dominated by two main features [14]; finite-size effects (single-domain, multi-domain structures and quantum confinement) and surface effects, which results from the symmetry breaking of the crystal structure at the surface of the particle, oxidation, dangling bonds, surface stain, *etc*. Surface effects become significant as the particle size decreases because the ratio of the number of surface atoms to the core atoms increases. It is well established that several magnetic properties such as magnetic anisotropy, magnetic moment per atom, Curie temperature, and the coercivity field of NPs can be different than those of a bulk material [14,15]. In most medical applications, the preferred size of the nanoparticles is typically around 10–50 nm. In this range of sizes, usually a nanoparticle becomes a single magnetic domain (for minimization of its magnetic energy) and shows superparamagnetic behavior when the temperature is above a certain temperature called the blocking temperature. In the superparamagnetic state, a nanoparticle possesses a large magnetic moment and behaves like a giant paramagnetic atom with a fast response to applied magnetic fields with negligible remanence and coercivity. For hyperthermia applications, MNPs must have high saturation magnetization, M_s values. High M_s values will result in large thermal energy dissipation in the tumor cells. On the other hand, large M_s values give more control on the movement of the MNPs in the blood using external magnetic field. However, it is important to understand that in order to apply MNPs in hyperthermia, the NPs must satisfy two main conditions: they should have large heating power, and they should have good stability. For good stability, MNPs are preferred to be superparamagnetic. In the absence of an applied magnetic field, the superparamagnetic NPs lose their magnetism at temperatures above the blocking temperature. This enables the particles to avoid aggregation and maintain their colloidal stability. On the other hand, dipolar interactions between MNPs become very small as the particles' sizes become very small. This is the case because the dipole-dipole interaction energy scale as r^6 (r is the inter-particle distance between particles). Reducing the dipolar interactions will minimize particle aggregation in the existence of applied magnetic field. However, with regard to the other condition, superparamagnetic NPs might not be the best choice. In [16], it was reported that heating power was maximized in large ferromagnetic NPs with low anisotropy. In addition, as reported in [17], the optimum size for the maximum power loss varies with the amplitude of the applied magnetic field. Hence, the choice between superparamagnetic and ferromagnetic NPs for hyperthermia is not a simple task where several experimental conditions should be considered.

There are five main factors that determine the magnetic properties of nanoparticles. These are: (a) the geometrical properties of the nanoparticles; (b) magnetic interactions that occur inside the nanoparticle (intra-particle interactions); (c) particle-particle magnetic interactions (inter-particle interactions); (d) magnetic interactions that occur between the nanoparticles and the matrix material; and (e) particle-applied magnetic field interactions. The geometrical properties of the nanoparticles include: (a) sizes of particles; (b) shapes of particles; (c) distributions of sizes; and (d) distributions of anisotropy axes. Interactions inside a magnetic nanoparticle (intra-particle interactions) include: (a) those inside the domain of a ferromagnetic particle, where the magnetic moments of atoms interact via the exchange interaction; (b) those inside the domain of a ferrimagnetic particle, where the magnetic moments of atoms interact via the super-exchange interaction; (c) interaction of moments inside a multi-domain particle with each other; and (d) interaction of moments at the surface of a nanoparticle with moments of the interior of the particle. Interactions between magnetic nanoparticles (inter-particle interactions) include: (a) dipolar (dipole-dipole) interactions between the net magnetic moments of the particles; (b) direct exchange interactions between moments at the surfaces of touching particles; and (c) super-exchange interactions between non-touching particles such as magnetic particles which are placed in an insulating matrix. The interactions between particles and the magnetic field include: (a) interaction between magnetic moments of the magnetic domains and the applied magnetic field; and (b) interactions between moments at the surface of a nanoparticle with the applied magnetic field. In any particular sample of nanoparticles, some or even all of these factors and interactions might exist simultaneously. It is not simple to separate the geometrical roles from the interaction roles. The most dominant interaction between particles is the dipolar interaction.

2. Physics of Heating of MNPs

2.1. Relaxation of Magnetic Moment

The spin-orbital interactions of the electrons in the NP produce magnetic anisotropy. For isolated systems, the magnetic anisotropy is responsible for keeping the spins in a particular direction. Because atomic orbitals mainly have non-spherical shapes, they prefer to align in a specific crystallographic direction which is called the easy direction. Because in materials with large magnetocrystalline anisotropy, the atomic spin and orbital angular moments are strongly coupled, magnetization prefers to align along the easy direction. Energy is needed in order to rotate the magnetization away from the easy direction. This energy is called the anisotropy energy. In the case with uniaxial anisotropy, the anisotropy energy per particle is given by [18]:

$$E = K V \sin^2\theta + higher\ order\ terms \tag{1}$$

where K is the anisotropy constant (it includes all sources of anisotropy), V is the volume of the particle, and θ is the angle between the particle magnetization and the easy magnetization axis of the particle. The higher order terms can be ignored since they are very small compared with the first term. This anisotropy energy with uniaxial anisotropy has one easy axis with two energy minima separated by the energy maximum, KV. As can be seen from Equation (1), the anisotropy energy directly depends on the particle size and on the anisotropy constant. For a fixed K, as V decreases, E decreases. At very small particle sizes, the particle will prefer to have only one magnetic domain and thus called single-domain NP. At this small size, the anisotropy energy might become smaller than the thermal energy, $E_{th} = k_B T$ (k_B is the Boltzmann constant). Once this happens, the particle magnetic moment starts to rotate freely in all directions leading to zero net magnetization in the absence of an external magnetic field. If the flipping of magnetic moment occurs while the particle orientation is fixed, then the relaxation time of the moment of a particle is called the Néel relaxation time, τ_N and is given by [8,9,19,20]:

$$\tau_N = \frac{\tau_o}{2} \sqrt{\frac{\pi k_B T}{K_{eff} V}} \exp\left(\frac{K_{eff} V}{k_B T}\right) \tag{2}$$

where K_{eff} is the effective anisotropy and the factor $\tau_0 \approx 10^{-13} - 10^{-9}$ s [4,21].

When measuring the magnetization of a superparamagnetic NP, we define τ_m to be the measurement time. If $\tau_m \gg \tau_N$, the magnetization of the NP will flip several times during the measurement giving zero average magnetization. In this case, the NP is said to be in the superparamagnetic state. If $\tau_m \ll \tau_N$, the magnetization will not have enough time to flip during the measurement and will be blocked at the initial non-zero value at the beginning of the measurement. In this case, the NP is said to be in the blocked state. The transition between the superparamagnetic state and the blocked state occurs when $\tau_m = \tau_N$ [18]. If in an experiment, the measurement time is kept constant while the temperature was varied, the transition between superparamagnetic and blocked states is obtained as a function of temperature. The temperature at which this transition occurs is called the blocking temperature, T_b. Thus, at T_b the measurement time will be equal to the Neel relaxation time, $\tau_m = \tau_N$. As mentioned above, the size of the particle is crucial in determining the blocking temperature. The blocking temperature also depends on other factors such as particle-particle interactions. Superparamagnetic behavior of non-interacting single-domain particles occurs at temperatures larger than the blocking temperature. It is important to note that mainly all superparamagnetic particles are single-domain particles, but not all single-domain particles are superparamagnetic. In the superparamagnetic state, magnetization disappears as long as no magnetic field is applied.

If the particle itself rotates while the flipping of the particle's moment occurs, moment relaxation is called the Brownian relaxation mechanism, τ_B and is given by [8,9,22]:

$$\tau_B = \frac{3 V_H \eta}{k_B T} \tag{3}$$

where η is the viscosity of the liquid containing the particles and V_H is the hydrodynamic volume of the particle. Because of particle coating, absorbed surfactants or interaction with the fluid, V_H is larger than the original volume of the particle, V. The effective magnetic relaxation time, τ_{eff} is then given by:

$$\frac{1}{\tau_{eff}} = \frac{1}{\tau_N} + \frac{1}{\tau_B} \tag{4}$$

It can be seen that it is the shorter relaxation time which controls the effective time. For MNPs with diameter smaller than 15 nm, τ_N is smaller than τ_B and hence τ_{eff} is dominated by τ_N [8]. On the other hand for NPs with diameter larger than 15 nm, τ_B is smaller than τ_N and hence τ_{eff} is dominated by τ_B [8]. Both of these mechanisms contribute towards magnetic hyperthermia of magnetic NP. In obtaining Equations (3) and (4) it is assumed that the particles are identical (same size and shape) and non-interacting single-domain particles. In addition, Equations (1) and (2) are valid for zero applied magnetic field. If the applied field is not zero, then Zeeman energy should be included [9].

2.2. Power Dissipation in MNPs

The internal energy of a magnetic system in an adiabatic process is equal to the magnetic work done on it [22]:

$$U = -\mu_0 \oint M dH \tag{5}$$

The power dissipation in the magnetic system, during a complete magnetic field cycle, is equal to internal energy divided by the time. Thus, the power dissipation, during several cycles, is equal to internal energy multiplied by the frequency:

$$P = Uf \tag{6}$$

The power dissipated in a MNP due to the application of an alternating magnetic field of maximum strength H, frequency f ($\omega = 2\pi f$) was proposed to depend on magnetic spin relaxations of superparamagnetic NPs and is given by [22]:

$$P(f,H) = Uf = \pi\mu_0\chi''H^2 f \tag{7}$$

where μ_0 is the permeability of free space and χ'' is the imaginary part of the susceptibility χ ($\chi = \chi' - i\chi''$). In the linear response theory (LRT), χ is assumed to remain constant with increasing H ($M = \chi H$). This approach was shown to be valid for very small magnetic fields. To be more specific, the LRT is valid in the superparamagnetic regime where $H_{max} < k_B T / \mu_0 M_s V$ and when the magnetization of NPs is linearly proportional to the applied magnetic field. This means that the applied fields should be much smaller than the saturation field of the NPs ($H_{max} \ll H_K$) where H_{max} is the amplitude of the alternating applied magnetic field and H_K is the anisotropy field [16]. The imaginary part of the susceptibility, χ'' is given by [23,24]:

$$\chi'' = \frac{\omega\tau}{1 + (\omega\tau)^2}\chi_0 \tag{8}$$

$$\chi_0 = \frac{\mu_0 M_s^2 V}{k_B T} \tag{9}$$

where τ is the effective magnetic relaxation time, V is the volume of the NP, and M_s is the saturation magnetization.

The heating efficiency is represented by the specific loss power (SLP) also referred to as the specific absorption rate (SAR). The specific loss power (SLP), which is measured in watts per gram, is given by [25]:

$$SLP(f,H) = \frac{P(f,H)}{\rho} = \frac{\pi\mu_0\chi''H^2 f}{\rho} \tag{10}$$

where ρ is the mass density of the magnetic material.

In living organs, the water-based medium around the cells absorb a lot of the heat generated by the NPs. Hence, for MNPs to have practical medical applications they should generate large SLP. It is clear that in the Rosensweig's theory or LRT, the heat dissipation of the MNPs depends on several factors such as: strength and frequency of the applied magnetic field, the solvent viscosity, the size of the particles, the saturation magnetization and the magnetic anisotropy of the MNP. The strength and frequency of the applied magnetic field cannot have any value for applications on living organs. It is well-known that eddy currents are induced in a conductor due to an alternating magnetic field. These currents cause heating in the conductor. In human body, water is a conductor and hence eddy currents can be induced in the body under an alternating magnetic field which could damaging effect. Hence, there must be a criterion which imposes an upper limit for the allowed magnetic field that can be applied to living organs. The allowed frequency and amplitude of the alternating magnetic field that can be considered safe is not completely agreed on. In [26], the authors discussed the origins of these safety limits and pointed out they are self-imposed limits which are the subject of some debate. These safety limits were based on the work of Atkinson in 1984 [27] who performed some clinical tolerance tests on a healthy volunteers. He conducted the test using a single-turn induction coil which was placed around the thorax of the volunteer. Atkinson found that field intensities up to 35.8 A·turns/m at a frequency of 13.56 MHz can be thermally tolerated for extended periods of time [27]. This clinical tolerance is not known to be repeated [26] and the results reported by Atkinson become accepted as a safety limit which is known now as "Brezovich criterion" [28] where the product $C = H \cdot f$ should not exceed 4.5×10^8 $Am^{-1}s^{-1}$. The Brezovich criterion is considered at best an upper limit for $H \cdot f$ when applying a uniform field over an entire thorax of an adult [26]. In practice, smaller coils are used with inhomogeneous fields and off-axis field directions which are significantly different conditions than those used by Atkinson. These factors are expected to reduce eddy current heating. In addition to that clinical tolerability to counteract cancer is expected to be higher than that of a healthy volunteer.

Hence, the Brezovich criterion should not be considered as the only criterion. In [25], the authors suggested another criterion $C = H \cdot f = 5 \times 10^9$ Am^{-1}s^{-1} which is 10 times larger than the Brezovich criterion. Hence, the two criteria for the product of the amplitude and the frequency $(H \cdot f)$ of the applied magnetic field are 4.85×10^8 Am^{-1} s^{-1} (6×10^6 Oe Hz) [26] and 5×10^9 Am^{-1} s^{-1} (6.25×10^7 Oe Hz) [25]. When the frequency is fixed at 100 kHz (which is very suitable for medical applications) H will be between 4.85×10^3 Am^{-1} (60 Oe) and 50×10^3 Am^{-1} (625 Oe) [29].

As mentioned above, the LRT is valid for very small applied fields. Hence, LRT will not be applicable for MNPs with low anisotropy energies where the magnetization is saturated at low applied fields. The alternative is to use Stoner-Wohlfarth model. However, the standard Stoner-Wohlfarth model is applied when $T = 0$ or in the limit of infinite frequency. Because thermal activation is not involved, the magnetization can switch direction between the two equilibrium positions (potential wells) by removing the energy barrier using an applied magnetic field. Thus, a modification of Stoner-Wohlfarth model which takes into account the thermal activation of magnetization and the sweeping rate of the alternating magnetic field was investigated in [16]. There, the role of finite temperature and frequency on the coercive field and areas of hysteresis loops, were studied using Stoner-Wohlfarth based theories. Analytical formulas for temperature and frequency dependent coercive field as well as for temperature and frequency dependent hysteresis loop area. A time-dependent magnetic field $H(t) = H_{max}\cos(\omega t)$ that is applied to the MNP along a direction that makes an angle ϕ with respect to the easy axis. The authors used a two-level approximation where the thermally activated reversals of moments occur between the metastable points (θ_1, E_1) and (θ_2, E_2) across the saddle point (θ_3, E_3). They then calculated the time dependence of the probabilities of the magnetization being in the first and second potential well, p_1 and $p_2 = (1 - p_1)$ respectively. The magnetization was calculated using the equation:

$$M = M_s(p_1\cos\theta_1 + (1 - p_2)\cos\theta_2) \tag{11}$$

where θ is the angle between the magnetization and the easy axis. The authors conducted a large number of simulations to investigate the dependence of the coercivity field on the frequency of the magnetic field. For the case with $\phi = 0$ (magnetic field is aligned along the easy direction), the results in [16] have good agreement with the results in [30] for κ smaller than 0.5 where they obtained the following equation:

$$\mu_0 H_C = \mu_0 H_K(1 - \kappa^{1/2}) \tag{12}$$

For the random orientation case, the authors obtained the following equation:

$$\mu_0 H_C = 0.48\mu_0 H_K(b - \kappa^n) \tag{13}$$

where $b = 1$ and $n = 0.8 \pm 0.05$, H_K is the anisotropy field, and κ is a dimensionless parameter for the variation of H_C that includes temperature and takes into account the sweeping rate of the magnetic field. The authors then conducted simulations to investigate the frequency and temperature dependence of the hysteresis loop area. The authors also deduced suitable formulas to calculate the areas of major hysteresis loops using Stoner-Wohlfarth based theories. It is worth mentioning that in [16], the easy axis of MNPs were considered to be fixed since rotation of the whole MNP was not considered in the analytical estimates [31,32].

In [32,33], the authors used numerical simulations to discuss the dynamics of rotatable superparamagnetic and ferromagnetic NPs in aqueous phase resembling the cytoplasm in a large alternating magnetic field. The authors considered monodisperse spheroidal magnetite nanoparticles with non-magnetic surfactant layers. Hence, the dipolar interactions among NPs were neglected and the NPs are considered to be uniformly dispersed and do not aggregate. In the calculations, the crystalline, and surface anisotropy energies were neglected compared to the uniaxial shape anisotropy. The authors used a two-level approximation, which considers thermally activated reversals between two meta-stable directions via a midway saddle point. In the simulation of reversal and rotation, Brownian dynamics simulation were considered where the inertia of the nanoparticle were neglected. The results of this numerical study could not be explained by the conventional models that consider a linear response of thermodynamic equilibrium states ($H_0 = 0, T \neq 0$) or magnetic field-driven reversals ($H_0 \neq 0, T = 0$). For rotatable superparamagnetic NPs, the relaxation loss was found to have two maxima; the primary one which is attributed to the rapid Neel relaxation, and a secondary one which is attributed to the slow rotation of the magnetic easy axis of each nanoparticle in the large field. For the rotatable ferromagnetic NPs, due to high-frequency alternating magnetic field, longitudinal and planar orientations were formed, irrespective of the free energy, as dissipative structures.

In magnetic hyperthermia, heating of MNPs mostly occurs in a liquid medium. In the LRT, the magnetic response of MNPs in a liquid is assumed to be characterized by the effective magnetic relaxation time, τ_{eff} (Equation (4)). However, the conventional LRT does not take into account the complex dynamics of MNPs in in a viscous liquid in an alternating external magnetic field of finite amplitude and hence it oversimplifies the real situation [34]. In [34], magnetic dynamics of an assembly of NPs dispersed in a viscous Liquid were theoretically studied using stochastic equations of motion. In this method, stochastic equations of motion were constructed and solved for two unit vectors; the unit magnetization vector and the director which is a unit vector that determines the space orientation of a MNP with uniaxial anisotropy. Two regimes of the stationary magnetization oscillations were obtained, depending on the amplitude of the alternating magnetic field. In the viscous regime, which occurs for low magnetic field amplitudes, $H_0 \ll H_k$, the two unit vectors move in unison and out of phase with respect to the phase of the alternating magnetic field. In the magnetic regime, which occurs for $H_0 \geq H_k$, the director oscillates slightly, while the unit magnetization vector jumps between the directions along and opposite to the direction of the external magnetic field. The transition between the oscillation regimes occurs within the range $0.5H_k \leq H_0 < H_k$, depending on the magnetic field frequency and the liquid viscosity. The authors described the behavior of the low-frequency hysteresis loops as a function of the liquid viscosity and the amplitude and frequency of alternating magnetic field. The authors showed that SAR of an assembly of MNPs in a liquid can be significantly increased by selecting a suitable mode of magnetization oscillations. The authors reported that for an assembly of MNPs in viscous liquid, large SAR can be obtained in the intermediate excitation regime, $H_0 \approx 0.5H_k$. For magnetic parameters typical for iron oxides, and for H_0 = 200–300 Oe, and f = 300–500 kHz, the estimated SAR values can be of the order of 1 kW/g. The results of this paper [34] clearly show that the magnetic dynamics (for low frequency hysteresis loops) of NPs dispersed in a viscous liquid is significantly different from the behavior of NPs immobilized in a solid matrix.

In [35], the authors studied superparamagnetic particles, with uniaxial anisotropy, suspended in a viscous fluid and subjected to an alternating magnetic field. Both dissipation mechanisms; the internal

(Néel) and the external (Brownian) magnetic relaxations were considered. The authors obtained simple expression for the dynamic susceptibility that takes into account both dissipation mechanisms. The energy absorption was compared to the conventional approach using a model polydisperse colloid containing maghemite nanoparticles. The viscous losses due to particle motion in the fluid were found to have important contribution to the full magnetic response of the particles and thus to the specific loss power. The authors suggested a modification to the conventional LRT where the field-independent Brownian relaxation time τ_B should be replaced by a field-dependent Brownian relaxation time $\tau_B(\mu H/k_B T)$ [36].

2.3. The Role of Anisotropy on Heating Efficiency

According to the LRT it is clear that the anisotropy is an important factor in enhancing the Néel relaxation time (Equation (2)). However, it should be emphasized that from Equations (8) and (10) the SLP is maximized when $\omega\tau = 1$. This means that the increase of relaxation time does not always yield an increase in SLP. The frequency of the applied magnetic field must be correlated with the relaxation time such that $\omega\tau = 1$ [8]. Hence increasing the anisotropy results in an increase in the relaxation time and allows for the use of lower frequencies of the magnetic field. In [16], the authors concluded that the anisotropy of MNPs is a key parameter in tuning magnetic hyperthermia. They suggested that magnetic anisotropy should become central in the experimental investigations of magnetic hyperthermia. However, it is important to realize that depending on other factors, the anisotropy can increase or decrease heating efficiency of NPs. For example, the authors in [16] reported that heating efficiency was maximized in low anisotropy ferromagnetic Fe NPs.

3. Types of Anisotropies

Most magnetic materials contain some type of anisotropy that affects their magnetic behavior [18]. The most common types of anisotropy are: (a) magnetocrystalline anisotropy (or magnetic anisotropy or crystalline anisotropy); (b) surface anisotropy; (c) shape anisotropy; (d) exchange anisotropy; and (e) induced anisotropy (for example, by stress). All these anisotropies have influence on the magnetic properties to certain extent. In nanoparticles, shape anisotropy and magnetocrystalline anisotropy are the most important. Magnetocrystalline anisotropy arises from spin-orbit interaction and energetically favors alignment of the magnetic moments along a specific crystallographic direction called the easy axis of the material. The magnetocrystalline anisotropy depends on the type of material, temperature and impurities and is independent of the sample shape and size. Shape anisotropy causes magnetization to depend on the shape of the sample. The magnetization of a long thin needle shaped sample is easier along its long axis compared with that along any of its short axes. For nanoparticles, shape anisotropy is the dominant form of anisotropy. Stress anisotropy implies that magnetization might change with stress. It was shown that magnetic anisotropy changes when the surfaces are modified or adsorb different molecules [37]. This means that surface structure significantly influence the magnetic anisotropy. Hence, due to their large ratio of surface to bulk atoms, the surface anisotropy of nanoparticles could be more significant than both the crystalline and shape anisotropy. Coating of nanoparticles can have an influence on their magnetic anisotropies and hence on their magnetic properties [38,39]. In this work we focus on surface and exchange anisotropies.

3.1. The Role of Exchange Anisotropy in Core-Shell Nanoparticles

In 1956, Meiklejohn and Beans discovered that the hysteresis loop of a sample of ferromagnetic cobalt (Co) nanoparticles that is surrounded by antiferromagnetic oxidized layer (CoO) was shifted along the field (horizontal) axis after cooling in an applied magnetic field, H [40,41]. An increase in coercivity, H_C is usually observed with the shift of hysteresis loop [41,42]. This new effect is called exchange bias effect and the amount of the horizontal loop shift is called the exchange bias field, H_{EB}. This new type of magnetic anisotropy is called the exchange anisotropy or exchange coupling. The exchange anisotropy is suggested to be due to the interaction between the antiferromagnetic and the ferromagnetic materials. The exchange coupling could occur at the core-shell interface of different magnetic phases such as at an interface of a FM and AFM or FIM [42]. Some researchers suggested that the existence of pinned uncompensated spins in the AFM shell [43,44] or at the interface between the FM core and the AFM shell [43–45] could be the source of the exchange bias coupling. Several experimental studies showed the existence of uncompensated spins [46–50] but with orientations relative to the ferromagnetic magnetization. A satisfactory understanding of the mechanism of the exchange anisotropy of core-shell NPs at the microscopic level is not achieved yet [51]. In [52,53], core (FIM)-shell (AFM) Mn_3O_4–MnO and Mn_3O_4–Mn NPs it was found that the atomic structure and strain at the interface determine the interfacial exchange coupling. In [54] microstructural properties of core (AFM) MnO-shell (FM) Mn_3O_4 NPs were investigated. This arrangement is opposite the usual FM-AFM core-shell arrangement. The interface was found to be ordered, implying a strong interfacial coupling. At temperatures below T_C of the FM Mn_3O_4 shell, large exchange field (H_{EB}) values were obtained. In [45], the exchange coupling (or exchange anisotropy) at the core (AFM)-shell (FIM) interface of FeO–Fe_3O_4 NPs determines H_C and H_E. Large effective interface area reults in large interface exchange anisotropy which leads to large values of H_C and H_E. In [52,53], defects at the core-shell interface resulted in a small interfacial exchange coupling. The interfacial defects were suggested to produce interfacial uncompensated spins. In [55], FePt (FM)–Fe_3O_4 (FIM) core-shell NPs were investigated. The intimate contact between the FePt core and Fe_3O_4 shell was reported to lead to an effective interface exchange coupling. Hence, tailoring of the magnetic properties of these NPs can be achieved by controlling the core-shell dimensions, and by varying the material parameters of the core and shell. In an interesting study [56], $CoFe_2O_4$ (core)–$MnFe_2O_4$ (shell) MNPs were investigated. The diameter of the $CoFe_2O_4$ core was 9 mm and the thickness of the $MnFe_2O_4$ shell was 3 nm. The coercivity, H_C was found to have values between those for the core and shell materials which reflects the magnetic coupling of the core-shell structure. The SLP of the core-shell structures were found to have nearly one order of magnitude larger than those of NPs of the core or shell materials alone. In addition to an enhanced SLP, the SLP was found to vary by varying the core or shell structure. It is suggested that the exchange coupling at the interface can be tuned to produce effective anisotropy, K and magnetization that could lead to enhanced SLP.

These studies show that the dimensions and structure of core-shell interface have an impact on the interface exchange anisotropy and hence, on the effective anisotropy of the core-shell NPs. Thus, by tuning the core–shell parameters, the effective anisotropy can be varied to result in large heating efficiency of core–shell MNPs.

3.2. Surface Anisotropy in Nanoparticles

Because of the increased ratio of surface atoms to core atoms in NPs, surface effects were suggested to have significant role on the properties of NPs. Surface effects in MNPs include lattice relaxation [57], charge transfer [58,59], oxidation [60], surface spin disorder which results in spin-glass-like structures and spin canting [61]. These effects and others could the cause of several observed magnetic properties of MNPs such as the enhancement of magnetic anisotropy [37,62–64] and the reduction in saturation magnetization [61]. The total magnetization of a MNP has two contributions; the magnetization due to the surface spins and the magnetization due to the core. The magnetization of the surface is suggested to be due to the surface effects [65]. Surface spin disorder was reported to occur in iron oxide nanoparticles and was though to lead to extremely high magnetic anisotropy [66,67]. Some of these observations were explained initially in terms of a dead magnetic layer at the surface of the NP [68]. However, others attributed these observations to the disordered surface spins that freeze in a spin glass-like state or to surface spin canting [61,69,70]. A frozen disordered surface spin structures could make it difficult to attain saturation even under the application of high magnetic fields [71,72]. It was suggested also that exchange bias effect occurs between the surface and core spins of antiferromagnetic NPs and resulted in shifts in the magnetization hysteresis loops [73]. Using molecular dynamic modeling, non-uniform strains in the surface layers with an average expansion of a few percent compared to bulk were predicted [74]. A stress-induced anisotropy field was suggested to result from this expansion. An increase in the effective magnetic anisotropy due to surface effects was reported in several studies [6,8,75,76].

With decreasing the size of NPs, the ratio of the number of surface atoms to that of the bulk atoms become larger yielding larger contribution of the surface magnetization. Hence, surface magnetic anisotropy is expected to contribute towards the total magnetic anisotropy of the MNP. The total magnetic anisotropy of MNPs that includes the contribution of surface and core of MNPs is given by the phenomenological expression [37,72,77]:

$$K = K_V + \frac{6K_S}{D} \tag{14}$$

where K_V is the magnetocrystalline anisotropy of the core, K_S is the surface anisotropy of the particle, and D is the diameter of the particle (which is assumed to be spherical). Equation (14) has been used for its simplicity, but it might not be accurate to combine the surface anisotropy with the core anisotropy in this simple additive way. It is clear from Equation (14) that the surface contribution to the effective anisotropy increases with decreasing the size of the particle. Modified version of Equation (14) was proposed by some researchers [63].

The role of surface anisotropy on the efficiency of heating in magnetic hyperthermia was studied [78]. Single-domain cubic iron oxide particles were found to have superior magnetic heating efficiency compared to spherical particles of similar sizes. Using Monte Carlo simulations at the atomic level [78] cubic particles were reported to have larger surface anisotropy compared with the spherical particles. These results show the beneficial role of surface anisotropy in the improved heating power. These results demonstrate the importance of both the crystal quality and surface bond on the magnetic properties of the ferrimagnetic nanoparticles. However, it should be kept in mind that increasing the anisotropy does not always increase heating efficiency. For example, as mentioned earlier, heating power was maximized in large ferromagnetic NPs with low anisotropy [16].

4. The Role of Inter-Particle Interactions on the Heating Efficiency

Inter-particle dipolar interaction energy is proportional to $1/r^6$, where r is the inter-particle distance between particles. Hence, dipolar interactions between MNPs increases with decreasing the inter-particle distance when particle concentration increases. Dipolar interactions are expected to influence the magnetic relaxations of MNPs and thus influence their heating efficiency in the existence of an alternating magnetic field. Although the role of dipolar interactions on heating efficiency in magnetic hyperthermia is important, it is not well understood.

The role of magnetic interactions between magnetic nanoparticles on the heating efficiency for hyperthermia has been studied experimentally and theoretically [79–83]. In [17], the authors conducted numerical simulations and experiments on a system of Fe NPs of sizes between 5.5 and 28 nm. By comparing SAR from numerical simulations with those from experiments, the authors suggested that magnetic interactions decrease heating efficiency. In [83], the influence of magnetic interactions on magnetic hyperthermia efficiency was investigated by conducting SAR and high-frequency hysteresis loop measurements on systems of MNPs with the same size (with diameter around 13.5 nm) but with a varying anisotropy. Both kind of measurements were performed at the same frequency $f = 54$ kHz. The samples investigated were colloidal solutions composed of Fe, Fe_xC_y, Fe (core)–Fe_xC_y (shell), and FeCo nanoparticles. The anisotropy was varied by changing their composition. High-frequency hysteresis loops were measured at maximum applied magnetic field $\mu_0 H_{max} = 42$ mT while in SAR measurements, it was varied between 0 and 60 mT. The authors suggested that the formation of chains of MNPs could be a key element to understand experimental data. They reported that the large particle–particle magnetic interactions (compared with the magneto-crystalline anisotropy) leads to the formation of chains of MNPs during hyperthermia experiments. The authors observed a correlation between the magnetic nanoparticle magnetocrystalline anisotropy and the squareness of their hysteresis loop in colloidal solution where particles with larger anisotropy displayed smaller squareness. The authors claimed that *"chains of MNPs with a uniaxial anisotropy are the only way to reach the maximum possible SAR with a given magnetic material"* [83]. These results could explain contradictory results in the literature on the influence of magnetic dipolar interactions on heating efficiency of MNPs for magnetic hyperthermia [84–90].

In an interesting work [91], the SAR of two series of aqueous magnetite (Fe_3O_4) NPs and polyacrylic acid (PAA)-coated Fe_3O_4 NPs based dispersions were investigated at different magnetite concentrations. Heat efficiency was found to decrease with magnetite concentration for the PAA–Fe_3O_4 NPs. On the other hand, the heating efficiency for the bare Fe_3O_4 NPs, was found to increase with increasing particle concentration. This behavior was attributed to dipolar interactions. It was suggested that with increasing NP concentration, dipolar interactions cause Neel relaxation times to increase resulting in decreasing SAR for the PAA–Fe_3O_4 NPs. The PAA coating also was suggested to change the hydrodynamic size of the particles and thus modifying the Brownian relaxation time. For the bare NPs it was suggested that dipolar interactions are significant even at low concentrations, while aggregations occur at high concentrations. This work summarizes the conflicting role MNP concentration on SAR.

In an excellent review [8], the authors tried to resolve these controversial reports where they calculated $\omega\tau$ in some conflicting reports and suggested that MNP concentration always suppresses the relaxation time in all situations. They clarified that this reduction in relaxation time has opposite effects on SAR depending on whether the value of $\omega\tau > 1$ or $\omega\tau < 1$. When $\omega\tau < 1$, SAR will decrease as the relaxation time, τ decreases while for $\omega\tau > 1$, SAR will increase as τ decreases. Although this work [8] is very interesting and provide a coherent explanation of the conflicting experimental work, we have to emphasize that in the calculations of $\omega\tau$, the authors used the formulas for Néel and Brownian relaxation times which are known to be valid for identical and non-interacting particles. However, dipolar interactions were suggested to change the effective anisotropy of the particles [92]. In addition, the applied magnetic field strength should be in the linear region of the Langevin curve [93].

The effect of inter-particle dipolar interactions on heating efficiency was studied in [92]. The influence of particle chain formation was investigated on the heating efficiency. The experimental part of the study was conducted on low-anisotropy (spherical) as well as high-anisotropy (parallelepiped) ferrite-based magnetic fluids. It was found that heating efficiency decreases with increasing dipolar interactions. Using a theoretical model (which is valid for linear response regime) for dipole interactions it was found that in general dipolar interactions decrease heating efficiency. The theoretical model is based on the fact that dipolar interactions in linear chain arrangements increase the effective magnetic anisotropy. The authors suggested that several factors play roles in this process, such as particle size, chain size and experimental conditions, need to be optimized to enhance heating. It is important to mention that the anisotropy can increase or decrease heating efficiency of NPs depending on other experimental factors. For example, in [16], heating efficiency was maximized in low anisotropy ferromagnetic Fe NPs.

In [81], numerical investigation of the role of dipolar interactions on the hyperthermia efficiency was conducted. The authors reported that dipolar interactions decrease heating efficacy of MNPs. When studying different sample shapes the authors reported that hysteresis might slightly increase by small dipolar interactions.

In [94], two separate sets of agglomerated and dispersed (non-agglomerated) Fe_2O_3 ferrite NPs were studied. The heating efficiency of intensely agglomerated 15 nm Fe_2O_3 ferrite nanoparticles, was investigated in alternating magnetic field with frequency, $f = 100$ kHz and maximum strength, $H_0 = 13$ kA/m. Although the inter-particle interactions are strong in such agglomerate, moderate SAR value was found. To determine the effect of the NP diameter on SAR, the authors also used a model which includes the dipolar interactions among MNPs in the agglomerate. Because the amplitude of the alternating magnetic field, H_0 was considered to be comparable to the anisotropy field, H_K, the model was based on the hysteresis losses that is valid for the non-linear region. Because of the large size of the agglomerates (hydrodynamic mean diameters larger than 200 nm), the mechanical movement of the particles in the fluid can be neglected. Hence, in the simulated model, Brown relaxations were neglected as a heat generation mechanism. For the dispersed sets of MNPs the authors showed that heating is dominated by Neel and Brown relaxations. The authors reported a clear dependence of SAR on MNP size in both the agglomerated and dispersed samples.

The inter- and intra-particle interactions were recently investigated in core–shell nanoparticles [95]. In that report [95] the authors reported the results of low-temperature magnetic measurements that were conducted on very small (3.3 nm in size) core-shell structured NPs. The core ($MnFe_2O_4$) was found to be well-ordered ferromagnetic. The shell (γ-Fe_2O_3) was found to display uniaxial anisotropy with disordered spins. The magnetic measurements were conducted on two NP samples; one with non-textured frozen dispersions and the other with disordered powder. The authors discussed three types of particle interactions; the dipolar, the intra-particle exchange bias, and the inter-particle exchange bias. The dipolar interaction is the magnetic interaction of the magnetic moments of the cores of the NPs. The intra-particle exchange bias interaction exists between the core and surface of each particle and results in horizontal shifts of the field-cooled hysteresis loops. The inter-particle exchange interaction exists between the surface spins of particles that are in contact with each other. The authors [95] found that for dilute frozen dispersions of NPs that are at a distance from each other (not in contact), the dipolar interactions and inter-particle exchange bias interactions are neglected while the intra-particle exchange bias interaction is the only existing interaction. On the other hand, in concentrated frozen dispersions of NPs at a distance, dipolar interactions become significant. In powder NPs that are in contact, the inter-particle exchange interactions were found to be dominant. This is an interesting study [95] since it allows for distinguishing between intra- and inter-particle exchange bias interactions by comparing the results of magnetic measurements on samples with non-textured frozen dispersions of NPs and powder NPs. These results enhances the knowledge about the factors that could contribute towards the heating efficiency of NPs.

In the interesting paper [96], a global view of the role of particle–particle magnetic interactions was given. To calculate hysteresis loops, the authors used a kinetic Monte–Carlo algorithm that correctly account for both time and temperature. SAR of MNPs dispersed inside spherical lysosomes was studied as a function of several parameters including volume concentration of the 20 nearest neighbors around a given NP. For large magnetic fields, magnetic interactions of NPs increase the coercive field, saturation field and hysteresis area of major loops. However, for small amplitude magnetic field such as those used in magnetic hyperthermia, the heating power as function of concentration can increase, decrease or display a bell shape, depending on the relationship between the applied magnetic field and the coercive/saturation fields of the NPs [96]. The volume concentration was found to strongly influence the heating properties of a given NP. Heating power was shown to be not homogeneous inside lysosomes and drastically changes with the position inside them. Hence, the local environment of a given NP was found to have a significant impact on its heating power. In certain conditions, the amplitude of variation of heating power with position inside lysosomes could be more than one order of magnitude. The NP diameter, anisotropy and the amplitude of the applied magnetic field were also found to significantly affect magnetic interactions.

5. Some Remarks about the Physics of Heating of Nanoparticles for Localized Magnetic Hyperthermia

5.1. The Role of Size Distribution

In determining the relaxation times and heat generated in MNPs, the particles were assumed to have the same size. However, in reality this assumption cannot be satisfied because usually there will be some size distribution of the NPs regardless of the synthesis method used. In practice, size distributions are broad and may extend from single domain to multi-domain NPs. Producing NP systems with sufficiently narrow size distribution, where only one defined reversal mechanism appears, is not a simple task. There are not enough studies to clearly understand the effect of the size distribution width on heating efficiency of NPs. In [97], the effect of size distribution of NPs (in the diameter range from 10 to 100 nm) on magnetic hysteresis losses was investigated using a phenomenological model. The authors derived an empirical expression for the dependence of hysteresis loss on field amplitude and particle size. It was shown that a useful choice of field amplitude and frequency depends strongly on the mean particle size and variance of the NPs that will be used for hyperthermia. The authors suggested that iron oxide NPs with narrow size distribution and with a mean diameter that corresponds to the maximum coercivity in the single domain size range could lead to maximum heating efficiency. Hence, in addition to the required narrow size distribution, the mean particle size should be adjusted in relation to the magnetic field amplitude in order to obtain maximum SLP. If the accurate mean particle size and the corresponding coercivity are not known, then the relation between the magnetic field amplitude and mean particle size might not be satisfied resulting in a situation where many particles will not be able to reach maximum SLP. In this case, the authors indicated that using NPs with a broader size distribution may be advantageous. In [22], polydispersity in the size distribution of MNPs was found to reduce the heating efficiency. When the size distribution was changed from highly monodisperse ($\sigma = 0$) to polydisperse ($\sigma = 2.5$) the heating rate was significantly decreased, where σ is standard deviation of the lognormal size distribution.

5.2. The Heating Curve

In magnetic hyperthermia experiments, an alternating magnetic field is applied to the nanoparticle sample and the variation of temperature is measured. The heating efficiency is represented by SAR or SLP is usually obtained from the initial slope of the measured data. This method ignores the entire heating curve and hence does not display the entire temperature dependence of SAR [98].

The authors in [99] discussed several analytical method that is used to obtain the SAR from calorimetric measurements and pointed out that SAR values depend also on the analytical method used. The commonly used "initial slope" method was found to sensitive to the experimental conditions and could underestimate values by up to 25%. The full-curve fit method was found to be better but also with underestimation by up to 10% can be expected. The "corrected slope" method which was derived by the authors [99] was found to be the most accurate method. The combined errors that associate the analytical methods with the experimental errors could lead to large variations between the actual SAR and the reported values.

5.3. Temperature Dependence of Saturation Magnetization and Magnetic Anisotropy

Usually, in magnetic hyperthermia models that are based on Rosensweig's theory [22] magnetic properties such as the saturation magnetization and the anisotropy are considered to be constant with temperature. However these properties and others change significantly with temperature [100].

The saturation magnetization as function of temperature in bulk ferromagnetic or ferrimagnetic materials at low temperatures, is governed by the Bloch's law [101]:

$$M_S(T) = M(0)\left[1 - \left(\frac{T}{T_0}\right)^{\alpha}\right] \tag{15}$$

Here, T_0 is the temperature at which M_s becomes zero and $M(0)$ is the saturation magnetization at 0 K. The Bloch's exponent $\alpha = 3/2$ for bulk materials. Bloch's $T^{3/2}$ law was based on magnon excitation of long wave-length spin-waves at low temperatures. However, due to finite size effects in nanoparticles, magnons could have wavelengths larger than the size of the particle leading to deviations from the $T^{3/2}$ law. Several studies discussed the temperature dependence of the saturation magnetization in nanoparticles and reported deviations from Bloch's law at low temperatures [102–111].

In the modified Bloch's law for nanoparticles at intermediate temperatures, the Bloch's exponent α was found to have values larger and smaller than 3/2 [10,112] and decays exponentially at low temperatures [106]. In [102], the temperature dependence of saturation magnetization in nickel ferrite nanoparticles was investigated where the surface spin and finite size effects in nanoparticles were found to have an important role. The deviations from Block's law could be due to inter-particle interactions [113] and the size distribution of the particles, significant and the disordered surface spins which influence the surface anisotropy and hence the effective anisotropy in the particles.

The magnetic anisotropy is known to vary with temperature in bulk magnetic materials and was expected to change with temperature also in nanoparticles [114–117]. A theoretical investigation of the influence of temperature on the magneto-crystalline anisotropy in Fe, Co and Ni nanoparticles showed a clear decrease with increasing temperature [118]. The dependence of effective magnetic anisotropy on temperature was recently investigated in $MnFe_2O_4$ nanoparticles with cubic anisotropy [119]. The magnetic anisotropy was found to decrease significantly with increasing temperature. Surface effects, which are more pronounce in particles with cubic anisotropy, were suggested to play a role in small particles.

Using Monte Carlo simulations surface effects were also found to have a dominant factor in determining the effective anisotropy at low temperatures resulting in an overall cubic effective anisotropy even in spherical nanoparticles with uniaxial anisotropy [120]. The contribution of cubic anisotropy was found to decrease with increasing temperature faster than that in particles with uniaxial anisotropy. The effective magnetic anisotropy constant of Fe_2O_3 nanoparticles was found to decrease significantly with temperature [121].

5.4. Experimental and Theoretical Limitations in the Determination of SAR

There are several methods where SAR or SLP can be measure. These methods involves magnetic or calorimetric measurements. However, there are always some inaccuracies in these measurements and results should be carefully discussed. In a recent and interesting review [122] the sources of uncertainties of several available methods in measuring SAR were analyzed. Comparison between magnetic methods and calorimetric methods were also discussed. It was shown that inaccuracies in magnetic measurements mainly result from the lack of experimental set-ups that are needed for the application of suitable strength and frequency of the alternating magnetic field in magnetic hyperthermia experiments. Inaccuracies in SAR when using the calorimetric methods result mainly from the lack of matching between measuring conditions, thermal models, and experimental setups.

In [99], the authors indicated that when SAR are determined by calorimetric measurements, it is preferred to conduct the measurements under adiabatic conditions where external heat transfer is minimized. However, because it is difficult to build adiabatic measurement systems and because the measurements in these systems are time-consuming, the SAR measurements are mainly conducted in non-adiabatic systems which lead to in accurate results. The authors pointed out that by using suitable experimental and analytical methods, accurate SAR measurements can be made using non-adiabatic conditions, as long as heat losses from the non-adiabatic setup are accounted for. The paper also discussed the different ways of heat loss which are due to conduction, convection, radiation at high temperatures, and evaporating or melting of the sample. Then possible sources of the inaccuracy in SAR measurement were discussed. One of these is the spatial inhomogeneity of temperature in the sample which makes the location of the thermal probe in the sample important. Other source of error is the delaying of heating where it takes some time for the heating curve to take off after the start of the heating process. A third source of error is the change of heat capacity with temperature. A fourth source of error is the inhomogeneity of the magnetic field. In addition to those, peripheral heating, which is due to the experimental setup itself and is expected to vary in different laboratories depending on the system used.

An interesting paper [16] discussed the three types of theories that can be used for describing hysteresis loops of MNPs. These are: equilibrium functions, theories based on Stoner–Wohlfarth model, and the linear response theory (LRT). Limitations and domains of validity were discussed. The authors proposed that the separation between "relaxation losses" and "hysteresis losses" is artificial and not correct. The authors showed that the LRT is only pertinent for MNPs with strong anisotropy and for particles with small anisotropy, theories based on Stoner–Wohlfarth model should be used. The authors also stressed that LRT including Brownian motion is only valid for small magnetic field [36].

5.5. Self-Regulated Hyperthermia

Heating cells to temperatures between 42 and 46 °C (315–319 K) results in killing only tumor cells [5]. Above this temperature, healthy cells might be affected resulting in necrosis. Hence, it is preferred that the temperature of MNPs does not exceed this level. In practice, it is not simple to determine the temperature of cells in accurate manner during hyperthermia. Thus, having MNPs with Curie temperature, T_C above 42 °C and below 46 °C is essential for self-regulated hyperthermia. Once T_C is reached, the MNPs lose their magnetization. Hence, heating stops without the need to remove the external magnetic field. In order to have self-regulated hyperthermia it is essential to investigate new materials and structures of MNPs. Several studies reported partial success in this regard [123]. Here we discuss some of the recent work on controlling T_C of MNPs. In [124], the role of shape, size and composition on T_C of ferromagnetic NPs was theoretically investigated. It was found that reducing the particle size could result in a decrease of T_C. The authors also investigated different ferromagnetic material compositions in combination with nonmagnetic materials such as Zn and Cu and magnetic ions such as Gd and Cr. It was suggested [124] that introducing these materials in the ferromagnetic material results in lower T_C due to the reduction of the exchange interaction between the magnetic ions in the NPs. In [125] Cu–Ni alloy NPs were found to have T_C in the range of 43–46 °C. In [126], manganese perovskite nanoparticles $La_{1-x}Sr_xMnO_3$ with $x = 0.25$ with size in the range of 30–49 nm were found to have T_C near 352 K with large SAR values. This indicates that this material could be a good candidate for self-regulated hyperthermia. In [127], Curie temperatures of $Mn_{0.5}Zn_{0.5}Gd_xFe_{(2-x)}O_4$ ferrite nanoparticles, with $x = 0, 0.5, 1.0,$ and 1.5 were investigated. T_C was found to vary with changing the Gd concentration indicating the possibility of tuning T_C of ferrite NPs. Doping Mn ferrite with Zn ($Mn_{1-x}Zn_xO$) and doping Zn ferrite with Gd ($ZnGd_xFe_{2-x}O_4$) was investigated to tune the Curie temperature in the range (42–43 °C) which is very suitable for hyperthermia applications [128]. In [129], the authors investigated the magnetic properties of $Mn_xZn_{1-x}Fe_2O_4$ nanoparticles with Mn concentrations $x = 0.8, 0.61, 0.5,$ and 0.2. They found that T_C increases with decreasing Mn concentration. Hence, the authors suggested that Curie temperature close to 42 °C (315 K) might be achieved by adjusting the Mn and Zn concentrations. In [130], magnetic measurements of $Zn_xGd_{1-x}Fe$ nanoparticles with $x = 0.02, 0.05, 0.1,$ and 0.2 were conducted. The authors found that T_C has a nonmonotonic behavior with increasing x. All samples were found to have T_C larger than 665 K. The lowest T_C of 665 K (392 C) was found for the sample with $x = 0.02$. The authors suggested that larger Zn concentrations (0.2–0.5) might result in T_C in the range suitable for self-controlled hyperthermia. In [131] $Ni_{1-x}Cr_x$ NPs were prepared by standard arc melting technique. Their magnetic properties were investigated using Vibrating Sample Magnetometer (VSM) and Superconducting Quantum Interference Device (SQUID) magnetometer. As the Cr concentration was increased from $x = 4.54$ wt% to $x = 5.90$ wt%, T_C was found to decrease almost linearly from 401 to 289 K. Hence $Ni_{1-x}Cr_x$ NPs were found to be suitable material for self-regulating magnetic hyperthermia. In [132], chromium–nickel alloy (Cr_xNi_{1-x}) NPs were prepared using water-in-oil microemulsion and mechanical milling and investigated for self-controlled magnetic hyperthermia. The T_C of the sample synthesized by microemulsion method ($x = 20$) was found to be 320 °C. For the series of NPs prepared by mechanical milling, some of the NPs ($x = 26, 27, 28, 29$) were found to have low T_C and some of them ($x = 10, 15, 20$) were found to have high T_C. The NPs with $x = 29$, were found to have T_C of 43 °C and for NPs with $x = 28$, T_C was found to be 44 °C. As the Cr content (x) decreases, T_C was found to increase. The results in [131] and in [132] clearly revealed that T_C of Cr_xNi_{1-x} NPs can tuned by varying the synthesis method and by varying the Cr/Ni molar ratio.

6. Conclusions

The factors that influence local magnetic hyperthermia using magnetic nanoparticles were discussed. We have shown that surface anisotropy and core-shell interface anisotropy have a noticeable impact on the relaxation time and hence, could be tuned to enhance the heating efficiency of MNPs. The role of dipolar interactions was emphasized as an important factor in heating efficiency where the concentration of MNPs was found to suppress the relaxation time. We have shown that the role of size distribution of the particles was not well-investigated with contradictory results. The magnetic anisotropy and magnetization of MNPs were shown to decrease with temperature but not considered in most heating efficiency studies. Self-regulated hyperthermia was shown to have limited success.

Acknowledgments

This work was financially supported by the National Research Foundation (NRF) under the Grant No. 31S087.

Author Contributions

All authors contributed equally to the reported research and writing of the paper.

Conflicts of Interest

The authors declare no conflict of interest.

References

1. Gilchrist, R.K.; Shorey, W.D.; Hanselman, R.C.; Parrott, J.C.; Taylor, C.B. Selective inductive heating of lymph nodes. *Ann. Surg.* **1957**, *146*, 596–606.

2. Jordan, A.; Scholz, R.; Wust, P.; Schirra, H.; Schiestel, T.; Schmidt, H.; Felix, R. Endocytosis of dextran and silan-coated magnetite nanoparticles and the effect of intracellular hyperthermia on human mammary carcinoma cells *in vitro*. *J. Magn. Magn. Mater.* **1999**, *194*, 185–196.

3. Moroz, P.; Jones, S.K.; Gray, B.N. Magnetically mediated hyperthermia: Current status and future directions. *Int. J. Hyperth.* **2002**, *18*, 267–284.

4. Laurent, S.; Dutz, S.; Häfeli, U.O.; Mahmoudi, M. Magnetic fluid hyperthermia: Focus on superparamagnetic iron oxide nanoparticles. *Adv. Colloid Interface Sci.* **2011**, *166*, 8–23.

5. Kumar, C.S.S.R.; Mohammad, F. Magnetic nanomaterials for hyperthermia-based therapy and controlled drug delivery. *Adv. Drug Deliv. Rev.* **2011**, *63*, 789–808.

6. Hergt, R.; Dutz, S.; Müller, R.; Zeisberger, M. Magnetic particle hyperthermia: Nanoparticle magnetism and materials development for cancer therapy. *J. Phys. Condens. Matter* **2006**, *18*, S2919–S2934.

7. Vallejo-Fernandez, G.; Whear, O.; Roca, A.G.; Hussain, S.; Timmis, J.; Patel, V.; O'Grady, K. Mechanisms of hyperthermia in magnetic nanoparticles. *J. Phys. D* **2013**, *46*, doi:10.1088/0022-3727/46/31/312001.

8. Deatsch, A.E.; Evans, B.A. Heating efficiency in magnetic nanoparticle hyperthermia. *J. Magn. Magn. Mater.* **2014**, *354*, 163–172.

9. Dennis, C.L.; Ivkov, R. Physics of heat generation using magnetic nanoparticles for hyperthermia. *Int. J. Hyperth.* **2013**, *29*, 715–729.

10. Sun, S.; Murray, C.B.; Weller, D.; Folks, L.; Moser, A. Monodisperse FePt nanoparticles and ferromagnetic FePt nanocrystal superlattices. *Science* **2000**, *287*, 1989–1992.

11. Rusponi, S.; Cren, T.; Weiss, N.; Epple, M.; Buluschek, P.; Claude, L.; Brune, H. The remarkable difference between surface and step atoms in the magnetic anisotropy of two-dimensional nanostructures. *Nat. Mater.* **2003**, *2*, 546–551.

12. Tartaj, P.; Morales, M.D.P.; Veintemillas-Verdaguer, S.; Gonz lez-Carreo, T.; Serna, C.J. The preparation of magnetic nanoparticles for applications in biomedicine. *J. Phys. D* **2003**, *36*, R182–R197.

13. Gupta, A.K.; Gupta, M. Synthesis and surface engineering of iron oxide nanoparticles for biomedical applications. *Biomaterials* **2005**, *26*, 3995–4021.

14. Koksharov, Y.A. Magnetism of Nanoparticles: Effects of Size, Shape, and Interactions. In *Magnetic Nanoparticles*, 1st ed.; Gubin, S.P., Ed.; Wiley-VCH: Berlin, Germany, 2009; pp. 228–229.

15. Issa, B.; Obaidat, I.M.; Albiss, B.A.; Haik, Y. Magnetic nanoparticles: Surface effects and properties related to biomedicine applications. *Int. J. Mol. Sci.* **2013**, *14*, 21266–21305.

16. Carrey, J.; Mehdaoui, B.; Respaud, M. Simple models for dynamic hysteresis loop calculations of magnetic single-domain nanoparticles: Application to magnetic hyperthermia optimization. *J. Appl. Phys.* **2011**, *109*, doi:10.1063/1.3551582.

17. Mehdaoui, B.; Meffre, A.; Carrey, J.; Lachaize, S.; Lacroix, L.-M.; Gougeon, M.; Chaudret, B.; Respaud, M. Optimal size of nanoparticles for magnetic hyperthermia: A combined theoretical and experimental study. *Adv. Funct. Mater.* **2011**, *21*, 4573–4581.

18. Guimarães, A.P. *Principles of Nanomagnetism*; Springer-Verlag: Berlin, Germany, 2009.

19. Néel, L. Théorie du trainage magnétique des ferromagné tiques en grains fins avec applications aux terres cuites. *Ann. Geophys.* **1949**, *5*, 99–136. (In French)

20. Brown, W.F. Thermal fluctuations of a single-domain particle. *Phys. Rev.* **1963**, *34*, 1677–1686.

21. Leslie-Pelecky, D.L.; Rieke, R.D. Magnetic properties of nanostructured materials. *Chem. Mater.* **1996**, *8*, 1770–1783.

22. Rosensweig, R.E. Heating magnetic fluid with alternating magnetic field. *J. Magn. Magn. Mater.* **2002**, *252*, 370–374.

23. Delaunay, L.; Neveu, S.; Noyel, G.; Monin, J. A new spectrometric method, using a magneto-optical effect, to study magnetic liquids. *J. Magn. Magn. Mater.* **1995**, *149*, L239–L245.

24. Glöckl, G.; Hergt, R.; Zeisberger, M.; Dutz, S.; Nagel, S.; Weitschies, W. The effect of field parameters, nanoparticle properties and immobilization on the specific heating power in magnetic particle hyperthermia. *J. Phys. Condens. Matter* **2006**, *18*, S2935–S2949.

25. Hergt, R.; Dutz, S. Magnetic particle hyperthermia—Biophysical limitations of a visionary tumour therapy. *J. Magn. Magn. Mater.* **2007**, *311*, 187–192.

26. Ortega, D.; Pankhurst, Q.A. Magnetic Hyperthermia. In *Nanoscience, Volume 1: Nanostructures through Chemistry*, O'Brien, P., Ed.; Royal Society of Chemistry: Cambridge, UK, 2013; pp. 60–88.

27. Atkinson, W.J.; Brezovich, I.A.; Chakraborty, D.P. Usable frequencies in hyperthermia with thermal seeds. *IEEE Trans. Biomed. Eng.* **1984**, *31*, 70–75.

28. Brezovich, I.A. Low frequency hyperthermia: Capacitive and ferromagnetic thermoseed methods. *Med. Phys. Monogr.* **1988**, *16*, 82–111.

29. Kita, E.; Oda, T.; Kayano, T.; Sato, S.; Minagawa, M.; Yanagihara, H.; Kishimoto, M.; Mitsumata, C.; Hashimoto, S.; Yamada, K.; *et al.* Ferromagnetic nanoparticles for magnetic hyperthermia and thermoablation therapy. *J. Phys. D* **2010**, *43*, doi:10.1088/0022-3727/43/47/474011.

30. Usov, N.A.; Grebenshchikov, Y.B. Hysteresis loops of an assembly of superparamagnetic nanoparticles with uniaxial anisotropy. *J. Appl. Phys.* **2009**, *106*, doi:10.1063/1.3173280.

31. Usov, N.A.; Liubimov, B.Y. Dynamics of magnetic nanoparticle in a viscous liquid: Application to magnetic nanoparticle hyperthermia. *J. Appl. Phys.* **2012**, *112*, doi:10.1063/1.4737126.

32. Mamiya, H. Recent advances in understanding magnetic nanoparticles in AC magnetic fields and optimal design for targeted hyperthermia. *J. Nanomater.* **2013**, doi:10.1155/2013/752973.

33. Mamiya, H.; Jeyadevan, B. Hyperthermic effects of dissipative structures of magnetic nanoparticles in large alternating magnetic fields. *Sci. Rep.* **2011**, *1*, doi:10.1038/srep00157.

34. Usov, N.A. Low frequency hysteresis loops of superparamagnetic nanoparticles with uniaxial anisotropy. *J. Appl. Phys.* **2010**, *107*, doi:10.1063/1.3445879.

35. Raikher, Y.L.; Stepanov, V.I. Physical aspects of magnetic hyperthermia: Low-frequency AC field absorption in a magnetic colloid. *J. Magn. Magn. Mater.* **2014**, *368*, 421–427.

36. Raikher, Y.L.; Stepanov, V.I. Absorption of AC field energy in a suspension of magnetic dipoles. *J. Magn. Magn. Mater.* **2008**, *320*, 2692–2695.

37. Bødker, F.; Mørup, S.; Linderoth, S. Surface effects in metallic iron nanoparticles. *Phys. Rev. Lett.* **1994**, *72*, 282–285.

38. Homola, A.; Lorenz, M.; Mastrangelo, C.; Tilbury, T. Novel magnetic dispersions using silica stabilized particles. *IEEE Trans. Magn.* **1986**, *22*, 716–719.

39. Hormes, J.; Modrow, H.; Bönnemann, H.; Kumar, C.S.S.R. The influence of various coatings on the electronic, magnetic, and geometric properties of cobalt nanoparticles. *J. Appl. Phys.* **2005**, *97*, doi:10.1063/1.1855191.

40. Meiklejohn, W.H.; Bean, C.P. New magnetic anisotropy. *Phys. Rev.* **1956**, *102*, 1413–1414.

41. Meiklejohn, W.H.; Bean, C.P. New magnetic anisotropy. *Phys. Rev.* **1957**, *105*, 904–913.

42. Nogués, J.; Sort, J.; Langlais, V.; Skumryev, V.; Suriñach, S.; Muñoz, J.S.; Baró, M.D. Exchange bias in nanostructures. *Phys. Rep.* **2005**, *422*, 65–117.

43. Berkowitz, A.E.; Takano, K. Exchange anisotropy—A review. *J. Magn. Magn. Mater.* **1999**, *200*, 552–570.

44. Nogués, J.; Schuller, I.K. Exchange bias. *J. Magn. Magn. Mater.* **1999**, *192*, 203–232.

45. Sun, X.; Huls, N.F.; Sigdel, A.; Sun, S. Tuning exchange bias in core/shell FeO/Fe_3O_4 nanoparticles. *Nano Lett.* **2012**, *12*, 246–251.

46. Takano, K.; Kodama, R.H.; Berkowitz, A.E.; Diego, S.; Jolla, L. Interfacial uncompensated antiferromagnetic spins: Role in unidirectional anisotropy in polycrystalline $Ni_{81}Fe_{19}/CoO$ bilayers. *Phys. Rev. Lett.* **1997**, *79*, 1130–1133.

47. Antel, W.J., Jr.; Perjeru, F.; Harp, G.R. Spin structure at the interface of exchange biased FeMn/Co bilayers. *Phys. Rev. Lett.* **1999**, *83*, 1439–1442.

48. Ohldag, H.; Regan, T.; Stöhr, J.; Scholl, A.; Nolting, F.; Lüning, J.; Stamm, C.; Anders, S.; White, R. Spectroscopic identification and direct imaging of interfacial magnetic spins. *Phys. Rev. Lett.* **2001**, *87*, doi:10.1103/PhysRevLett.87.247201.

49. Ohldag, H.; Scholl, A.; Nolting, F.; Arenholz, E.; Maat, S.; Young, A.; Carey, M.; Stöhr, J. Correlation between exchange bias and pinned interfacial spins. *Phys. Rev. Lett.* **2003**, *91*, doi:10.1103/PhysRevLett.91.017203.

50. Eimüller, T.; Kato, T.; Mizuno, T.; Tsunashima, S.; Quitmann, C.; Ramsvik, T.; Iwata, S.; Schütz, G. Uncompensated spins in a micro-patterned CoFeB/MnIr exchange bias system. *Appl. Phys. Lett.* **2004**, *85*, 2310–2312.

51. Kiwi, M. Exchange bias theory. *J. Magn. Magn. Mater.* **2001**, *234*, 584–595.

52. Si, P.Z.; Li, D.; Lee, J.W.; Choi, C.J.; Zhang, Z.D.; Geng, D.Y.; Brück, E. Unconventional exchange bias in oxide-coated manganese nanoparticles. *Appl. Phys. Lett.* **2005**, *87*, doi:10.1063/1.2072807.

53. Si, P.Z.; Li, D.; Choi, C.J.; Li, Y.B.; Geng, D.Y.; Zhang, Z.D. Large coercivity and small exchange bias in Mn_3O_4/MnO nanoparticles. *Solid State Commun.* **2007**, *142*, 723–726.

54. Berkowitz, A.E.; Rodriguez, G.F.; Hong, J.I.; An, K.; Hyeon, T.; Agarwal, N.; Smith, D.J.; Fullerton, E.E. Monodispersed MnO nanoparticles with epitaxial Mn_3O_4 shells. *J. Phys. D* **2008**, *41*, 134007.

55. Zeng, H.; Sun, S.; Li, J.; Wang, Z.L.; Liu, J.P. Tailoring magnetic properties of core/shell nanoparticles. *Appl. Phys. Lett.* **2004**, *85*, 792–795.

56. Lee, J.-H.; Jang, J.-T.; Choi, J.-S.; Moon, S.H.; Noh, S.-H.; Kim, J.-W.; Kim, J.-G.; Kim, I.-S.; Park, K.I.; Cheon, J. Exchange-coupled magnetic nanoparticles for efficient heat induction. *Nat. Nanotechnol.* **2011**, *6*, 418–422.

57. Evans, R.; Nowak, U.; Dorfbauer, F.; Shrefl, T.; Mryasov, O.; Chantrell, R.W.; Grochola, G. The influence of shape and structure on the Curie temperature of Fe and Co nanoparticles. *J. Appl. Phys.* **2006**, *99*, doi:10.1063/1.2167636.

58. Coey, J.M.D.; Wongsaprom, K.; Alaria, J.; Venkatesan, M. Charge-transfer ferromagnetism in oxide nanoparticles. *J. Phys. D* **2008**, *41*, doi:10.1088/0022-3727/41/13/134012.

59. Cótica, L.F.; Santos, I.A.; Girotto, E.M.; Ferri, E.V.; Coelho, A.A. Surface spin disorder effects in magnetite and poly(thiophene)-coated magnetite nanoparticles. *J. Appl. Phys.* **2010**, *108*, doi:10.1063/1.3488634.

60. Makhlouf, S.A.; Al-attar, H.; Kodama, R.H. Particle size and temperature dependence of exchange bias in NiO nanoparticles. *Solid State Commun.* **2008**, *145*, 1–4.

61. Kodama, R.H. Magnetic nanoparticles. *J. Magn. Magn. Mater.* **1999**, *200*, 359–372.

62. Luis, F.; Torres, J.M.; García, L.M.; Bartolomé, J.; Stankiewicz, J.; Petroff, F.; Fettar, F.; Maurice, J.-L.; Vaurès, A. Enhancement of the magnetic anisotropy of nanometer-sized Co clusters: Influence of the surface and of interparticle interactions. *Phys. Rev. B* **2002**, *65*, doi:10.1103/PhysRevB.65.094409.

63. Luis, F.; Bartolomé, F.; Petroff, F.; Bartolomé, J.; García, L.M.; Deranlot, C.; Jaffrès, H.; Martínez, M.J.; Bencok, P.; Wilhelm, F.; *et al.* Tuning the magnetic anisotropy of Co nanoparticles by metal capping. *Europhys. Lett.* **2006**, *76*, 142–148.

64. Lima, E.; de Biasi, E.; Mansilla, M.V.; Saleta, M.E.; Effenberg, F.; Rossi, L.M.; Cohen, R.; Rechenberg, H.R.; Zysler, R.D. Surface effects in the magnetic properties of crystalline 3 nm ferrite nanoparticles chemically synthesized. *J. Appl. Phys.* **2010**, *108*, doi:10.1063/1.3514585.

65. Ho, C.-H.; Lai, C.-H. Size-dependent magnetic properties of PtMn nanoparticles. *IEEE Trans. Magn.* **2006**, *42*, 3069–3071.

66. Del Muro, M.G.; Batlle, X.; Labarta, A. Erasing the glassy state in magnetic fine particles. *Phys. Rev. B* **1999**, *59*, 584–587.

67. Batlle, X.; del Muro, M.G.; Labarta, A. Interaction effects and energy barrier distribution on the magnetic relaxation of nanocrystalline hexagonal ferrites. *Phys. Rev. B* **1997**, *55*, 6440–6445.

68. Berkowitz, A.E.; Schuele, W.J.; Flanders, P.J. Influence of crystallite size on the magnetic properties of acicular-Fe_2O_3 particles. *J. Appl. Phys.* **1968**, *39*, 1261–1263.

69. Batlle, X.; Pérez, N.; Guardia, P.; Iglesias, O.; Labarta, A.; Bartolomé, F.; Garcia, L.M.; Bartolomé, J.; Roca, A.G.; Morales, M.P.; *et al.* Magnetic nanoparticles with bulk-like properties. *J. Appl. Phys.* **2011**, *109*, doi:10.1063/1.3559504.

70. Bedanta, S.; Kleemann, W. Supermagnetism. *J. Phys. D* **2009**, *42*, doi:10.1088/0022-3727/42/1/013001.

71. Papaefthymiou, G.C. Nanoparticle magnetism. *Nano Today* **2009**, *4*, 438–447.

72. Ahmed, S.R.; Ogale, S.B.; Papaefthymiou, G.C.; Ramesh, R.; Kofinas, P. Magnetic properties of $CoFe_2O_4$ nanoparticles synthesized through a block copolymer nanoreactor route. *Appl. Phys. Lett.* **2002**, *80*, 1616–1618.

73. Park, T.-J.; Papaefthymiou, G.C.; Viescas, A.J.; Moodenbaugh, A.R.; Wong, S.S. Size-dependent magnetic properties of single-crystalline multiferroic $BiFeO_3$ nanoparticles. *Nano Lett.* **2007**, *7*, 766–772.

74. Nunes, A.C.; Yang, L. Calculated ferrite nanocrystal relaxation and its magnetic implications. *Surf. Sci.* **1998**, *399*, 225–233.

75. Jamet, M.; Wernsdorfer, W.; Thirion, C.; Mailly, D.; Dupuis, V.; Mélinon, P.; Pérez, A. Magnetic anisotropy of a single cobalt nanocluster. *Phys. Rev. Lett.* **2001**, *86*, 4676–4679.

76. Jamet, M.; Wernsdorfer, W.; Thirion, C.; Dupuis, V.; Mélinon, P.; Pérez, L.; Mailly, D. Magnetic anisotropy in single clusters. *Phys. Rev. B* **2004**, *69*, doi:10.1103/PhysRevB.69.024401.

77. Pérez, N.; Guardia, P.; Roca, A.G.; Morales, M.P.; Serna, C.J.; Iglesias, O.; Bartolomé, F.; García, L.M.; Batlle, X.; Labarta, A. Surface anisotropy broadening of the energy barrier distribution in magnetic nanoparticles. *Nanotechnology* **2008**, *19*, doi:10.1088/0957-4484/19/47/475704.

78. Martinez-Boubeta, C.; Simeonidis, K.; Makridis, A.; Angelakeris, M.; Iglesias, O.; Guardia, P.; Cabot, A.; Yedra, L.; Estradé, S.; Peiró, F.; *et al.* Learning from nature to improve the heat generation of iron-oxide nanoparticles for magnetic hyperthermia applications. *Sci. Rep.* **2013**, *3*, doi:10.1038/srep01652.

79. Burrows, F.; Parker, C.; Evans, R.F.L.; Hancock, Y.; Hovorka, O.; Chantrell, R.W. Energy losses in interacting fine-particle magnetic composites. *J. Phys. D* **2010**, *43*, doi:10.1088/0022-3727/43/47/474010.

80. Gudoshnikov, S.A.; Liubimov, B.Y.; Usov, N.A. Hysteresis losses in a dense superparamagnetic nanoparticle assembly. *AIP Adv.* **2012**, *2*, doi:10.1063/1.3688084.

81. Haase, C.; Nowak, U. Role of dipole-dipole interactions for hyperthermia heating of magnetic nanoparticle ensembles. *Phys. Rev. B* **2012**, *85*, doi:10.1103/PhysRevB.85.045435.

82. Martinez-Boubeta, C.; Simeonidis, K.; Serantes, D.; Conde-Leborán, I.; Kazakis, I.; Stefanou, G.; Peña, L.; Galceran, R.; Balcells, L.; Monty, C.; *et al*. Adjustable hyperthermia response of self-assembled ferromagnetic Fe-MgO core-shell nanoparticles by tuning dipole-dipole interactions. *Adv. Funct. Mater.* **2012**, *22*, 3737–3744.

83. Mehdaoui, B.; Tan, R.P.; Meffre, A.; Carrey, J.; Lachaize, S.; Chaudret, B.; Respaud, M. Increase of magnetic hyperthermia efficiency due to dipolar interactions in low-anisotropy magnetic nanoparticles: Theoretical and experimental results. *Phys. Rev. B* **2013**, *87*, doi:10.1103/PhysRevB.87.174419.

84. Hansen, M.F.; Mørup, S. Models for the dynamics of interacting magnetic nanoparticles. *J. Magn. Magn. Mater.* **1998**, *184*, 262–274.

85. Urtizberea, A.; Natividad, E.; Arizaga, A.; Castro, M.; Mediano, A. Specific absorption rates and magnetic properties of ferrofluids with interaction effects at low concentrations. *J. Phys. Chem. C* **2010**, *114*, 4916–4922.

86. Dennis, C.L.; Jackson, A.J.; Borchers, J.A.; Ivkov, R.; Foreman, A.R.; Lau, J.W.; Goernitz, E.; Gruettner, C. The influence of collective behavior on the magnetic and heating properties of iron oxide nanoparticles. *J. Appl. Phys.* **2008**, *103*, doi:10.1063/1.2837647.

87. Mørup, S.; Tronc, E. Superparamagnetic relaxation of weakly interacting particles. *Phys. Rev. Lett.* **1994**, *72*, 3278–3281.

88. Dormann, J.; Fiorani, D.; Tronc, E. On the models for interparticle interactions in nanoparticle assemblies: Comparison with experimental results. *J. Magn. Magn. Mater.* **1999**, *202*, 251–267.

89. Dormann, J.; D'Orazio, F.; Lucari, F.; Tronc, E.; Prené, P.; Jolivet, J.; Fiorani, D.; Cherkaoui, R.; Noguès, M. Thermal variation of the relaxation time of the magnetic moment of γ-Fe$_2$O$_3$ nanoparticles with interparticle interactions of various strengths. *Phys. Rev. B* **1996**, *53*, 14291–14297.

90. Mørup, S.; Hansen, M.H.; Frandsen, C. Magnetic interactions between nanoparticles. *Beilstein J. Nanotechnol.* **2010**, *1*, 182–190.

91. Piñeiro-Redondo, Y.; Bañobre-López, M.; Pardiñas-Blanco, I.; Goya, G.; López-Quintela, M.A.; Rivas, J. The influence of colloidal parameters on the specific power absorption of PAA-coated magnetite nanoparticles. *Nanoscale Res. Lett.* **2011**, *6*, 383–389.

92. Branquinho, L.C.; Carrião, M.S.; Costa, A.S.; Zufelato, N.; Sousa, M.H.; Miotto, R.; Ivkov, R.; Bakuzis, A.F. Effect of magnetic dipolar interactions on nanoparticle heating efficiency: Implications for cancer hyperthermia. *Sci. Rep.* **2013**, *3*, 2887–2896.

93. Hergt, R.; Dutz, S.; Zeisberger, M. Validity limits of the Néel relaxation model of magnetic nanoparticles for hyperthermia. *Nanotechnology* **2010**, *21*, doi:10.1088/0957-4484/21/1/015706.

94. Lima, E.; de Biasi, E.; Mansilla, M.V.; Saleta, M.E.; Granada, M.; Troiani, H.E.; Effenberger, F.B.; Rossi, L.M.; Rechenberg, H.R.; Zysler, R.D. Heat generation in agglomerated ferrite nanoparticles in an alternating magnetic field. *J. Phys. D* **2013**, *46*, doi:10.1088/0022-3727/46/4/045002.

95. Silva, F.G.; Aquino, R.; Tourinho, F.A.; Stepanov, V.I.; Raikher, Y.L.; Perzynski, R.; Depeyrot, J. The role of magnetic interactions in exchange bias properties of MnFe$_2$O$_4$@γ-Fe$_2$O$_3$ core/shell nanoparticles. *J. Phys. D* **2013**, *46*, doi:10.1088/0022-3727/46/28/285003.

96. Tan, R.P.; Carrey, J.; Respaud, M. Magnetic hyperthermia properties of nanoparticles inside lysosomes using kinetic Monte–Carlo simulations: Influence of key parameters, of dipolar interactions and spatial variations of heating power. *Phys. Rev. B* **2014**, *90*, doi:10.1103/PhysRevB.90.214421.

97. Hergt, R.; Dutz, S.; Röder, M. Effects of size distribution on hysteresis losses of magnetic nanoparticles for hyperthermia. *J. Phys. Condens. Matter* **2008**, *20*, doi:10.1088/0953-8984/20/38/385214.

98. Landi, G.T. Simple models for the heating curve in magnetic hyperthermia experiments. *J. Magn. Magn. Mater.* **2013**, *326*, 14–21.

99. Wildeboer, R.R.; Southern, P.; Pankhurst, Q.A. On the reliable measurement of specific absorption rates and intrinsic loss parameters in magnetic hyperthermia materials. *J. Phys. D* **2014**, *47*, doi:10.1088/0022-3727/47/49/495003.

100. Ondeck, C.L.; Habib, A.H.; Ohodnicki, P.; Miller, K.; Sawyer, C.A.; Chaudhary, P.; McHenry, M.E. Theory of magnetic fluid heating with an alternating magnetic field with temperature dependent materials properties for self-regulated heating. *J. Appl. Phys.* **2009**, *105*, doi:10.1063/1.3076043.

101. Bloch, F. Zur Theorie des Ferromagnetismus. *Z. Phys.* **1930**, *61*, 206–219. (In German)

102. Maaz, K.; Mumtaz, A.; Hasanain, S.K.; Bertino, M.F. Temperature dependent coercivity and magnetization of nickel ferrite nanoparticles. *J. Magn. Magn. Mater.* **2010**, *322*, 2199–2202.

103. Della Torre, E.; Bennett, L.H.; Watson, R.E. Extension of the Bloch $T^{3/2}$ law to magnetic nanostructures: Bose-Einstein condensation. *Phys. Rev. Lett.* **2005**, *94*, doi:10.1103/PhysRevLett.94.147210.

104. Senz, V.; Röhlsberger, R.; Bansmann, J.; Leupold, O.; Meiwes-Broer, K.-H. Temperature dependence of the magnetization in Fe islands on W(110): Evidence for spin-wave quantization. *New J. Phys.* **2003**, *5*, doi:10.1088/1367-2630/5/1/347.

105. Hendriksen, P.V.; Linderoth, S.; Lindgard, P.A. Finite-size effects in the magnetic properties of ferromagnetic clusters. *J. Magn. Magn. Mater.* **1992**, *104–107*, 1577–1579.

106. Hendriksen, P.V.; Linderoth, S.; Lindgard, P.A. Finite-size modifications of the magnetic properties of clusters. *Phys. Rev. B* **1993**, *48*, 7259–7273.

107. Linderoth, S.; Balcells, L.; Labarta, A.; Tejada, J.; Hendriksen, P.V.; Sethi, S.A. Magnetization and Mössbauer studies of ultrafine Fe–C particles. *J. Magn. Magn. Mater.* **1993**, *124*, 269–276.

108. Eggeman, A.S.; Petford-Long, A.K.; Dobson, P.J.; Wiggins, J.; Bromwich, T.; Dunin-Borkowski, R.; Kasam, T. Synthesis and characterisation of silica encapsulated cobalt nanoparticles and nanoparticle chains. *J. Magn. Magn. Mater.* **2006**, *301*, 336–342.

109. Alves, C.R.; Aquino, R.; Sousa, M.H.; Rechenberg, H.R.; Goya, G.F.; Tourinho, F.A.; Depeyrot, J. Low temperature experimental investigation of finite-size and surface effects in $CuFe_2O_4$ nanoparticles of ferrofluids. *J. Metastable Nanocryst. Mater.* **2004**, *20–21*, 694–699.

110. Mandal, K.; Mitra, S.; Kumar, P.A. Deviation from Bloch $T^{3/2}$ law in ferrite nanoparticles. *Europhys. Lett.* **2006**, *75*, 618–623.

111. Morup, S. Comment on "Deviation from the Bloch $T^{3/2}$ law in ferrite nanoparticles" by K. Mandal *et al. Europhys. Lett.* **2007**, *77*, 27003–27004.

112. Ortega, D.; Vélez-Fort, E.; García, D.A.; García, R.; Litrán, R.; Barrera-Solano, C.; Ramírez-del-Solar, M.; Domínguez, M. Size and surface effects in the magnetic properties of maghemite and magnetite coated nanoparticles. *Phil. Trans. R. Soc. A* **2010**, *368*, 4407–4418.

113. Gubin, S.P.; Koksharov, Y.A.; Khomutov, G.B.; Yurkov, G.Y. Magnetic nanoparticles: Preparation, structure and properties. *Russ. Chem. Rev.* **2005**, *74*, 489–520.

114. Fonseca, F.; Goya, G.; Jardim, R.; Muccillo, R.; Carreño, N.; Longo, E.; Leite, E. Superparamagnetism and magnetic properties of Ni nanoparticles embedded in SiO_2. *Phys. Rev. B* **2002**, *66*, doi:10.1103/PhysRevB.66.104406.

115. Tronc, E.; Fiorani, D.; Noguès, M.; Testa, A.; Lucari, F.; D'Orazio, F.; Grenèche, J.; Wernsdorfer, W.; Galvez, N.; Chanéac, C.; *et al.* Surface effects in noninteracting and interacting γ-Fe_2O_3 nanoparticles. *J. Magn. Magn. Mater.* **2003**, *262*, 6–14.

116. Tronc, E.; Noguès, M.; Chanéac, C.; Lucari, F.; D'Orazio, F.; Grenèche, J.; Jolivet, J.; Fiorani, D.; Testa, A. Magnetic properties of γ-Fe_2O_3 dispersed particles: Size and matrix effects. *J. Magn. Magn. Mater.* **2004**, *272–276*, 1474–1475.

117. De Julián Fernández, C.; Mattei, G.; Sangregorio, C.; Battaglin, C.; Gatteschi, D.; Mazzoldi, P. Superparamagnetism and coercivity in HCP-Co nanoparticles dispersed in silica matrix. *J. Magn. Magn. Mater.* **2004**, *272–276*, E1235–E1236.

118. De Julián Fernández, C. Influence of the temperature dependence of anisotropy on the magnetic behavior of nanoparticles. *Phys. Rev. B* **2005**, *72*, doi:10.1103/PhysRevB.72.054438.

119. Yoon, S.; Krishnan, K.M. Temperature dependence of magnetic anisotropy constant in manganese ferrite nanoparticles at low temperature. *J. Appl. Phys.* **2011**, *109*, doi:10.1063/1.3563068.

120. Yanes, R.; Chubykalo-Fesenko, O.; Evans, R.F.L.; Chantrell, R.W. Temperature dependence of the effective anisotropies in magnetic nanoparticles with Néel surface anisotropy. *J. Phys. D* **2010**, *43*, doi:10.1088/0022-3727/43/47/474009.

121. Yoon, S. Determination of the temperature dependence of the magnetic anisotropy constant in magnetite nanoparticles. *J. Korean Phys. Soc.* **2011**, *59*, 3069–3073.

122. Andreu, I.; Natividad, E. Accuracy of available methods for quantifying the heat power generation of nanoparticles for magnetic hyperthermia. *Int. J. Hyperth.* **2013**, *29*, 739–751.

123. Martirosyan, K.S. Thermosensitive magnetic nanoparticles for self-controlled hyperthermia cancer treatment. *J. Nanomed. Nanotechol.* **2012**, *3*, 1–2.

124. Apostolova, I.; Wesselinowa, J.M. Possible low-T_C nanoparticles for use in magnetic hyperthermia treatments. *Solid State Commun.* **2009**, *149*, 986–990.

125. Chatterjee, J.; Bettge, M.; Haik, Y.; Chen, C.J. Synthesis and characterization of polymer encapsulated Cu–Ni magnetic nanoparticles for hyperthermia applications. *J. Magn. Magn. Mater.* **2005**, *293*, 303–309.

126. Pollert, E.; Knížek, K.; Maryško, M.; Kašpar, P.; Vasseur, S.; Duguet, E. New-tuned magnetic nanoparticles for self-controlled hyperthermia. *J. Magn. Magn. Mater.* **2007**, *316*, 122–125.

127. Obaidat, I.M.; Mohite, V.; Issa, B.; Tit, N.; Haik, Y. Predicting a major role of surface spins in the magnetic properties of ferrite nanoparticles. *Cryst. Res. Technol.* **2009**, *44*, 489–494.

128. Apostolov, A.T.; Apostolova, I.N.; Wesselinowa, J.M. $MO.Fe_2O_3$ nanoparticles for self-controlled magnetic hyperthermia. *J. Appl. Phys.* **2011**, *109*, doi:10.1063/1.3580476.

129. Hejase, H.; Hayek, S.S.; Qadri, S.; Haik, Y. MnZnFe nanoparticles for self-controlled magnetic hyperthermia. *J. Magn. Magn. Mater.* **2012**, *324*, 3620–3628.

130. Hejase, H.A.; Hayek, S.S.; Qadri, S.M.; Haik, Y.S. Self-controlled hyperthermia characteristics of ZnGdFe nanoparticles. *IEEE Trans. Magn.* **2012**, *48*, 2430–2439.

131. Akin, Y.; Obaidat, I.M.; Issa, B.; Haik, Y. $Ni_{1-x}Cr_x$ alloy for self-controlled magnetic hyperthermia. *Cryst. Res. Technol.* **2009**, *44*, 386–390.

132. Ban, I.; Stergar, J.; Drofenik, M.; Ferk, G.; Makovec, D. Synthesis of chromium-nickel nanoparticles prepared by a microemulsion method and mechanical milling. *Acta Chim. Slov.* **2013**, *60*, 750–755.

Gene Expression, Protein Function and Pathways of *Arabidopsis thaliana* Responding to Silver Nanoparticles in Comparison to Silver Ions, Cold, Salt, Drought, and Heat

Eisa Kohan-Baghkheirati [1,2] **and Jane Geisler-Lee** [1,*]

[1]   Department of Plant Biology, Southern Illinois University Carbondale, Carbondale, IL 62901, USA;
      E-Mail: eisa_kohan@yahoo.com

[2]   Department of Biology, Golestan University, Gorgan 49138-15739, Iran

*   Author to whom correspondence should be addressed; E-Mail: geislerlee@siu.edu

Academic Editor: Robert Tanguay

Abstract: Silver nanoparticles (AgNPs) have been widely used in industry due to their unique physical and chemical properties. However, AgNPs have caused environmental concerns. To understand the risks of AgNPs, *Arabidopsis* microarray data for AgNP, Ag^+, cold, salt, heat and drought stresses were analyzed. Up- and down-regulated genes of more than two-fold expression change were compared, while the encoded proteins of shared and unique genes between stresses were subjected to differential enrichment analyses. AgNPs affected the fewest genes (575) in the *Arabidopsis* genome, followed by Ag^+ (1010), heat (1374), drought (1435), salt (4133) and cold (6536). More genes were up-regulated than down-regulated in AgNPs and Ag^+ (438 and 780, respectively) while cold down-regulated the most genes (4022). Responses to AgNPs were more similar to those of Ag^+ (464 shared genes), cold (202), and salt (163) than to drought (50) or heat (30); the genes in the first four stresses were enriched with 32 PFAM domains and 44 InterPro protein classes. Moreover, 111 genes were unique in AgNPs and they were enriched in three biological functions: response to fungal infection, anion transport, and cell wall/plasma membrane related. Despite shared similarity to Ag^+, cold and salt stresses, AgNPs are a new stressor to *Arabidopsis*.

Keywords: silver nanoparticles; silver ions; abiotic stresses; gene expression; protein functions; pathways; *Arabidopsis thaliana*

1. Introduction

Nanoparticles of 1–100 nm in size [1,2] have been used in different sectors of industry [3]. In 2010, it was reported that 63%–91% of the 260,000–309,000 metric tons of worldwide products containing nanoparticles ended up in landfills while 8%–28% of them went into soil [4]. Of all nanoparticles, silver nanoparticles (AgNPs) have wide and successful applications in clothing, coatings on domestic products, food packaging, pesticides, electronics, photonics, medical drug delivery and biological tagging medicine [5–10].

Human health, food safety and environmental impacts are of prime concern regarding the usage of AgNPs [11–14]. A recent study showed that application of sewage biosolid with a low concentration of 21 ± 17 nm AgNPs (0.14 mg Ag kg^{-1} soil) to a field produced only one third of the original biomass in plants and soil microbes [15]. If AgNPs are released to the environment, they can be taken up and internalized into cells, tissues and systems. AgNPs in human, plant and microbial cells can result in adverse effects, including oxidative stress (imbalance between free radicals and their containments), cytotoxicity and genotoxicity (ability to damage the genetic information within a cell) [14,16–18].

AgNPs are a novel abiotic stressor and an emerging environmental contaminant to plants [19–21]. Uptake and accumulation of AgNPs in root caps and columella cells and transport of AgNPs through intercellular space (*i.e.*, short distance transport) and via vascular tissue (*i.e.*, long distance transport) were reported in *Arabidopsis thaliana* (herein, *Arabidopsis*) [19,22–24]. AgNPs accumulate in the cell walls of *Arabidopsis* and rice (*Oryza sativa* L.) [19,25]. Exposure of roots to AgNPs produced conflicting results, either inhibiting or promoting root growth [26,27]. But a recent study of the effects of AgNPs noted that lateral root initiation and development was promoted after the primary root apical meristem was abolished and the primary root growth was inhibited [22].

The causes of silver nanotoxicity are still in debate. One school of thoughts is that silver ions (Ag^+) are released by AgNPs, causing chemical damage [28,29], while the other school considers the nano size AgNPs cause physical/mechanical damage [19]. Chemical silver specification in plant physiology due to physical nano silver uptake in plant tissue is also considered [30,31]. For example, ethylene is a plant hormone in various stress responses that involve Ag^+. In the presence of such ethylene biosynthesis inhibitors, such as Ag^+ (as silver thiosulfate, $[Ag(S_2O_3)_2]^{3-}$), in the hydroponic nutrient solution, the Fe-deficiency stress responses were inhibited in the roots of cucumber (*Cucumis sativus L. cv Ashley*) [32]. Within plant cells, more AgNPs will pose more physical harm while greater surface area of AgNPs will release more Ag^+ to drive more toxicity. However, a recent expression study in *Arabidopsis* showed that gene expression profiles in AgNP and Ag^+ treatments are shared and thus, concluded phytotoxicity (toxicity to plants) between the two stresses are similar [29].

Plants, being sessile, have adapted to abiotic stresses such as cold, salt, drought and heat. Cellular and molecular responses of plants to these four abiotic stresses have been studied extensively [33–35]. The initial responses to abiotic stresses include a transient increase of cytoplasmic Ca^{2+}, elevated intracellular secondary messengers, such as inositol polyphosphate, reactive oxygen species (ROS, such as oxygen ions and peroxides) and Abscisic acid (ABA, a plant hormone), and increase in mitogen-activated protein kinase (MAPK) pathways [36–41]. The next level of stress response involves regulatory proteins that are directly involved in protection from cellular damage, and up- and down-regulation of stress-specific genes [42,43]. Secondary metabolites are also important for plants in response to abiotic stress. They are involved in structure stabilization, photoprotection, protection from antioxidants and antiradicals, signal transducing, and accumulation of polyamines; some are precursors of plant hormones and contribute to signal transduction of hormones [44–47].

When exposed to abiotic and biotic stresses, plant cell wall is the first mechanical layer of stress perception and plays a dynamic and structural role in plant adaptation [48]. Extracellular peroxidases act as modifiers of cell wall and produce superoxide, hydrogen peroxidase and oxidative burst when encountering stresses [49–51]. Oxidative burst triggers production of ROS, accumulation of phenylpropanoid (a type of secondary metabolites) biosynthesis enzymes, and changes of gene expression in plant defense response [50,52]. Plasmodesmata are pores of 50–60 nm in diameter and connect adjacent neighboring plant cells. Plasmodesmata can carry out trafficking and transport of proteins, mRNAs and small molecules between cells [53]. When plants are in stress, small RNAs are found in plasmodesmata [54,55]. AgNPs were found to aggregate in the cell walls and plasmodesmata in *Arabidopsis* [19] and gold nanoparticles were found to transport through plasmodesmata in poplar [56].

In contrast to commonly known abiotic stresses, the understanding of AgNP stress or silver nanotoxicity in plants is still in its infancy and remains elusive [12,15,19,31]. This study aimed to understand whether AgNP stress is similar to other abiotic stresses in plants. Four well-studied abiotic stresses (cold, salt, drought, heat) and silver ion (Ag^+) stress were comprehensively compared with AgNP stress in *Arabidopsis*. Gene expression, protein function and pathways were used to elucidate similarities and differences in the six abiotic stresses.

2. Results

2.1. Overview of the Affected Genes by the Six Abiotic Stresses

Six sets of publically available microarray data from GEO and Array Express were used. Based on M-values generated from these collective data, the genes with either $M \geq 1$ or $M \leq -1$ were listed separately for the six abiotic stresses (Table S1). The list of differentially expressed genes showed that different number of genes in the *Arabidopsis thaliana* genome were affected by the six different abiotic stresses: between 575 and 6536 genes were differentially expressed, with AgNPs and Ag^+ having the least (575 and 1010, respectively) and cold and salt stresses having the most (6536 and 4133, respectively) numbers of affected genes (Table 1). Drought and heat stresses have similar numbers of affected genes (1435 and 1374, respectively) (Table 1). In addition, cold stress changed the expression of 23.84% of genes (total 6536) in the *Arabidopsis* genome (27416 protein-coding nuclear genes based on the TAIR 10 release) and exhibited a predominantly down-regulating effect on gene expression. In terms of gene numbers in the AgNP, Ag^+ and drought stresses, there were more up-regulated than down-regulated genes. The salt and heat stresses had approximately similar numbers of up- and down-regulated genes. The total number of genes affected by Ag^+ (1010) is more than that by AgNPs (575); however both stresses induced more genes than they suppressed by a 3:1 ratio.

Table 1. The number list of differentially expressed genes that have more than two-fold differences (*i.e.*, $M \geq 1$ or $M \leq -1$) in *Arabidopsis thaliana* affected by six abiotic stresses, silver nanoparticles (AgNPs), silver ions (Ag^+), cold, salt, drought and heat. % [a] = regulated gene number/total affected genes.

Stress	Number of up regulated genes (% [a])	Number of down regulated genes (% [a])	Number of total affected genes	Percentage of total affected genes in genome
AgNPs	439 (76.34)	136 (23.65)	575	2.10
Ag^+	780 (77.22)	230 (22.77)	1010	3.68
Cold	2514 (38.46)	4022 (61.54)	6536	23.84
Salt	2057 (49.77)	2076 (50.23)	4133	15.08
Drought	814 (56.72)	621 (43.28)	1435	5.23
Heat	694 (50.50)	680 (49.50)	1374	5.01

Overviews of metabolic/regulatory pathway and cellular compartments were displayed for all the expressed genes in the six abiotic stresses in Figure 1. The displays allowed the first glimpse of global comparison among the six abiotic stresses: no stresses shared identical expression patterns. In the six stresses, cold stress mainly suppressed the genes in major primary and secondary metabolism (Figure 1B); salt induced the genes in both primary and secondary metabolism (Figure 1F). Drought and heat stresses showed differential patterns though shared some similarity (Figure 1C,E); heat also induced more genes than drought in both primary and secondary metabolism. AgNP and Ag^+ stresses exhibited a similar, but not identical, pattern (Figure 1A,D). Moreover, Ag^+ suppressed more genes in photosynthesis and sugar metabolism than AgNPs did, while AgNPs induced more genes in cell wall biosynthesis than Ag^+.

More than 30 metabolic/regulatory pathways and cell compartments were compared to further understand the differences and similarities in the differential gene expression patterns between AgNP and Ag^+ stresses (Figure S1). Reactive oxygen species (ROS) associated genes were up-regulated by both AgNPs and Ag^+; this agreed with previous results [57–60]. Although it has been reported that DNA repair might be involved in the AgNP stress in animal and human cell culture studies [61–64], there was no difference in this *Arabidopsis* study (Figure S1). In the secondary metabolism, AgNPs demonstrated more up-regulated genes of lignin and lignans than Ag^+. In nitrogen metabolism, nitrate reductase gene was up-regulated in the Ag^+ stress; this was probably due to the source of NO_3^- from $AgNO_3$. Some ion transport genes were up-regulated in the AgNP stress but not present in Ag^+ as shown in transport overview (Figure S1). The genes of sulfate (SO_4^{2-}) carbonic anhydrase pathway were up-regulated (*i.e.*, induced) by AgNPs but not by Ag^+ (Figure S1).

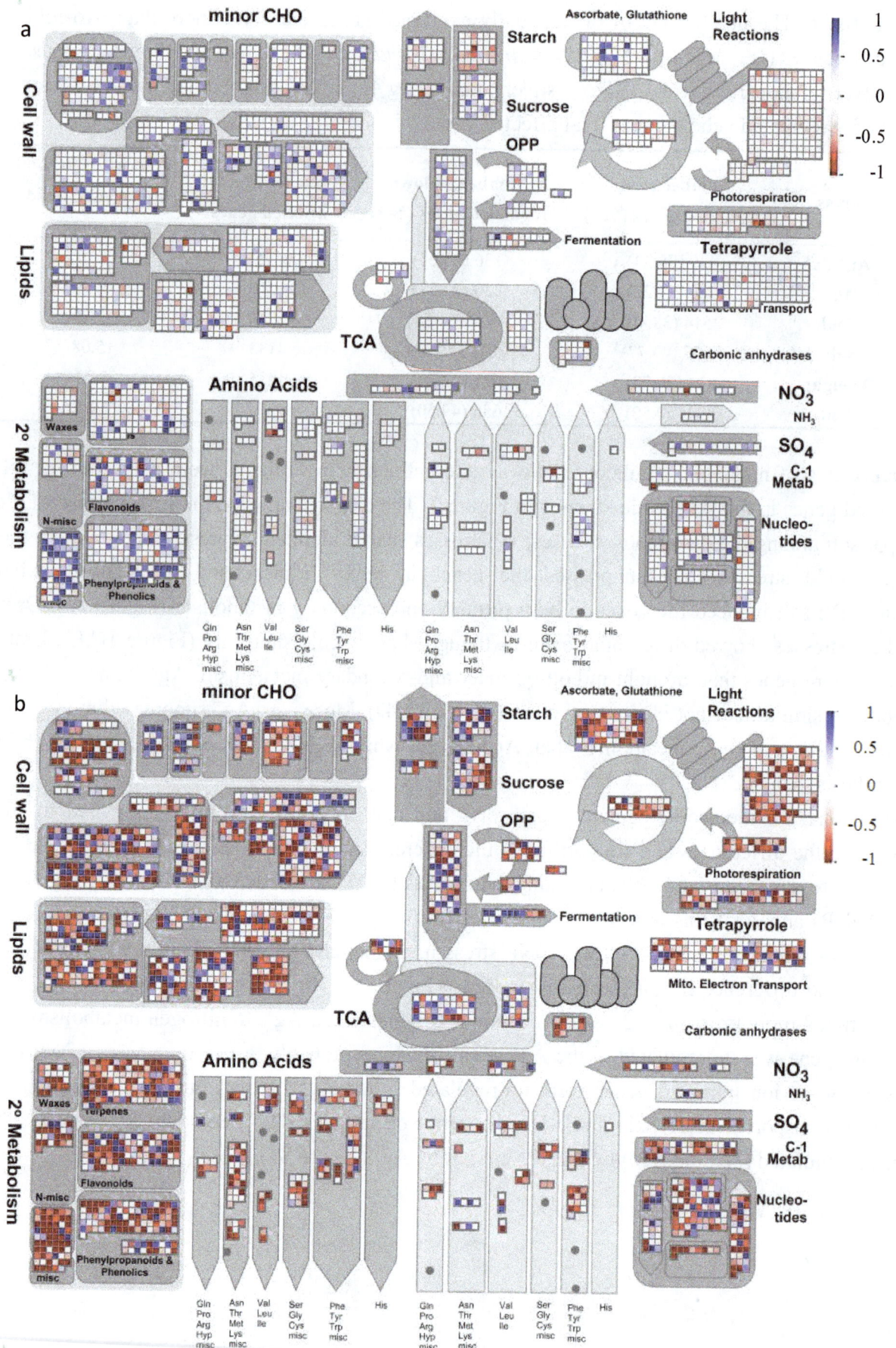

Figure 1. *Cont.*

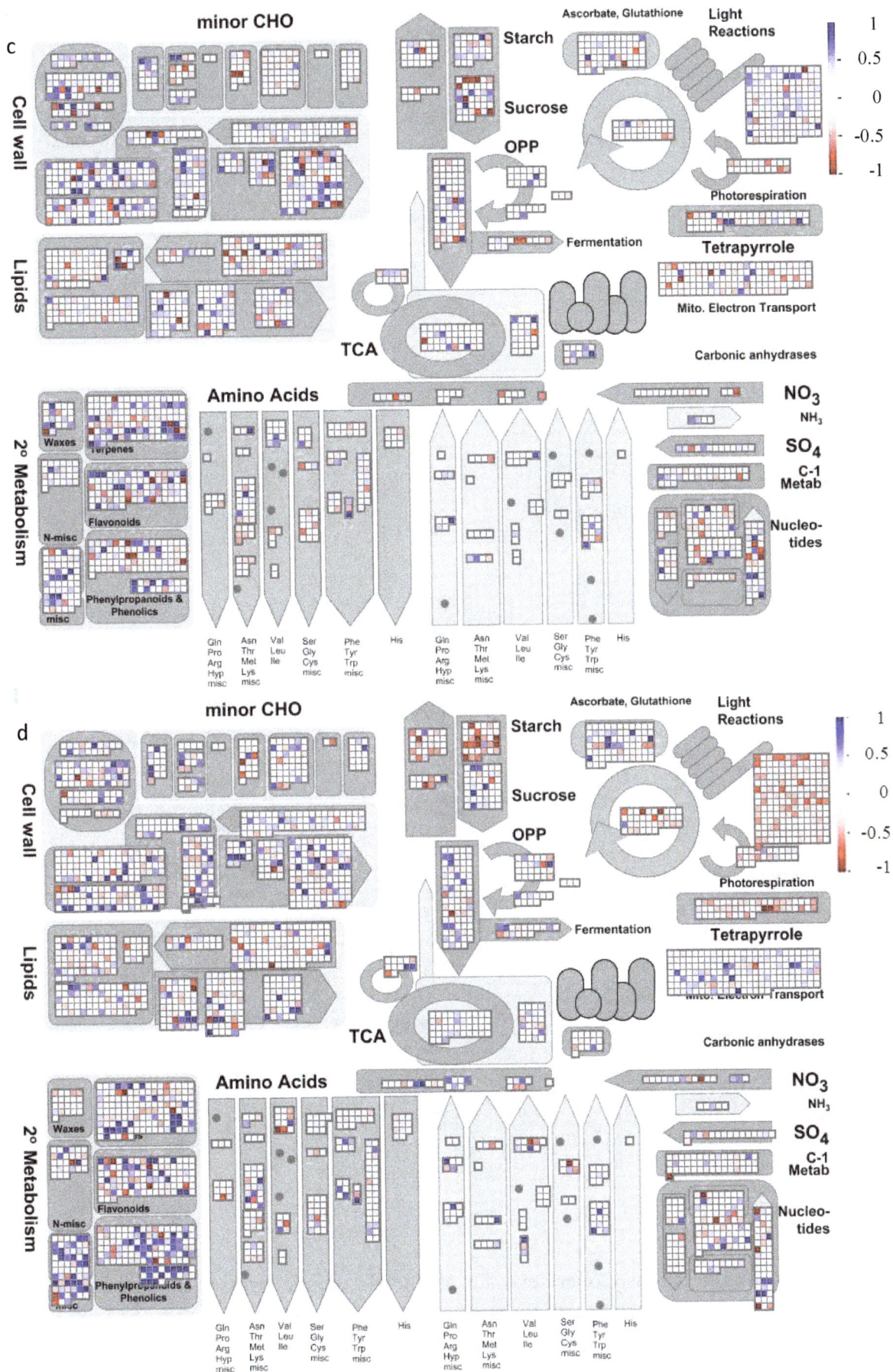

Figure 1. *Cont.*

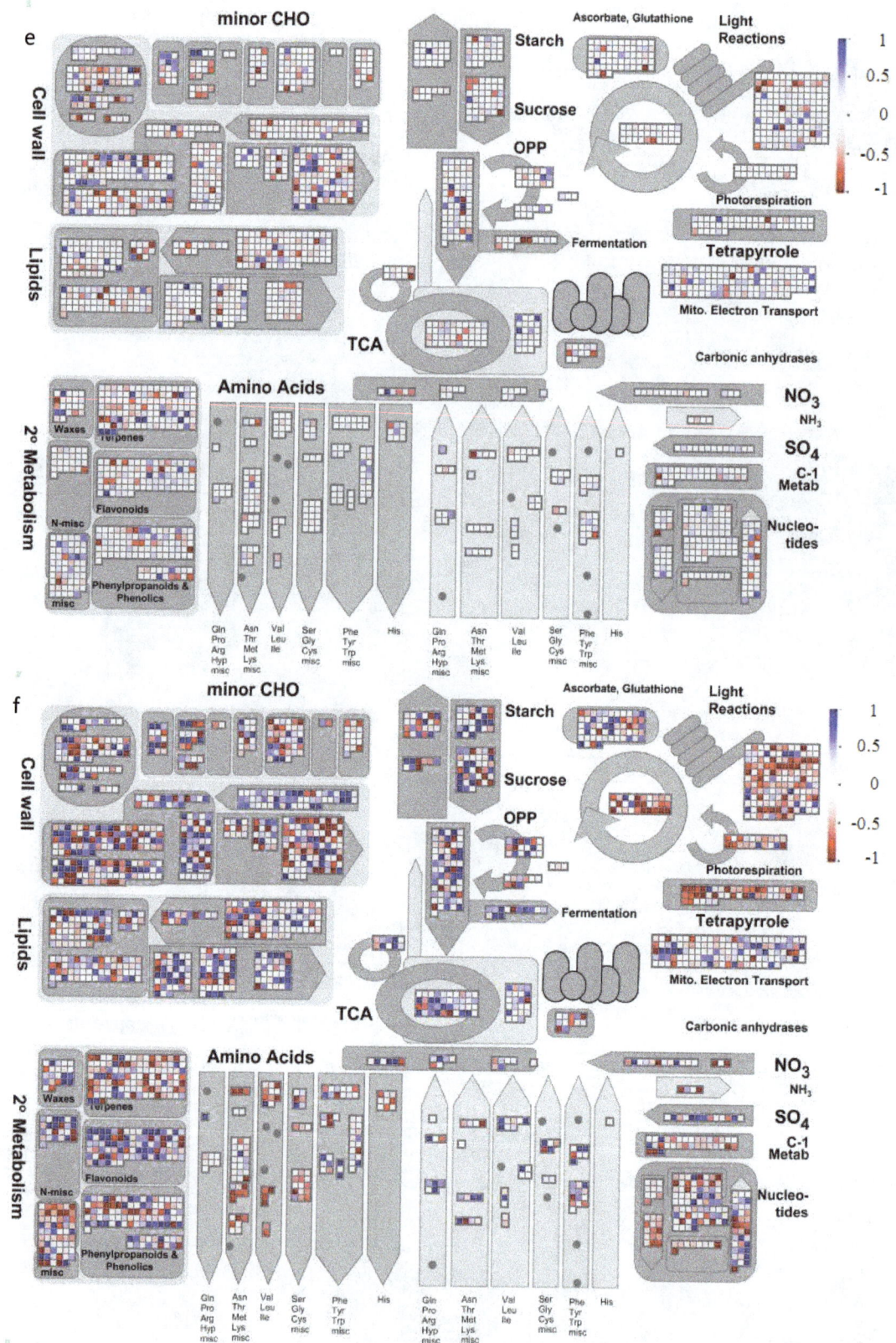

Figure 1. Metabolic pathway overviews for all the six abiotic stresses. *M*-value data in Table S1 for all the identified *Arabidopsis* genes were used to display in MapMan Image Annotator. Two color scale schemes were used; blue was to denote genes that were induced and red was to denote genes that were suppressed by (**a**) AgNPs; (**b**) Cold; (**c**) Drought; (**d**) Ag⁺ (AgNO₃); (**e**) Heat; and (**f**) Salt.

2.2. Gene Ontology Term Enrichment

No difference was found in gene ontology (GO) term enrichments of the total up-regulated genes by AgNP and Ag$^+$ stresses (Figure 2). In addition, there was no enrichment for the down-regulated genes by AgNP and/or Ag$^+$ stresses. The up-regulated genes in both stresses were enriched in lipid transport and transition metal ion in the category of biological process, peroxidase activity in the category of molecular function, and extracellular region in the category of cellular component (Figure 2A–C; Table S2).

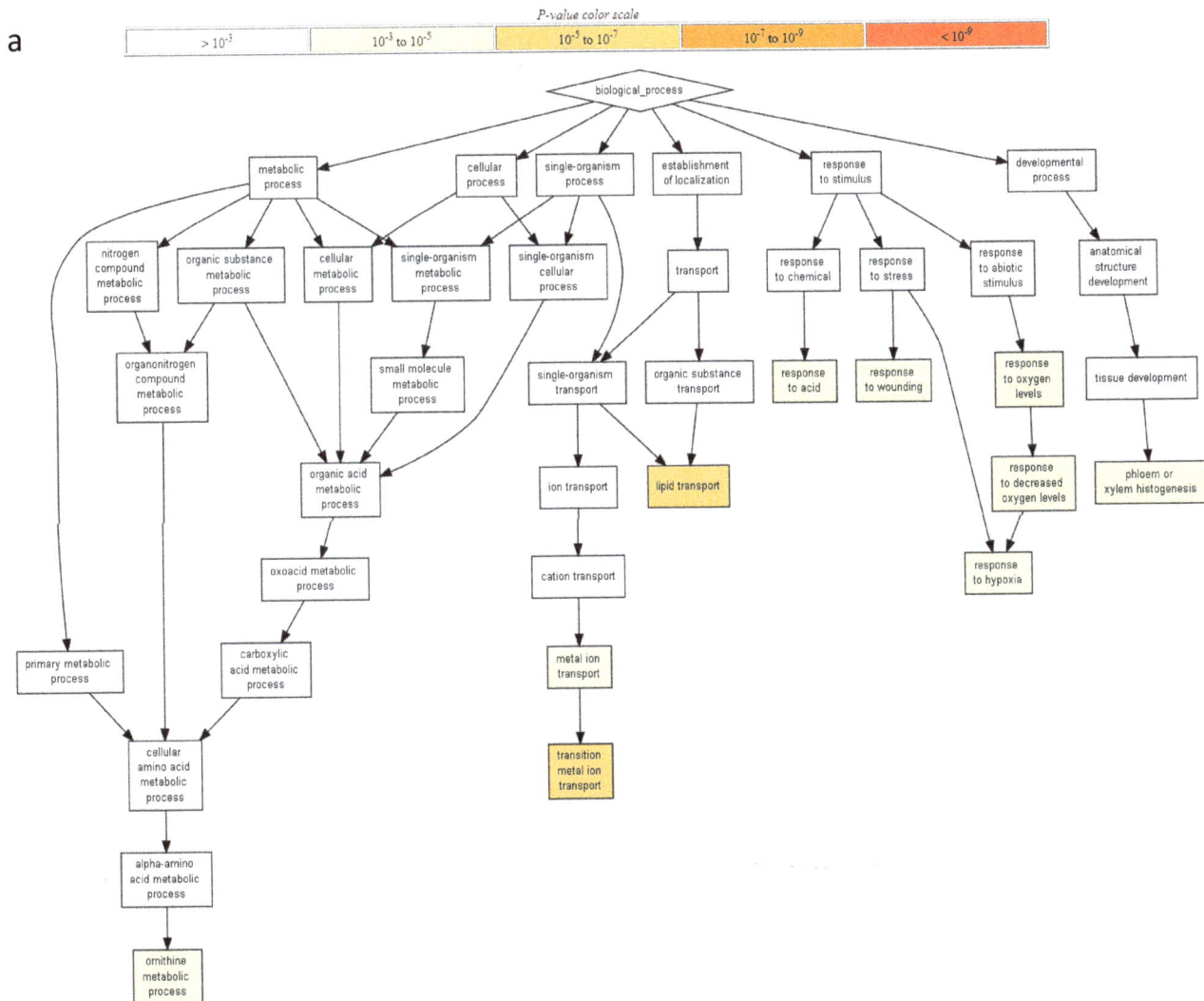

Figure 2. *Cont.*

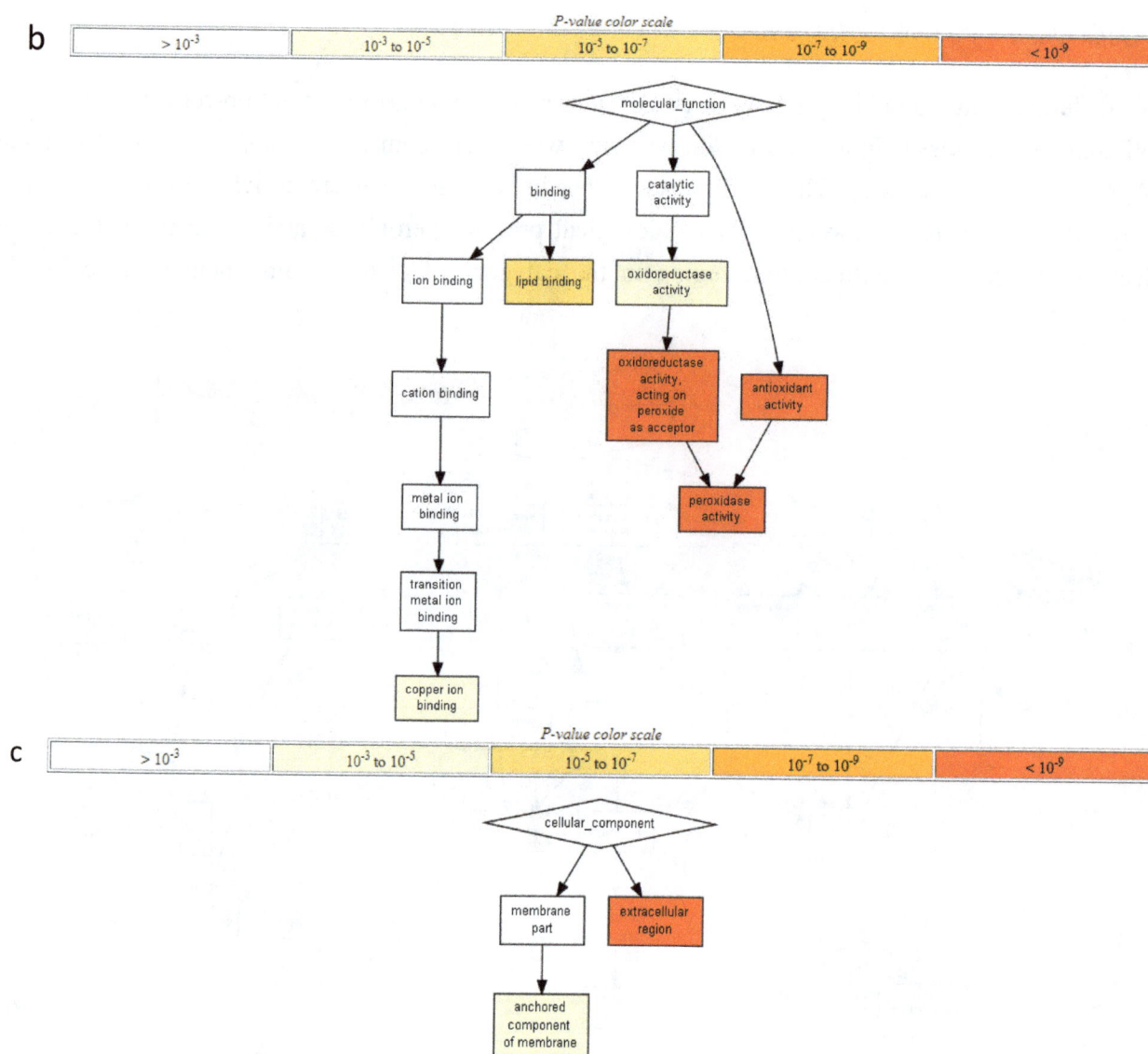

Figure 2. Gene Ontology (GO) term enrichment based on GO terms for total up-regulated genes by AgNP and Ag^+ stresses. GO term enrichment results for (**a**) Biological processes; (**b**) Molecular function; and (**c**) Cellular components were presented. All colored boxes are enriched with q-value (FDR) less than 0.05 ($q < 0.05$) and the density of color shows the degree of enrichment, *i.e.*, red (p-value $< 10^{-9}$), dark orange (p-value 10^{-7} to 10^{-9}), orange (p-value 10^{-5} to 10^{-7}), yellow (p-value 10^{-3} to 10^{-5}) and white (p-value $> 10^{-3}$).

To understand the similarities in AgNP and Ag^+ stresses, GO term enrichment analysis of the shared genes in both the stresses was compared. The Venn diagram data showed a total of 464 genes were shared by AgNP and Ag^+ stresses (Figure 3A; these genes are listed in Table S3). These genes were enriched in lipid transport (GO:0006869) and transition metal ion transport II (GO:0000041) in the category of biological process (Figure S2A); antioxidant activity (GO:0016684) and peroxidase activity (GO:0004601) in the category of molecular function (Figure S2B); the extracellular regions in the category of cellular components (GO:0005576) (Figure S2C).

To understand the differences in AgNP and Ag^+ stresses, GO enrichment analysis was compared for the specific genes in either AgNP or Ag^+ stress. A total of 546 Ag^+-specific genes (Figure 3A, listed in

Table S3) were enriched for more than 30 biological processes (Figure S3A). For example, nitrate transport (GO:0015706), transition metal ion transport I (GO:0000041), response to nitrate (GO:0010167). In contrast, 111 AgNP-specific genes (Figure 3A, listed in Table S3) were slightly enriched for only one biological process, response to fungus (GO:0009620) (Figure S3B).

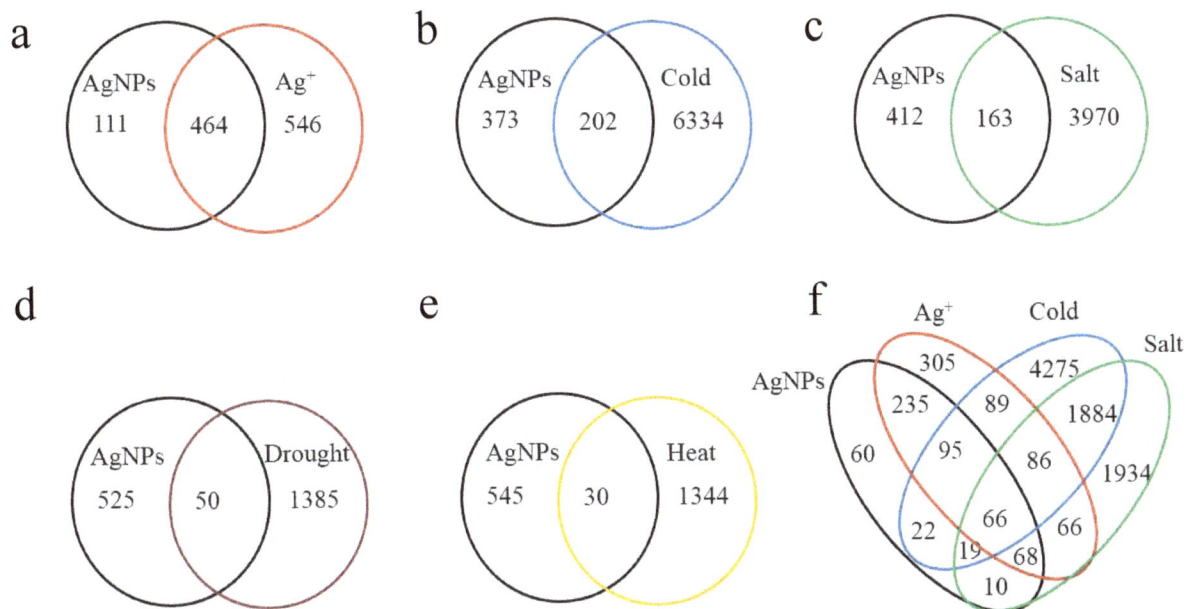

Figure 3. Venn diagrams of the genes with more than two fold expression changes and shared among the six stresses. (**a**–**e**) were two way comparison and (**f**) was four way. Overlapped areas were shared genes while non-overlapped areas were specific/unique genes for individual stress. (**a**) Between AgNPs and Ag$^+$; (**b**) Between AgNPs and cold; (**c**) Between AgNPs and salt; (**d**) Between AgNPs and drought; (**e**) Between AgNPs and heat; (**f**) Among AgNPs, Ag$^+$, cold and salt.

2.3. Protein Domain Enrichment

Protein domains curated by PFAM are categorized based on the similarity of global sequence alignments [65,66]. The coded proteins of induced and suppressed genes by the six abiotic stresses were subjected to PFAM protein domain enrichment analysis. A total of 32 uniquely enriched PFAM protein domains were identified across the four abiotic stresses, cold, salt, AgNPs, Ag$^+$ (see Table S2). This implies that these four stresses differ from the other two stresses, drought and heat. Four enriched domains, PF01419:Jacalin, PF00141:peroxidase, PF00234:Tryp_alpha_amyl and PF00067:p450, were shared in AgNP and/or Ag$^+$ stress. PF01419:Jacalin, Jacalin-like lectin domain, is a mannose/galactose-binding lectin domain with three beta-sheets [67,68]. Jacalin-like lectin domain containing proteins include Jacalin, which is seed lectin and agglutinin from jackfruit (*Artocarpus heterophyllus*) [69]. The peroxidases containing PF00141:peroxidase domain use hydrogen peroxide (H$_2$O$_2$) to accept electrons and produce water when catalyzing oxidative reactions [70]. One class of plant-specific peroxidases is involved in tissue-specific reactions; two of their notable reactions are ethylene production and defense against wounding [71]. The proteins containing PF00234:Tryp_alpha_amyl domain is a group of plant lipid transfer proteins (LTPs) and is involved in plant defense mechanisms [72,73]. LTPs transfer lipids in membranes. The proteins containing the PF00067:p450 domain belong to a superfamily of cytochrome p450 (p450), which catalyze the final reactions [RH + O_2 + NADPH + H^+ → ROH + H_2O + $NADP^+$] in biological electron transfer chains [74]. Plant p450s are involved in diverse reactions, especially in plant defense and secondary metabolite production [75–77]. Among these four enriched domains, the genes to encode the proteins containing PF00067:p450 domains were also associated with down-regulated genes by cold [78,79]. In addition, PF03106:WRKY and PF00847:AP2 were shared by the upregulated protein-encoding genes in Ag^+ and salt stresses and in salt and cold stresses, respectively. PF03106:WRKY domains belong to DNA-binding transcription factors which are one of the largest signaling/regulatory protein families in plants [80,81]. WRKYs could integrate with such signaling cascades as mitogen-activated protein kinase (MAPK), MAPK kinases and defense proteins. The proteins containing the PF00847:AP2 domain are transcription factors Apetala 2 in the large family of AP2/EREBP [82]. EREBP is ethylene-responsive element binding protein. It implied that the signaling pathways in Ag^+ and salt stresses are involved in ethylene and WRKY transcription factors. The 32 enriched protein domains with their related stresses could be visualized in Cytoscape in Figure 4A.

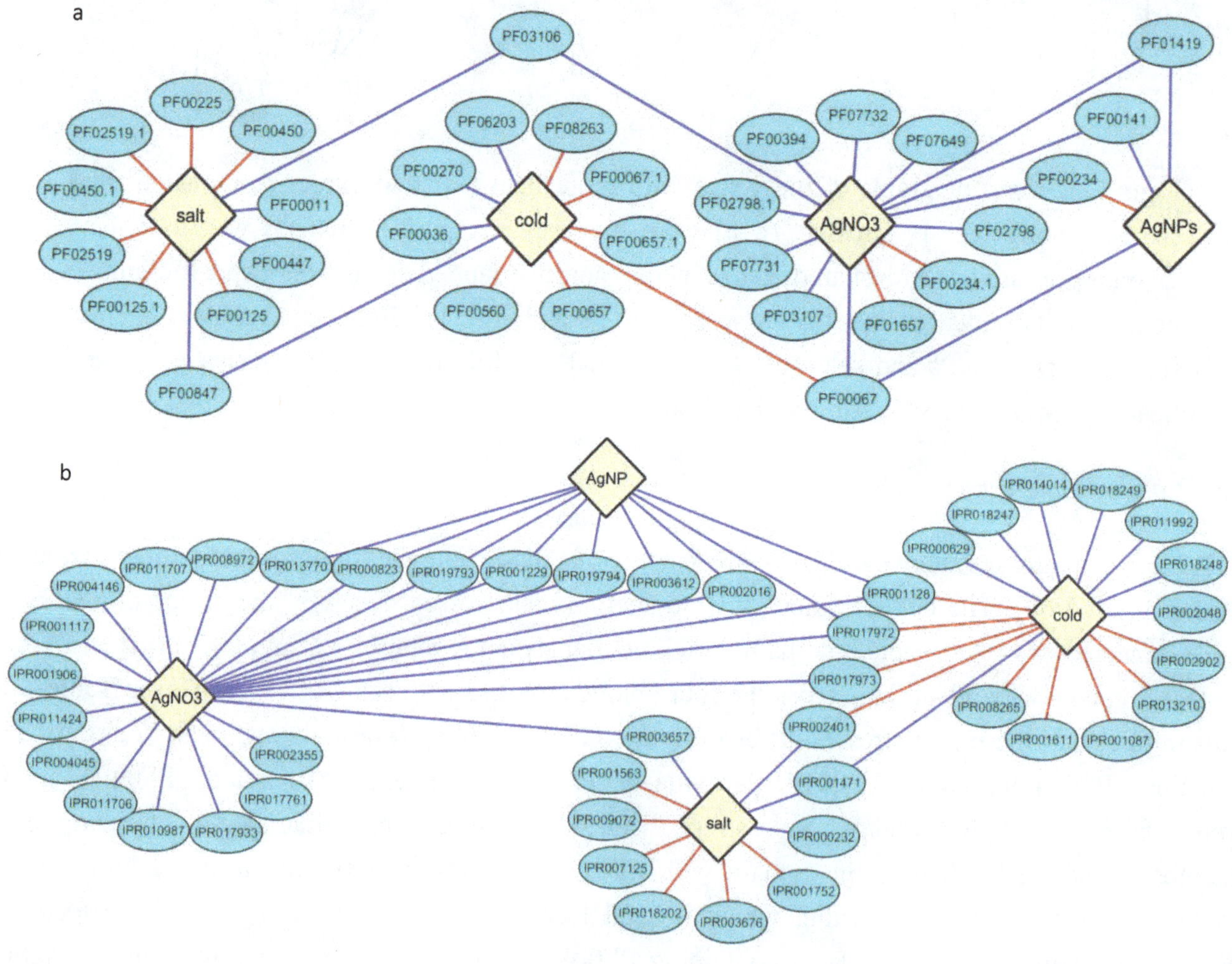

Figure 4. *Cont.*

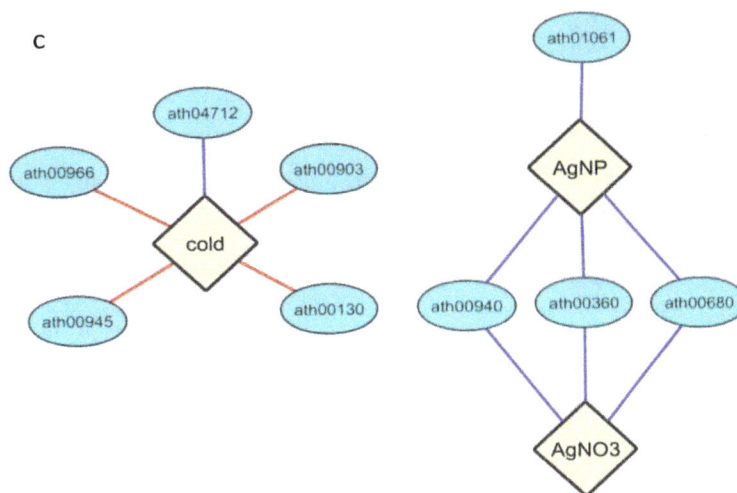

Figure 4. PFAM protein domain, InterPro protein class, and KEGG pathway enrichment of the genes with more than two fold expression changes for all the six abiotic stresses. (**a**) Visualization of 32 unique enriched PFAM protein domain across the six abiotic stresses; (**b**) Display of 44 definite enriched InterPro classes associated with the six stresses; (**c**) Nine enriched KEGG pathways in six stresses were shown. The enrichment results were visualized using Cytoscape 3.1.0, where blue edges denote enrichment for up-regulated genes and red edges denote enrichment for down-regulated genes. The description of PFAM protein domain, InterPro protein class, and KEGG pathway were in Tables S2, S4 and S5.

2.4. Enrichment of InterPro Protein Classes

InterPro [83,84] classifies proteins into families and predicts domains and reaction sites by providing functional analysis of proteins [85,86]. InterPro classified protein (herein, InterPro protein class) enrichment was based on predictive models as protein signatures, which were annotated in the InterPro database. There would be a similarity between PFAM protein domain analysis and InterPro protein class analysis; but the enrichment analysis by the latter could provide more specific data about interested proteins, due to protein signatures. No enriched InterPro protein classes were found related to drought and heat; this result matched PFAM protein domain enrichment (Tables S2 and S4). Among the four abiotic stresses studied (cold, salt, AgNPs, Ag^+), forty-four definite InterPro protein classes were found associated with one or multiple stresses (Figure 4B and Table S4). Both the enrichment analyses of PFAM protein domains and InterPro protein classes demonstrated that AgNP stress induced more peroxidase (including domain, signature and function) encoding genes than Ag^+ stress did. For example, IPR000823:Plant peroxidase; IPR002016:Heme peroxidase, plant/fungal/bacterial; IPR019794:Peroxidase, active site; PF00141:peroxidase in PFAM protein domain.

Between PFAM domain and InterPro protein class enrichment analyses, as predicted, an overall similarity was found in the four abiotic stresses (AgNPs, Ag^+, cold, salt; Figure 4A,B). However, there were two major differences in these two enrichment analyses. The first difference was differential occurrences of p450 domain-containing proteins in the four stresses (Figure 4B and Table S4). Based on the PFAM domain enrichment (Figure 4A), only PF00067:p450 domain was associated with the up-regulated genes by cold, and the down-regulated genes by AgNPs and/or Ag^+. But the InterPro protein class enrichment presented differential results in four different classes of p450s. IPR002401:Cytochrome p450 (E-class, group I) was shared between the up-regulated genes by salt stress and down-regulated genes by cold. IPR017973:Cytochrome p450 (C-terminal region) was shared between the up-regulated genes by Ag^+ and the down-regulated genes by cold. Two other InterPro protein classes, IPR001128:Cytochrome p450 and IPR017972:Cytochrome p450 conserved site, were shared between the up-regulated genes by cold and down-regulated genes by AgNPs and/or Ag^+ (Figure 4B and Table S4).

The second difference in the two protein enrichment analyses was lipid transfer proteins (LTPs) (Figure 4A,B). LTPs shuttle phospholipids and other fatty acid groups between cell membranes to build cell walls [72]. Phospholipids are major components in cell membrane, including inositol phosphate (InoP). Despite the fact that several LTPs were shown in both the enrichment analyses, PF00234 was only in the PFAM protein domain analysis but was not in the InterPro protein class analysis (Figure 4A,B). PF00234 is protease inhibitor/seed storage/LTP family domain [87].

2.5. Enrichment within KEGG Pathways

Kyoto Encyclopedia of Genes and Genomes (KEGG) annotation was used to show biological pathway enrichment of up- and down-regulated genes of the six abiotic stresses. The connectivity of each pathway related to the six stresses studied was displayed in Figure 4C and listed in Table S5. Nine unique KEGG pathways in *Arabidopsis* (*i.e.*, prefix with "ath") were found in differentially expressed genes induced by AgNP, Ag^+ and cold stresses; but no enriched KEGG pathway was found by salt, drought, and heat stresses. These nine KEGG pathways were characterized into three groups, (1) five for cold stress; (2) three for both AgNP and Ag^+ stresses; and (3) one for only AgNPs, which was ath01061:Biosynthesis of phenylpropanoids.

The three shared KEGG pathways between AgNP and Ag^+ stresses-regulated genes were in secondary metabolism and methane metabolism. Ath00360:Phenylalanine metabolism is involved in metabolism of terpenoids and polyketides. The pathway of Ath00940:Phenylpropanoid biosynthesis starts with phenylalanine and produces a variety of secondary metabolites as precursors for signaling (such as phenolic volatiles, coumarin, flavonoids) and structure (such as lignin, suberin, wall-bound phenolics) [88,89]. Ath00680:Methane metabolism can reduce $NADP^+$ to NADPH and convert glycine to serine.

The single KEGG pathway of ath01061:Biosynthesis of phenylpropanoids was enriched in the only AgNP up-regulated genes. The ath01061 pathway starts with the products of primary metabolism (*i.e.*, glycolysis and the tricarboxylic acid cycle/the Krebs cycle) and ends up phenylpropanoids [88–91]. Phenylpropanoids are precursors to diverse secondary metabolites, such as tannins, lignans and flavonoids.

2.6. Comparison of Shared and Specific Genes among Six Abiotic Stresses

Figure 3 showed that the number of the shared genes between two stresses (AgNP *vs.* Ag^+, AgNP *vs.* cold, AgNP *vs.* salt, AgNP *vs.* drought, AgNP *vs.* heat) and among four stresses (cold, salt, AgNP and Ag^+). The gene number shared between AgNP and Ag^+ stresses (464) was much higher than those between AgNP and with the other four stresses (202, 163, 50, and 30, respectively). The high number of shared genes might partially attribute to the potential release of silver ion (Ag^+) from AgNPs [29,92]. Nevertheless, 111 genes were AgNP-specific but not Ag^+-specific (Figure 3A). This may be in agreement with our previous publication that indicated the effects of AgNPs were different from Ag^+ [19,31].

Among the other four abiotic stresses (cold, salt, drought, heat) studied, AgNP stress shared the most genes affected with cold, followed by salt, then drought and finally, heat (Figure 3B–E). In the category of biological processes, gene ontology (GO) term enrichment for AgNP-cold shared genes were involved in response to acid (GO:0001101), and in response to oxygen containing compounds (GO:1901700); in the category of molecular functions, involved in catalytic activity; and in the category of cellular components, involved in extracellular region (Figure S4). Based on the GO enrichment analysis, the similarity of AgNP and cold stresses may be due to their mechanical damages on membrane/cell wall and induction of oxidative stress [93–95]. The four-way Venn diagram showed that sixty-six genes were shared in response to AgNP, Ag^+, cold and salt stresses. These 66 genes were enriched in response to oxygen containing compounds and regulation of reactive oxygen species (ROS) metabolism processes (GO:2000377) (Figure S5). Shared genes across three, four or six different stresses were also provided in Figure S6 and Table S6.

There were another 60 genes specific to only AgNP stress but not to Ag^+, cold, salt, and even to drought and heat (Table S7). These genes were enriched in ion transport process, especially anion transport (GO:0006820). This implies that *Arabidopsis* plants in the AgNP stress may have utilized anion transporters to maintain ion homeostasis (or charge equilibrium) from unknown mechanism(s) induced by AgNPs. The release of Ag^+ by AgNPs cannot explain this phenomenon.

Only four genes (At5g10040, At4g17470, At1g01130, and At1g69500) were shared by all the six abiotic stresses—AgNP, Ag^+, cold, salt, drought, and heat stresses (Table S6). At5g10040 encodes one unknown protein involved in anaerobic respiration; At1g01130 one unknown calcium/calmodulin-dependent protein kinase-like [96]; At1g69500 cytochrome p450 [97]; and At4g17470 alpha/beta-hydrolases superfamily protein involved in changes in the endoplasmic reticulum lipid properties when experiencing low temperature [72].

2.7. Protein-Protein Interaction Networks of Affected Genes by Six Abiotic Stresses

The protein-protein interaction (PPi) networks of the affected encoding genes (*i.e.*, *M*-value ≥ 1 or ≤ -1) for all the six stresses were created (Figure 5). The PPi network of the cold stress was most densely connected, followed by that of the salt stress. The other four stresses showed sparsely connected with few protein hubs. The PPi network of the cold stress included 6536 gene-encoded proteins and they could build the biggest network (among the six stresses) with an average connectivity of 1.94269 (Table S8). An average connectivity of more than 1 indicates that the number of edges (*i.e.*, interactions) is more than the number of nodes in the network. This means each protein has averaged more than one connection with other proteins.

A PPi network was also created for shared and specific gene-encoded proteins that were induced/suppressed by the stress of AgNPs and/or Ag^+ (Figure S7A; Table S9). This network contained 368 nodes (derived from 1127 AgNP and/or Ag^+ affected genes) and 129 edges. The majority of the nodes (70%) had no connectivity. But in the AgNP stress, there was a major hub of heptahelical transmembrane protein (HHP2; encoded by At4g30850) (Figure S7B). HHP2 was reported to be involved in membrane transport [98,99]. At4g30850 gene was up-regulated only by the AgNP stress but not by the other five stresses. Interestingly, the hub of HHP2 had 20 edges (*i.e.*, interactions) with Ag^+-stress specific gene-encoded proteins and 11 edges with AgNP and Ag^+ shared gene-encoded proteins. Some of the edges include such transporters as ABC transporter family proteins, oligopeptides transporter, nucleotide/sugar transporter family protein and copper transporter. The list of HHP2 connected nodes and their connectivity's were presented in Table S10.

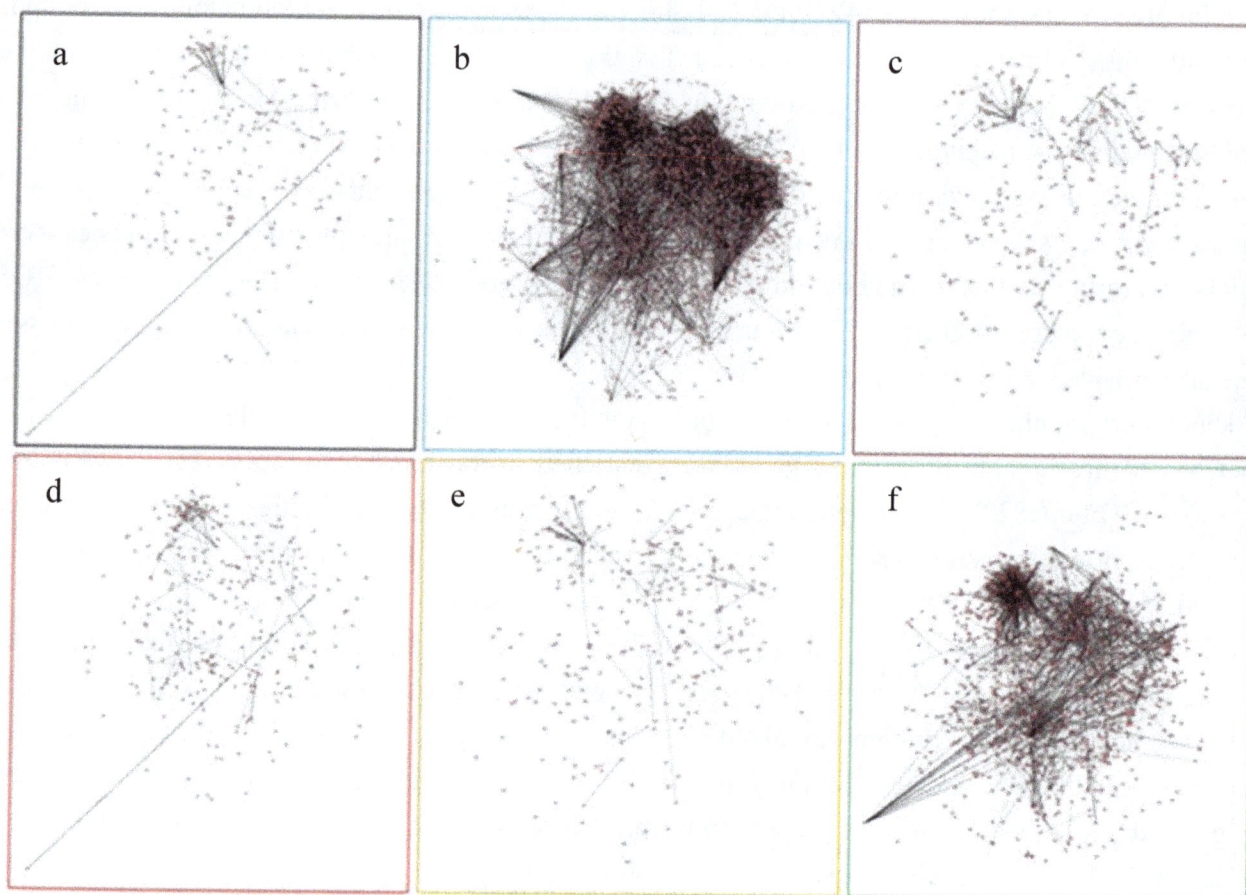

Figure 5. Protein-protein interaction (PPi) networks of affected genes (*i.e.*, *M*-value ≥ 1 or ≤ -1) for all the six stresses. (**a**) AgNPs; (**b**) Cold; (**c**) Drought; (**d**) Ag^+; (**e**) Heat; (**f**) Salt. Nodes represented proteins and edges showed the interaction between proteins.

3. Discussion

3.1. Similarities and Differences of AgNP Stress and Five Other Abiotic Stresses

Plants respond to abiotic and biotic stresses by changing their gene expression and metabolism in order to adapt the stresses [100,101]. *Arabidopsis* plants responded to cold and salt stresses by changing expression of large numbers of genes, 23.84% and 15.08% of their genome; however, AgNP stress did by only 2.10%, the lowest of all stresses examined (Table 1). It implies *Arabidopsis* plants has a much reduced response to AgNP stress by up-regulating/down-regulating fewer genes and producing/decreasing less of their encoded products than the other five stresses (Ag^+, cold, salt, drought and heat). This indicates that AgNPs are a new different stressor for *Arabidopsis* plants and in different plants and crop species [19,26,31,102]. However, the genetic differences elucidated in this study could be qualitative results that cannot be statistically evaluated nor in consideration of gene interactions.

Some of abiotic and biotic stresses trigger reaction oxygen species (ROS) responses [36,37,103]. The ROS reaction cascade triggered by stresses occurs in the membrane of the plant cells by generation of such secondary messengers as calcium and ROS, and then follows by phosphorylation of downstream proteins. This study showed that ROS-regulated genes (shown in Figure S5) and GO:2000377 were shared by AgNP, Ag^+, cold, and salt stresses. Although there is no direct evidence of secondary messenger calcium accumulation nor AgNP receptors found in *Arabidopsis* cell membranes, several studies reported induction of ROS in plants exposed to AgNPs [59,104]. In addition, the enrichment of antioxidant activity for the genes affected by AgNPs (Table S11) was in agreement with those studies. Upon the increase of ROS against stress, plants also produce antioxidants to remove ROS from damaging cells [38]. At the same time, ROS are also intermediate signals (*i.e.*, secondary messengers) to induce Abscisic acid (ABA) and calcium cascade [105]. ABA regulates approximate 10% of protein-coding genes in the *Arabidopsis* genome, the highest percentage among all the plant hormones [106]. Animal and human cell line studies showed generation of ROS and use of mitogen-activated protein kinase (MAPK) pathway to transduce signals of AgNPs [61,107]. Plants may also utilize oxidative stress signaling for AgNPs by using MAPK cascade modules.

3.2. Similarity and Difference of AgNP and Ag^+ Stresses

This study showed no major difference in the enrichment analyses of GO term, PFAM protein domain, InterPro protein classification, and KEGG pathways of AgNP and Ag^+ affected genes (Figure 2, Tables S2, S4, S5 and S11). However, most of their enrichments were related to oxygen level and ROS, which are also regulated by cold and salt stresses. Enrichment of ornithine metabolism process for AgNP/Ag^+ affected genes illustrated that AgNP and Ag^+ induced osmotic stress, which consequently changed ornithine metabolism to synthesize more osmolytes, such as polyamines and proline [108,109]. Osmotic stress is a rapid change in the solute concentration around a cell.

Another considerable enrichment in the both AgNP and Ag^+ stresses was for phloem or xylem histogenesis. This enrichment could be related to inhibition of primary root growth by AgNPs or Ag^+; thus, it implies possible production of lateral roots [21,22,25,27]. It was reported that AgNPs inhibited root growth by directly destroying meristematic cells (able to divide) in root apical meristem (RAM) [19] and indirectly promoted lateral root growth in *Arabidopsis* [22]. Although some lateral root primordia were destroyed by AgNPs, the others could have survived to take over the responsibility of nutrient and water uptake from primary roots [22]. Ag^+ (of $AgNO_3$) improved rooting of vanilla (*Vanilla planifolia*) explants [110]. Nevertheless, improved root growth by Ag^+ cannot explain why *Arabidopsis* RAM was abolished by AgNPs.

On the other hand, the phytotoxicity of AgNPs has been shown to be much worse than their released Ag^+ [19,31]. AgNPs could contribute their toxicity in both the nanoparticles themselves (*i.e.*, physical nano size) and their dissolved and released Ag^+ to their surroundings (*i.e.*, chemical Ag^+ factor) [14,19,31]. The *Arabidopsis* root phenotypes in AgNP stress differ from those in the identical concentrations of the released Ag^+ by AgNPs. In addition, AgNPs presented size- and concentration-dependent toxicity [19,63]. Any study using only Ag^+ (e.g., an $AgNO_3$ solution) could not answer size-dependent toxicity of AgNPs.

Cationic (or positive-charged) nanoparticles can pass through cell membranes by creating transitory holes in membranes [111]. This process, thus, induces cytotoxicity. If Ag^+ could penetrate plasma membrane fast, then cytotoxicity would be severe. Thus, it is hypothesized that fast penetration of Ag^+ across plasma membrane could affect photosynthetic electron transport and slow down primary metabolic pathways sooner [112,113]. Once primary metabolic pathways were slowed down, affected genes would be up- and down-regulated to allow plants to adapt into their Ag^+ stress.

GO term enrichment analysis presented unique differences between AgNP and Ag^+ stresses (Figure S3). The genes specifically regulated by Ag^+ were enriched for response to nitrate and related processes. This probably attributed to the addition of NO_3^- (in $AgNO_3$), a by-product of Ag^+ stress. Enrichment of nitrate related metabolism pathways could be corresponding with Ag^+ mediated responses such as in polyamine biosynthesis, ethylene- and calcium-mediated pathways [114]. PPi networks of the AgNP and Ag^+ affected genes-encoded proteins were similar to each other (Figure 5A,D, respectively); but the network of Ag^+ stress has slightly more connectivity than that of AgNP stress.

3.3. Comparison of AgNP and Cold Stresses

Cold stress changed the expression of approximately a quarter of total genes in the *Arabidopsis* genome and exhibited a predominantly suppressive effect on gene expression and most metabolic pathways (Table 1, Figure 1B). Based on the Venn diagram analysis among the four abiotic stresses (cold, drought, heat, salt) studied, AgNP stress shared the most genes affected by cold (Figure 3B), up to 35% of AgNP regulated genes were also regulated by cold. Among the genes shared by AgNP and cold stresses, 49 of them (including cold responsive gene, COR) were regulated by *DREB1A* gene-encoded protein DREB1A (dehydration responsive element binding factor 1A). DREB1A is also called CBF3 (C-repeat binding factor 3) and acts as a main regulon (a group of genes regulated by the same regulatory protein) in cold response [115,116]. Particularly in this regulon, the ICE1-CBF-COR signaling pathway has been known in regulating plant response to cold stress [117–122]. ICE is inducer of CBF expression 1. CBF (*i.e.*, DREB1) acts as a major player of the *Arabidopsis* regulatory network in response to cold stress; this could imply a possible signaling crosstalk between CBF-regulated cold response pathway [123] and other non-temperature signaling transduction pathways such as AgNPs.

Membrane leakage is the primary damage to cells upon cold stress [124], while ROS results in initial signaling of cold stress [42]. Thus, cold acclimation by plants includes stabilization of cell membrane integrity, production of ROS signals and antioxidative pathways, elevated levels in sugar and osmolytes, such as polyamines [108,125]. The similar ROS signaling and antioxidant pathways have been reported in the studies of rat cortical cell cultures and human murine dendritic cell lines when treated with AgNPs [60,126]. Despite the fact that no direct evidence of AgNP entry/transport to membrane is found in plant cells, aggregation of AgNPs in vacuoles and at plasmodesmata were recently reported [19,31,127] as well as gold and carbon coated iron nanoparticles [56,128]. It indirectly implies that AgNPs, like cold stress, may induce ROS generation and consequently, change the physical state of membranes.

GO term enrichment analysis also confirmed that both cold and AgNP stresses were enriched in the molecular functions of response to ROS. In addition, both stresses were enriched in the molecular function of response to fungus. In cold stress, ice formation was reported to cause a mechanical strain on cell wall and membrane leading to cell rupture in winter wheat (*Triticum aestivum*) [129]. Rupture of cells and their cell walls might have released some oligosaccharides similar to the elicitors induced by fungal infection [130]. Moreover, PFAM domain enrichment analysis showed PF00067:p450 domain associated with AgNP and/or Ag$^+$ up-regulated, and cold down-regulated genes (Table S2). It was reported that there were more than 270 cytochrome p450 genes in the *Arabidopsis* genome and they all played important roles in development and responses to abiotic and biotic stress [131]. However, most of stress-induced p450 genes could be triggered by multiple stresses but each response was regulated according to individual stress [79]. This concurred the PFAM enrichment analysis in the comparison of cold and AgNP stresses; Pf00067 was enriched for the down-regulated gene-encoded proteins in cold stress but it was enriched for the up-regulated gene-encoded proteins in AgNP stress (Figure 4A and Table S2).

3.4. AgNP-Specific Responses in Genes and Functions

AgNPs have been commonly used in human society for their unique antimicrobial properties [5,8]. They have been studied in assays, transport and accumulation and microarray studies to confirm their phytotoxicity (toxicity to plants) [19,29,31]. Although the controversy between AgNPs and Ag$^+$ continues, this current study could provide new insights and shed light to this controversy. Despite the fact that AgNP and the other five abiotic stresses (Ag$^+$, cold, drought, heat and salt) affected similar metabolic pathways, AgNPs had some unique effects on *Arabidopsis* plants. First, the gene ontology (GO) term enrichment analysis demonstrated that AgNP specific gene-encoded proteins were enriched in two biological processes; one was enriched in Response to fungus (*i.e.*, enriched beta-1,3-endoglucanase domain) and the other was enriched in Anion transport. Response to fungus demonstrates a similarity of AgNPs to biotic stresses (fungal infection specifically) and wounding. Anion transport implies that the AgNP stress regulated different ion transporters from Ag$^+$ or salt (Na$^+$) did. Second, among all the 60 AgNP-specific genes, they could be sorted into two categories, protection from oxidative burst and involvement in cell wall and/or plasma membrane. The category of protection from oxidative burst includes glutathione S-transferase and p450s [132]. The second category was beta carbonic anhydrase 3, cellulose synthase, glycosyl hydrolase superfamily protein, alpha/beta-hydrolases, hydroxyproline-rich glycoprotein, beta glucosidase, glycosyl hydrolase, and some related to proteolysis processes such as serine carboxypeptidase-like 30.

4. Experimental Section

4.1. Microarray Data and Data Processing

Microarray data of six abiotic stresses in *Arabidopsis thaliana* were obtained from Gene Expression Omnibus (GEO) [133,134] and from Array Express in the European Molecular Biology

Laboratory [135,136]. They are silver nanoparticles (herein, AgNPs), silver nitrate (AgNO₃; herein, Ag⁺), cold, salt, drought and heat. The microarray data were listed as below.

E-MEXP-3950.   AgNP and Ag⁺ stresses after 10-day treatment [29].
GSE5620.   Control after 24 h treatment [137].
GSE5621.   Cold stress after 24 h treatment [137].
GSE5623.   Salt (NaCl) stress after 24 h treatment [137].
GSE5624.   Drought stress after 24 h treatment [137].
GSE5628.   Heat stress after 24 h treatment [137].

E-MEXP-3950 data came from whole seedlings after growing for 10 days in the presence of 5mg/L AgNPs (of 20 nm) or Ag⁺ (*i.e.*, AgNO₃). Normalized log-2 transformed transcriptomic data [29] were used to find the genes with more than two fold expression changes. Three biological replicates for each treatment/control were averaged. The treatment average minus control average was taken as the final value for each gene. Since normalized log 2 transformed data, *i.e.*, *M*-values,

$$\left[M-value = log_2 \left(\frac{treatment}{control} \right) \right] \tag{1}$$

were used, the final values of more than 1 or less than −1 present a more than two fold change in gene expression. Genes with *M*-values ≥ 1 and *M*-value ≤ −1 mean more than two fold up-regulated and down-regulated, respectively.

The comprehensive data set at AtGenExpress [137] was used to identify genes with more than two fold expression changes under four diverse abiotic stress conditions: cold, salt, drought, heat. The AtGenExpress data came from shoots and roots separately while Kaveh's data [29] came from whole seedlings. Thus, the former's data would be proportionally scaled in order to be comparable with the latter's data. In doing so, a fresh weight biomass shoot-root ratio (S/R) was utilized, based on a similar growth stage and growing in a comparable medium [138]. The formula to convert shoot and root signal to whole plant signal is:

$$[whole\ plant\ signal = S/R\ ratio\ *\ shoot\ signal + root\ signal] \tag{2}$$

Once whole plant signal was calculated for each biological replicate, *M*-value was calculated for each replicate. Next, two biological replicates (in AtGenExpress data) were averaged. An initial list of the genes with more than two fold changes in the expression of six abiotic stresses was prepared for further analyses (Table S12). Based on Table S12, Venn diagrams were also created to display numbers of genes which were shared by the six abiotic stresses and which were unique for specific stresses for further analyses.

4.2. Visualization of Affected Genes in Metabolic Pathways

Microarray data of the six abiotic stresses were parsed into their respective metabolic pathways and cell compartments using MapMan software [139]. MapMan (version 3.5.1R2) was employed to display microarray data of the six stresses in a variety of metabolic and signaling pathways. *M*-value data in Table S1 were used for all the *Arabidopsis* identified genes (based on TAIR10 annotation) [140]. They were displayed in MapMan Image Annotator in two color scale schemes: blue is used to denote induced genes and red to denote suppressed genes.

4.3. Coded Proteins of Affected Genes by the Stresses in Protein-Protein Interaction Networks

Both the *Arabidopsis* predicted interactome 2.0 [141] and an experimentally verified interactome [142] were used as reference sets, based on protein orthologues (*i.e.*, proteins from divergence of a common gene), to create a PPi network for the coded proteins of the affected genes by the six abiotic stresses. The genes with M-value ≥ 1 or ≤ -1 (see Table S12) were first used as a coded protein query set to search their interacting protein partners. This query set of proteins (*i.e.*, coded gene products) became an initial reference network to find their edges (*i.e.*, interacting proteins). Next, these edges were used to identify protein analogues (*i.e.*, proteins from convergence of different genes but of the same function) and to expand a PPi network. The set of paired proteins from the query set and their analogues was then exported as a new PPi network. The combination of the initial reference network and its expanded networks became the final PPi network of the affected gene-coded proteins from the six abiotic stresses. All the proteins in the final PPi network were displayed in the Cytoscape 3.1.0 [143].

4.4. Enrichment Analyses of Differentially Expressed Genes in Six Abiotic Stresses

Two web-based applications, GOrilla [144] and DAVID 6.7 [145], were used in enrichment analyses to characterize the underlying biological processes, molecular functions and cellular components for the differentially expressed genes in the six abiotic stresses (*i.e.*, Table S12). The analyses investigated the coherence of the data across different mechanisms of *Arabidopsis* responses to the six abiotic stresses. Enrichment analyses included gene ontology (GO) [146], PFAM for protein domains [66], InterPro for protein signatures and functions [83,147], and Kyoto Encyclopedia of Genes and Genomes (KEGG) pathways [148]. GO term enrichment for biological process, molecular function and cellular component were performed by GOrilla. Annotated and characterized genes in *Arabidopsis* (TAIR10) were a "background gene list". GOrilla used a list of up- and down-regulated genes (from each stress) as a "target gene list" to search for GO enriched terms in this "target gene list" in comparison to the background gene list. PFAM domains came from global (amino acid) sequence alignment while InterPro classes came from local shorter aligned sequences (*i.e.*, signatures) and catalytic sites (*i.e.*, functions). *Arabidopsis* gene IDs of TAIR 10 as background list and the target gene list (Table S12) were subjected to DAVID 6.7 when enrichment analyses of PFAM domains, Interpro protein classes and KEGG pathways were performed. The output data by GOrilla and DAVID 6.7 were then filtered, using the q-values (*i.e.*, False Discovery Rate; it was adjusted from p-value) less than 0.05 (*i.e.*, $q < 0.05$). p-value is the probability of the observed results on the null hypothesis which is true. Enrichment analyses were also done for shared and/or specific genes that were derived from Venn diagram analysis (see below) for AgNP when compared with the other five stresses.

4.5. Comparison of Shared and Specific Genes in Venn Diagrams

A graphical Venn diagram helps visualize complex biological data sets and illustrate the overlap in genes shared by different conditions. One calculator and drawing Venn diagram's web-tool [149] was employed to compare genes with more than two fold expression difference (*i.e.*, M-value ≥ 1 or ≤ -1) that were shared by the six abiotic stresses. The list of genes with more than two fold expression changes (Table S12) was uploaded to the site and output data were used to draw the diagrams. Two-way Venn diagrams were used to compare AgNP with the other five abiotic stresses. Three-, four- or six-way Venn diagrams were also used to compare shared genes across three, four or six stresses.

4.6. Plasmodesmata Related Genes Expressed in AgNP and Ag⁺ Stresses

Two approaches were employed to identify how many genes were related to plasmodesmata and also affected by AgNPs and/or Ag⁺. First, a search was performed using "plasmodesmata" in the gene description and GO terms of AgNP and/or Ag⁺ affected genes, which were obtained from BioMart [150]. Second, a list of genes that are directly related to plasmodesmata was prepared based on GO terms in AmiGO2 [146,151–154], GONUTS (the Gene Ontology Normal Usage Tracking System [155], and literature search [156–170]. Altogether, a list of the 26 plasmodesmata related genes was collected and it was provided in Table S13.

5. Conclusions

Despite the similarities of regulated genes by AgNP stress and five other stresses, there are distinct differences by AgNPs. There are 60 AgNP-specific genes that are not affected/regulated by the other five stresses. The shared properties of Ag⁺ and AgNP stresses were due to chemical Ag⁺ ions; but AgNP stress differed from Ag⁺ stress, probably resulting from physical/mechanical damage due to nano-size of AgNPs. The similarities of AgNP and cold stresses could result from their mechanical damages and induction of ROS; but the two stresses were different. In sum, despite the shared similarity in gene expression and metabolic pathways to the three abiotic stresses (Ag⁺, cold, salt), AgNPs are also novel abiotic stressors that pose different toxicity risks to *Arabidopsis* plants.

Supplementary Materials

Supplementary materials can be accessed at: http://www.mdpi.com/2079-4991/5/2/436/s1.

Acknowledgments

The authors wish to thank Van Aken, B. for the microarray data, Matt Geisler for the discussion and Karen Renzaglia and Richard Thomas for their critical review. E.K.B. acknowledged the writing assistance from David Foutch. E.K.B. was supported by a scholarship from Ministry of Science, Research and Technology of Iran.

Author Contributions

E.K.B.: performing experiments, figures and tables making, writing; J.G.L.: concepts, experimental design, writing and discussion.

Conflicts of Interest

The authors declare no conflict of interest.

References

1. Oberdörster, G.; Oberdörster, E.; Oberdörster, J. Nanotoxicology: An emerging discipline evolving from studies of ultrafine particles. *Environ. Health Perspect.* **2005**, *113*, 823–839.

2. EPA. Module 3: Characteristics of particles-particle size categories. Available online: http://web.archive.org/web/20101203205130/http://www.epa.gov/apti/bces/module3/category/cat egory.htm (accessed on 6 December 2014).

3. Taylor, R.; Coulombe, S.; Otanicar, T.; Phelan, P.; Gunawan, A.; Lv, W.; Rosengarten, G.; Prasher, R.; Tyagi, H. Small particles, big impacts: A review of the diverse applications of nanofluids. *J. Appl. Phys.* **2013**, *113*, 011301.

4. Keller, A.; McFerran, S.; Lazareva, A.; Suh, S. Global life cycle releases of engineered nanomaterials. *J. Nanopart. Res.* **2013**, *15*, 1–17.

5. Ahamed, M.; AlSalhi, M.S.; Siddiqui, M. Silver nanoparticle applications and human health. *Clin. Chim. Acta* **2010**, *411*, 1841–1848.

6. Jo, Y.-K.; Kim, B.H.; Jung, G. Antifungal activity of silver ions and nanoparticles on phytopathogenic fungi. *Plant Dis.* **2009**, *93*, 1037–1043.

7. Kim, S.W.; Jung, J.H.; Lamsal, K.; Kim, Y.S.; Min, J.S.; Lee, Y.S. Antifungal effects of silver nanoparticles (AgNPs) against various plant pathogenic fungi. *Mycobiology* **2012**, *40*, 53–58.

8. Bechert, T.; Böswald, M.; Lugauer, S.; Regenfus, A.; Greil, J.; Guggenbichler, J.P. The erlanger silver catheter: *In vitro* results for antimicrobial activity. *Infection* **1999**, *27*, S24–S29.

9. Liong, M.; Lu, J.; Kovochich, M.; Xia, T.; Ruehm, S.G.; Nel, A.E.; Tamanoi, F.; Zink, J.I. Multifunctional inorganic nanoparticles for imaging, targeting, and drug delivery. *ACS Nano* **2008**, *2*, 889–896.

10. Korkin, A.; Rosei, F. *Nanoelectronics and Photonics: From Atoms to Materials, Devices, and Architecture*; Springer Science & Business Media: Berlin, Germany, 2008.

11. Geranio, L.; Heuberger, M.; Nowack, B. The behavior of silver nanotextiles during washing. *Environ. Sci. Technol.* **2009**, *43*, 8113–8118.

12. Gardea-Torresdey, J.L.; Rico, C.M.; White, J.C. Trophic transfer, transformation, and impact of engineered nanomaterials in terrestrial environments. *Environ. Sci. Technol.* **2014**, *48*, 2526–2540.

13. Tomczyk, M. *Nanoinnovation: What Every Manager Needs to Know*; John Wiley & Sons: Hoboken, NJ, USA, 2014.

14. Watson, C.; Ge, J.; Cohen, J.; Pyrgiotakis, G.; Engelward, B.P.; Demokritou, P. High-throughput screening platform for engineered nanoparticle-mediated genotoxicity using cometchip technology. *ACS Nano* **2014**, *8*, 2118–2133.

15. Colman, B.P.; Arnaout, C.L.; Anciaux, S.; Gunsch, C.K.; Hochella, M.F., Jr.; Kim, B.; Lowry, G.V.; McGill, B.M.; Reinsch, B.C.; Richardson, C.J.; *et al.* Low concentrations of silver nanoparticles in biosolids cause adverse ecosystem responses under realistic field scenario. *PLoS ONE* **2013**, *8*, doi:10.1371/journal.pone.0057189.

16. Sur, I.; Cam, D.; Kahraman, M.; Baysal, A.; Culha, M. Interaction of multi-functional silver nanoparticles with living cells. *Nanotechnology* **2010**, *21*, 175104.

17. Kim, S.; Choi, J.E.; Choi, J.; Chung, K.-H.; Park, K.; Yi, J.; Ryu, D.-Y. Oxidative stress-dependent toxicity of silver nanoparticles in human hepatoma cells. *Toxicol. In Vitro* **2009**, *23*, 1076–1084.

18. Park, E.-J.; Yi, J.; Kim, Y.; Choi, K.; Park, K. Silver nanoparticles induce cytotoxicity by a trojan-horse type mechanism. *Toxicol. In Vitro* **2010**, *24*, 872–878.

19. Geisler-Lee, J.; Wang, Q.; Yao, Y.; Zhang, W.; Geisler, M.; Li, K.; Huang, Y.; Chen, Y.; Kolmakov, A.; Ma, X. Phytotoxicity, accumulation and transport of silver nanoparticles by *Arabidopsis thaliana*. *Nanotoxicology* **2013**, *7*, 323–337.

20. Lee, W.-M.; Kwak, J.I.; An, Y.-J. Effect of silver nanoparticles in crop plants phaseolus radiatus and sorghum bicolor: Media effect on phytotoxicity. *Chemosphere* **2012**, *86*, 491–499.

21. Qian, H.; Peng, X.; Han, X.; Ren, J.; Sun, L.; Fu, Z. Comparison of the toxicity of silver nanoparticles and silver ions on the growth of terrestrial plant model *Arabidopsis thaliana*. *J. Environ. Sci.* **2013**, *25*, 1947–1956.

22. Geisler-Lee, J.; Brooks, M.; Gerfen, J.; Wang, Q.; Fotis, C.; Sparer, A.; Ma, X.; Berg, R.; Geisler, M. Reproductive toxicity and life history study of silver nanoparticle effect, uptake and transport in *Arabidopsis thaliana*. *Nanomaterials* **2014**, *4*, 301–318.

23. Ma, X.; Geisler-Lee, J.; Deng, Y.; Kolmakov, A. Interactions between engineered nanoparticles (ENPs) and plants: Phytotoxicity, uptake and accumulation. *Sci. Total Environ.* **2010**, *408*, 3053–3061.

24. Miralles, P.; Church, T.L.; Harris, A.T. Toxicity, uptake, and translocation of engineered nanomaterials in vascular plants. *Environ. Sci. Technol.* **2012**, *46*, 9224–9239.

25. Mirzajani, F.; Askari, H.; Hamzelou, S.; Farzaneh, M.; Ghassempour, A. Effect of silver nanoparticles on *Oryza sativa* l. And its rhizosphere bacteria. *Ecotoxicol. Environ. Saf.* **2013**, *88*, 48–54.

26. Yin, L.; Colman, B.P.; McGill, B.M.; Wright, J.P.; Bernhardt, E.S. Effects of silver nanoparticle exposure on germination and early growth of eleven wetland plants. *PLoS ONE* **2012**, *7*, e47674.

27. Dimkpa, C.O.; McLean, J.E.; Martineau, N.; Britt, D.W.; Haverkamp, R.; Anderson, A.J. Silver nanoparticles disrupt wheat (*Triticum aestivum* l.) growth in a sand matrix. *Environ. Sci. Technol.* **2012**, *47*, 1082–1090.

28. Vannini, C.; Domingo, G.; Onelli, E.; de Mattia, F.; Bruni, I.; Marsoni, M.; Bracale, M. Phytotoxic and genotoxic effects of silver nanoparticles exposure on germinating wheat seedlings. *J. Plant Physiol.* **2014**, *171*, 1142–1148.

29. Kaveh, R.; Li, Y.-S.; Ranjbar, S.; Tehrani, R.; Brueck, C.L.; van Aken, B. Changes in *Arabidopsis thaliana* gene expression in response to silver nanoparticles and silver ions. *Environ. Sci. Technol.* **2013**, *47*, 10637–10644.

30. Larue, C.; Castillo-Michel, H.; Sobanska, S.; Cécillon, L.; Bureau, S.; Barthès, V.; Ouerdane, L.; Carrière, M.; Sarret, G. Foliar exposure of the crop *Lactuca sativa* to silver nanoparticles: Evidence for internalization and changes in Ag speciation. *J. Hazard. Mater.* **2014**, *264*, 98–106.

31. Yin, L.; Cheng, Y.; Espinasse, B.; Colman, B.P.; Auffan, M.; Wiesner, M.; Rose, J.; Liu, J.; Bernhardt, E.S. More than the ions: The effects of silver nanoparticles on *Lolium multiflorum*. *Environ. Sci. Technol.* **2011**, *45*, 2360–2367.

32. Romera, F.J.; Alcantara, E. Iron-deficiency stress responses in cucumber (*Cucumis sativus* l.) roots (a possible role for ethylene?). *Plant Physiol.* **1994**, *105*, 1133–1138.

33. Yoshioka, K.; Shinozaki, K. *Signal Crosstalk in Plant Stress Responses*; Wiley: Hoboken, Germany, 2009.

34. Hirayama, T.; Shinozaki, K. Research on plant abiotic stress responses in the post-genome era: Past, present and future. *Plant J.* **2010**, *61*, 1041–1052.

35. Duque, A.S.; de Almeida, A.M.; da Silva, A.B.; da Silva, J.M.; Farinha, A.P.; Santos, D.; Fevereiro, P.; de Sousa Araújo, S. Abiotic Stress Responses in Plants: Unraveling the Complexity of Genes and Networks to Survive. In *Abiotic Stress-Plant Responses and Applications in Agriculture*; Vahdati, K., Leslie, C., Eds.; INTECH: Rijeka, Croatia, 2013.

36. Bailey-Serres, J.; Mittler, R. The roles of reactive oxygen species in plant cells. *Plant Physiol.* **2006**, *141*, 311.

37. Baxter, A.; Mittler, R.; Suzuki, N. Ros as key players in plant stress signalling. *J. Exp. Botany* **2013**, *65*, 1229–1240.

38. Gill, S.S.; Tuteja, N. Reactive oxygen species and antioxidant machinery in abiotic stress tolerance in crop plants. *Plant Physiol. Biochem.* **2010**, *48*, 909–930.

39. Nakagami, H.; Pitzschke, A.; Hirt, H. Emerging map kinase pathways in plant stress signalling. *Trends Plant Sci.* **2005**, *10*, 339–346.

40. Hirt, H. Multiple roles of map kinases in plant signal transduction. *Trends Plant Sci.* **1997**, *2*, 11–15.

41. Alcázar-Román, A.; Wente, S. Inositol polyphosphates: A new frontier for regulating gene expression. *Chromosoma* **2008**, *117*, 1–13.

42. Xiong, L.; Schumaker, K.S.; Zhu, J.-K. Cell signaling during cold, drought, and salt stress. *Plant Cell Online* **2002**, *14*, S165–S183.

43. Mahalingam, R.; Fedoroff, N. Stress response, cell death and signalling: The many faces of reactive oxygen species. *Physiol. Plant.* **2003**, *119*, 56–68.

44. Dixon, R.A.; Achnine, L.; Kota, P.; Liu, C.J.; Reddy, M.; Wang, L. The phenylpropanoid pathway and plant defence—A genomics perspective. *Mol. Plant Pathol.* **2002**, *3*, 371–390.

45. Edreva, A.; Velikova, V.; Tsonev, T.; Dagnon, S.; Gürel, A.; Aktaş, L.; Gesheva, E. Stress-protective role of secondary metabolites: Diversity of functions and mechanisms. *Gen. Appl. Plant Physiol.* **2008**, *34*, 67–78.

46. Oh, M.-M.; Trick, H.N.; Rajashekar, C. Secondary metabolism and antioxidants are involved in environmental adaptation and stress tolerance in lettuce. *J. Plant Physiol.* **2009**, *166*, 180–191.

47. Dixon, R.A.; Paiva, N.L. Stress-induced phenylpropanoid metabolism. *Plant Cell* **1995**, *7*, 1085–1097.

48. Degenhardt, B.; Gimmler, H. Cell wall adaptations to multiple environmental stresses in maize roots. *J. Exp. Bot.* **2000**, *51*, 595–603.

49. Rouet, M.; Mathieu, Y.; Barbier-Brygoo, H.; Laurière, C. Characterization of active oxygen-producing proteins in response to hypo-osmolarity in tobacco and Arabidopsis cell suspensions: Identification of a cell wall peroxidase. *J. Exp. Bot.* **2006**, *57*, 1323–1332.

50. Daudi, A.; Cheng, Z.; O'Brien, J.A.; Mammarella, N.; Khan, S.; Ausubel, F.M.; Bolwell, G.P. The apoplastic oxidative burst peroxidase in Arabidopsis is a major component of pattern-triggered immunity. *Plant Cell Online* **2012**, *24*, 275–287.

51. Passardi, F.; Penel, C.; Dunand, C. Performing the paradoxical: How plant peroxidases modify the cell wall. *Trends Plant Sci.* **2004**, *9*, 534–540.

52. Minibayeva, F.; Kolesnikov, O.; Chasov, A.; Beckett, R.; Lüthje, S.; Vylegzhanina, N.; Buck, F.; Böttger, M. Wound-induced apoplastic peroxidase activities: Their roles in the production and detoxification of reactive oxygen species. *Plant Cell Environ.* **2009**, *32*, 497–508.

53. Lucas, W.J.; Ham, B.-K.; Kim, J.-Y. Plasmodesmata–bridging the gap between neighboring plant cells. *Trends Cell Biol.* **2009**, *19*, 495–503.

54. Furuta, K.; Lichtenberger, R.; Helariutta, Y. The role of mobile small RNA species during root growth and development. *Curr. Opin. Cell Biol.* **2012**, *24*, 211–216.

55. Ruiz-Ferrer, V.; Voinnet, O. Roles of plant small RNAs in biotic stress responses. *Annu. Rev. Plant Biol.* **2009**, *60*, 485–510.

56. Zhai, G.; Walters, K.S.; Peate, D.W.; Alvarez, P.J.J.; Schnoor, J.L. Transport of gold nanoparticles through plasmodesmata and precipitation of gold ions in woody poplar. *Environ. Sci. Technol. Lett.* **2014**, *1*, 146–151.

57. He, D.; Jones, A.M.; Garg, S.; Pham, A.N.; Waite, T.D. Silver nanoparticle–reactive oxygen species interactions: Application of a charging–discharging model. *J. Phys. Chem. C* **2011**, *115*, 5461–5468.

58. Xu, H.; Qu, F.; Xu, H.; Lai, W.; Andrew Wang, Y.; Aguilar, Z.; Wei, H. Role of reactive oxygen species in the antibacterial mechanism of silver nanoparticles on *Escherichia coli* O157:H7. *Biometals* **2012**, *25*, 45–53.

59. Jiang, H.-S.; Qiu, X.-N.; Li, G.-B.; Li, W.; Yin, L.-Y. Silver nanoparticles induced accumulation of reactive oxygen species and alteration of antioxidant systems in the aquatic plant *Spirodela polyrhiza*. *Environ. Toxicol. Chem.* **2014**, *33*, 1398–1405.

60. Kang, K.; Jung, H.; Lim, J.-S. Cell death by polyvinylpyrrolidine-coated silver nanoparticles is mediated by ROS-dependent signaling. *Biomol. Ther.* **2012**, *20*, 399–405.

61. Lim, D.; Roh, J.Y.; Eom, H.J.; Choi, J.Y.; Hyun, J.; Choi, J. Oxidative stress-related PMK-1 P38 MAPK activation as a mechanism for toxicity of silver nanoparticles to reproduction in the nematode *Caenorhabditis elegans*. *Environ. Toxicol. Chem.* **2012**, *31*, 585–592.

62. Lim, H.K.; Asharani, P.V.; Hande, M.P. Enhanced genotoxicity of silver nanoparticles in DNA repair deficient mammalian cells. *Front. Genet.* **2012**, *3*, 104.

63. Gliga, A.R.; Skoglund, S.; Wallinder, I.O.; Fadeel, B.; Karlsson, H.L. Size-dependent cytotoxicity of silver nanoparticles in human lung cells: The role of cellular uptake, agglomeration and Ag release. *Part. Fibre Toxicol.* **2014**, *11*, doi:10.1186/1743-8977-11-11.

64. AshaRani, P.; Sethu, S.; Lim, H.; Balaji, G.; Valiyaveettil, S.; Hande, M.P. Differential regulation of intracellular factors mediating cell cycle, DNA repair and inflammation following exposure to silver nanoparticles in human cells. *Genome Integr.* **2012**, *3*, doi:10.1186/2041-9414-3-2.

65. Finn, R.D.; Bateman, A.; Clements, J.; Coggill, P.; Eberhardt, R.Y.; Eddy, S.R.; Heger, A.; Hetherington, K.; Holm, L.; Mistry, J.; *et al.* Pfam: The protein families database. *Nucleic Acids Res.* **2014**, *42*, D222–D230.

66. Bateman, A.; Coin, L.; Durbin, R.; Finn, R.D.; Hollich, V.; Griffiths-Jones, S.; Khanna, A.; Marshall, M.; Moxon, S.; Sonnhammer, E.L. The pfam protein families database. *Nucleic Acids Res.* **2004**, *32*, D138–D141.

67. Jeyaprakash, A.A.; Rani, P.G.; Reddy, G.B.; Banumathi, S.; Betzel, C.; Sekar, K.; Surolia, A.; Vijayan, M. Crystal structure of the jacalin-t-antigen complex and a comparative study of lectin-T-antigen complexes. *J. Mol. Biol.* **2002**, *321*, 637–645.

68. Raval, S.; Gowda, S.B.; Singh, D.D.; Chandra, N.R. A database analysis of jacalin-like lectins: Sequence-structure-function relationships. *Glycobiology* **2004**, *14*, 1247–1263.

69. Sankaranarayanan, R.; Sekar, K.; Banerjee, R.; Sharma, V.; Surolia, A.; Vijayan, M. A novel mode of carbohydrate recognition in jacalin, a *Moraceae* plant lectin with a β-prism fold. *Nat. Struct. Mol. Biol.* **1996**, *3*, 596–603.

70. Welinder, K.G. Superfamily of plant, fungal and bacterial peroxidases. *Curr. Opin. Struct. Biol.* **1992**, *2*, 388–393.

71. Campa, A. Biological Roles of Plant Peroxidases: Known and Potential Function. In *Peroxidases in Chemistry and Biology*; Everse, J., Everse, K.E., Grisham, M.B., Eds.; CRC Press, Taylor & Francis Group, Inc.: London, UK, 1990; Volume 2, pp. 25–50.

72. Kader, J.-C. Lipid-transfer proteins in plants. *Annu. Rev. Plant Physiol. Plant Mol. Biol.* **1996**, *47*, 627–654.

73. Kader, J.-C. Lipid-transfer proteins: A puzzling family of plant proteins. *Trends Plant Sci.* **1997**, *2*, 66–70.

74. Lamb, D.C.; Lei, L.; Warrilow, A.G.S.; Lepesheva, G.I.; Mullins, J.G.L.; Waterman, M.R.; Kelly, S.L. The first virally encoded cytochrome p450. *J. Virol.* **2009**, *83*, 8266–8269.

75. Werck-Reichhart, D.; Feyereisen, R. Cytochromes p450: A success story. *Genome Biol.* **2000**, *1*, REVIEWS3003.

76. Pinot, F.; Beisson, F. Cytochrome p450 metabolizing fatty acids in plants: Characterization and physiological roles. *FEBS J.* **2011**, *278*, 195–205.

77. Schuler, M.A. The role of cytochrome p450 monooxygenases in plant-insect interactions. *Plant Physiol.* **1996**, *112*, 1411–1419.

78. Seki, M.; Narusaka, M.; Ishida, J.; Nanjo, T.; Fujita, M.; Oono, Y.; Kamiya, A.; Nakajima, M.; Enju, A.; Sakurai, T.; *et al.* Monitoring the expression profiles of 7000 Arabidopsis genes under drought, cold and high-salinity stresses using a full-length cdna microarray. *Plant J.* **2002**, *31*, 279–292.

79. Narusaka, Y.; Narusaka, M.; Seki, M.; Umezawa, T.; Ishida, J.; Nakajima, M.; Enju, A.; Shinozaki, K. Crosstalk in the responses to abiotic and biotic stresses in Arabidopsis: Analysis of gene expression in cytochrome p450 gene superfamily by cdna microarray. *Plant Mol. Biol.* **2004**, *55*, 327–342.

80. Rushton, P.J.; Somssich, I.E.; Ringler, P.; Shen, Q.J. Wrky transcription factors. *Trends Plant Sci.* **2010**, *15*, 247–258.

81. Rushton, P.J.; Torres, J.T.; Parniske, M.; Wernert, P.; Hahlbrock, K.; Somssich, I.E. Interaction of elicitor-induced DNA-binding proteins with elicitor response elements in the promoters of parsley PR1 genes. *EMBO J.* **1996**, *15*, 5690–5700.

82. Riechmann, J.L.; Meyerowitz, E.M. The AP2/EREBP family of plant transcription factors. *Biol. Chem.* **1998**, *379*, 633–646.

83. Mitchell, A.; Chang, H.Y.; Daugherty, L.; Fraser, M.; Hunter, S.; Lopez, R.; McAnulla, C.; McMenamin, C.; Nuka, G.; Pesseat, S.; *et al.* The InterPro protein families database: The classification resource after 15 years. *Nucleic Acids Res.* **2015**, *43*, D213–D221.

84. The Homepage of InterPro. Available online: http://www.ebi.ac.uk/interpro (accessed on 30 March 2014).

85. Hunter, S.; Apweiler, R.; Attwood, T.K.; Bairoch, A.; Bateman, A.; Binns, D.; Bork, P.; Das, U.; Daugherty, L.; Duquenne, L. InterPro: The integrative protein signature database. *Nucleic Acids Res.* **2009**, *37*, D211–D215.

86. Hunter, S.; Jones, P.; Mitchell, A.; Apweiler, R.; Attwood, T.K.; Bateman, A.; Bernard, T.; Binns, D.; Bork, P.; Burge, S.; *et al.* InterPro in 2011: New developments in the family and domain prediction database. *Nucleic Acids Res.* **2011**, *40*, D306–D312.

87. Rico, M.; Bruix, M.; González, C.; Monsalve, R.I.; Rodríguez, R. 1H NMR assignment and global fold of napin Bnib, a representative 2S albumin seed protein. *Biochemistry* **1996**, *35*, 15672–15682.

88. Vogt, T. Phenylpropanoid biosynthesis. *Mol. Plant* **2010**, *3*, 2–20.

89. Petersen, M.; Hans, J.; Matern, U. Biosynthesis of Phenylpropanoids and Related Compounds. In *Annual Plant Reviews Volume 40: Biochemistry of Plant Secondary Metabolism*; Wiley-Blackwell: Hoboken, JN, USA, 2010; pp. 182–257.

90. Ferrer, J.L.; Austin, M.B.; Stewart, C.; Noel, J.P. Structure and function of enzymes involved in the biosynthesis of phenylpropanoids. *Plant Physiol. Biochem.* **2008**, *46*, 356–370.

91. Fraser, C.M.; Chapple, C. The phenylpropanoid pathway in Arabidopsis. *Arabidopsis Book* **2011**, *9*, e0152.

92. Lee, Y.J.; Kim, J.; Oh, J.; Bae, S.; Lee, S.; Hong, I.S.; Kim, S.H. Ion-release kinetics and ecotoxicity effects of silver nanoparticles. *Environ. Toxicol. Chem.* **2012**, *31*, 155–159.

93. Gubbins, E.J.; Batty, L.C.; Lead, J.R. Phytotoxicity of silver nanoparticles to *Lemna minor* l. *Environ. Pollut.* **2011**, *159*, 1551–1559.

94. Jaspers, P.; Kangasjärvi, J. Reactive oxygen species in abiotic stress signaling. *Physiol. Plant.* **2010**, *138*, 405–413.

95. Dietz, K.-J.; Herth, S. Plant nanotoxicology. *Trends Plant Sci.* **2011**, *16*, 582–589.

96. Wuest, S.E.; Vijverberg, K.; Schmidt, A.; Weiss, M.; Gheyselinck, J.; Lohr, M.; Wellmer, F.; Rahnenführer, J.; von Mering, C.; Grossniklaus, U. Arabidopsis female gametophyte gene expression map reveals similarities between plant and animal gametes. *Curr. Biol.* **2010**, *20*, 506–512.

97. Jiang, J.; Zhang, Z.; Cao, J. Pollen wall development: The associated enzymes and metabolic pathways. *Plant Biol.* **2013**, *15*, 249–263.

98. Bock, K.W.; Honys, D.; Ward, J.M.; Padmanaban, S.; Nawrocki, E.P.; Hirschi, K.D.; Twell, D.; Sze, H. Integrating membrane transport with male gametophyte development and function through transcriptomics. *Plant Physiol.* **2006**, *140*, 1151–1168.

99. Hsieh, M.H.; Goodman, H.M. A novel gene family in Arabidopsis encoding putative heptahelical transmembrane proteins homologous to human adiponectin receptors and progestin receptors. *J. Exp. Bot.* **2005**, *56*, 3137–3147.

100. Sanghera, G.S.; Wani, S.H.; Hussain, W.; Singh, N.B. Engineering cold stress tolerance in crop plants. *Curr. Genomics* **2011**, *12*, 30–43.

101. Lata, C.; Prasad, M. Role of DREBs in regulation of abiotic stress responses in plants. *J. Exp. Bot.* **2011**, *62*, 4731–4748.

102. Musante, C. *Nanoparticle Contamination of Agricultural Crops*; Chemistry, D.O.A., Ed.; The Connecticut Agricultural Experiment Station: New Haven, CT, USA, 2011; p. 32.

103. Atkinson, N.J.; Urwin, P.E. The interaction of plant biotic and abiotic stresses: From genes to the field. *J. Exp. Bot.* **2012**, *63*, 3523–3543.

104. Oukarroum, A.; Barhoumi, L.; Pirastru, L.; Dewez, D. Silver nanoparticle toxicity effect on growth and cellular viability of the aquatic plant *Lemna gibba*. *Environ. Toxicol. Chem.* **2013**, *32*, 902–907.

105. Apel, K.; Hirt, H. Reactive oxygen species: Metabolism, oxidative stress, and signal transduction. *Annu. Rev. Plant Biol.* **2004**, *55*, 373–399.

106. Nemhauser, J.L.; Hong, F.; Chory, J. Different plant hormones regulate similar processes through largely nonoverlapping transcriptional responses. *Cell* **2006**, *126*, 467–475.

107. Eom, H.-J.; Choi, J. p38 MAPK activation, DNA damage, cell cycle arrest and apoptosis as mechanisms of toxicity of silver nanoparticles in Jurkat T cells. *Environ. Sci. Technol.* **2010**, *44*, 8337–8342.

108. Hare, P.; Cress, W.; van Staden, J. Dissecting the roles of osmolyte accumulation during stress. *Plant Cell Environ.* **1998**, *21*, 535–553.

109. Kalamaki, M.S.; Merkouropoulos, G.; Kanellis, A.K. Can ornithine accumulation modulate abiotic stress tolerance in Arabidopsis. *Plant Signal. Behav.* **2009**, *4*, 1099–1101.

110. Nayak, L.; Raval, M.; Biswal, B.; Biswal, U. Silver nitrate influences *in vitro* shoot multiplication and root formation in *Vanilla planifolia Andr. Curr. Sci. India* **2001**, *14*, 1166.

111. Verma, A.; Uzun, O.; Hu, Y.; Hu, Y.; Han, H.-S.; Watson, N.; Chen, S.; Irvine, D.J.; Stellacci, F. Surface-structure-regulated cell-membrane penetration by monolayer-protected nanoparticles. *Nat. Mater.* **2008**, *7*, 588–595.

112. Wang, J.; Koo, Y.; Alexander, A.; Yang, Y.; Westerhof, S.; Zhang, Q.; Schnoor, J.L.; Colvin, V.L.; Braam, J.; Alvarez, P.J. Phytostimulation of poplars and Arabidopsis exposed to silver nanoparticles and Ag^+ at sublethal concentrations. *Environ. Sci. Technol.* **2013**, *47*, 5442–5449.

113. Singh, J.; Prasad, S.; Rai, L. Inhibition of photosynthetic electron transport in *Nostoc muscorum* by Ni2+ and Ag+. *J. Gen. Appl. Microbiol.* **1991**, *37*, 167–174.

114. Kumar, V.; Parvatam, G.; Ravishankar, G.A. AgNO₃: A potential regulator of ethylene activity and plant growth modulator. *Electron. J. Biotechnol.* **2009**, *12*, 8–9.

115. Shinwari, Z.K.; Nakashima, K.; Miura, S.; Kasuga, M.; Seki, M.; Yamaguchi-Shinozaki, K.; Shinozaki, K. An Arabidopsis gene family encoding DRE/CRT binding proteins involved in low-temperature-responsive gene expression. *Biochem. Biophys. Res. Commun.* **1998**, *250*, 161–170.

116. Kohan, E.; Bagherieh-Najjar, M.B. DRE-binding transcription factor (DREB1A) as a master regulator induced a broad range of abiotic stress tolerance in plant. *Afr. J. Biotechnol.* **2013**, *10*, 15100–15108.

117. Maruyama, K.; Sakuma, Y.; Kasuga, M.; Ito, Y.; Seki, M.; Goda, H.; Shimada, Y.; Yoshida, S.; Shinozaki, K.; Yamaguchi-Shinozaki, K. Identification of cold-inducible downstream genes of the Arabidopsis DREB1A/CBF3 transcriptional factor using two microarray systems. *Plant J.* **2004**, *38*, 982–993.

118. Seki, M.; Narusaka, M.; Abe, H.; Kasuga, M.; Yamaguchi-Shinozaki, K.; Carninci, P.; Hayashizaki, Y.; Shinozaki, K. Monitoring the expression pattern of 1300 Arabidopsis genes under drought and cold stresses by using a full-length cDNA microarray. *Plant Cell Online* **2001**, *13*, 61–72.

119. Chinnusamy, V.; Zhu, J.; Zhu, J.-K. Cold stress regulation of gene expression in plants. *Trends Plant Sci.* **2007**, *12*, 444–451.

120. Chinnusamy, V.; Zhu, J.-K.; Sunkar, R. Gene regulation during cold stress acclimation in plants. *Methods Mol. Biol.* **2010**, *639*, 39–55.

121. Chinnusamy, V.; Ohta, M.; Kanrar, S.; Lee, B.-H.; Hong, X.; Agarwal, M.; Zhu, J.-K. ICE1: A regulator of cold-induced transcriptome and freezing tolerance in Arabidopsis. *Genes Dev.* **2003**, *17*, 1043–1054.

122. Gilmour, S.J.; Zarka, D.G.; Stockinger, E.J.; Salazar, M.P.; Houghton, J.M.; Thomashow, M.F. Low temperature regulation of the Arabidopsis CBF family of AP2 transcriptional activators as an early step in cold-inducedcorgene expression. *Plant J.* **1998**, *16*, 433–442.

123. Yang, T.; Zhang, L.; Zhang, T.; Zhang, H.; Xu, S.; An, L. Transcriptional regulation network of cold-responsive genes in higher plants. *Plant Sci.* **2005**, *169*, 987–995.

124. Uemura, M.; Tominaga, Y.; Nakagawara, C.; Shigematsu, S.; Minami, A.; Kawamura, Y. Responses of the plasma membrane to low temperatures. *Physiol. Plant.* **2006**, *126*, 81–89.

125. Mahajan, S.; Tuteja, N. Cold, salinity and drought stresses: An overview. *Arch. Biochem. Biophys.* **2005**, *444*, 139–158.

126. Xu, F.; Piett, C.; Farkas, S.; Qazzaz, M.; Syed, N. Silver nanoparticles (AgNPs) cause degeneration of cytoskeleton and disrupt synaptic machinery of cultured cortical neurons. *Mol. Brain* **2013**, *6*, 1–15.

127. Amanda, S.M.; Laura, B.-S.K.; John, S.J.; Liming, D.; Saber, H.M. Can silver nanoparticles be useful as potential biological labels? *Nanotechnology* **2008**, *19*, 235104.

128. Corredor, E.; Testillano, P.; Coronado, M.-J.; Gonzalez-Melendi, P.; Fernandez-Pacheco, R.; Marquina, C.; Ibarra, M.R.; de la Fuente, J.; Rubiales, D.; Perez-de-Luque, A.; *et al.* Nanoparticle penetration and transport in living pumpkin plants: *In situ* subcellular identification. *BMC Plant Biol.* **2009**, *9*, doi:10.1186/1471-2229-9-45.

129. Kendall, E.J.; McKersie, B.D. Free radical and freezing injury to cell membranes of winter wheat. *Physiol. Plant.* **1989**, *76*, 86–94.

130. Ceron-Garcia, A.; Vargas-Arispuro, I.; Aispuro-Hernandez, E.; Martinez-Tellez, M.A. Oligoglucan Elicitor Effects during Plant Oxidative Stress, Cell Metabolism. In *Cell Homeostasis and Stress Response*; Bubulya, P., Ed.; InTech: Rijeka, Croatia, 2012.

131. Bak, S.; Beisson, F.; Bishop, G.; Hamberger, B.; Höfer, R.; Paquette, S.; Werck-Reichhart, D. *Cytochrome p450*; The American Society of Plant Biologists: Rockville, MD, USA, 2011; Volume 9, p. e0144.

132. Vranova, E.; Inzé, D.; van Breusegem, F. Signal transduction during oxidative stress. *J. Exp. Bot.* **2002**, *53*, 1227–1236.

133. Gene Expression Omnibus (GEO). Available online: http://www.ncbi.nlm.nih.gov/geo/ (accessed on 28 February 2014).

134. Barrett, T.; Wilhite, S.E.; Ledoux, P.; Evangelista, C.; Kim, I.F.; Tomashevsky, M.; Marshall, K.A.; Phillippy, K.H.; Sherman, P.M.; Holko, M.; *et al.* NCBI GEO: Archive for functional genomics data sets—update. *Nucleic Acids Res.* **2013**, *41*, D991–D995.

135. ArrayExpress. Available online: https://www.ebi.ac.uk/arrayexpress/ (accessed on 28 February 2014).

136. Kolesnikov, N.; Hastings, E.; Keays, M.; Melnichuk, O.; Tang, Y.A.; Williams, E.; Dylag, M.; Kurbatova, N.; Brandizi, M.; Burdett, T.; *et al.* ArrayExpress update—Simplifying data submissions. *Nucleic Acids Res.* **2015**, *43*, D1113–D1116.

137. Kilian, J.; Whitehead, D.; Horak, J.; Wanke, D.; Weinl, S.; Batistic, O.; D'Angelo, C.; Bornberg-Bauer, E.; Kudla, J.; Harter, K. The atgenexpress global stress expression data set: Protocols, evaluation and model data analysis of UV-B light, drought and cold stress responses. *Plant J.* **2007**, *50*, 347–363.

138. Hetu, M.; Tremblay, L.J.; Lefebvre, D.D. High root biomass production in anchored Arabidopsis plants grown in axenic sucrose supplemented liquid culture. *Biotechniques* **2005**, *39*, 345.

139. Thimm, O.; Bläsing, O.; Gibon, Y.; Nagel, A.; Meyer, S.; Krüger, P.; Selbig, J.; Müller, L.A.; Rhee, S.Y.; Stitt, M. Mapman: A user-driven tool to display genomics data sets onto diagrams of metabolic pathways and other biological processes. *Plant J.* **2004**, *37*, 914–939.

140. Lamesch, P.; Berardini, T.Z.; Li, D.; Swarbreck, D.; Wilks, C.; Sasidharan, R.; Muller, R.; Dreher, K.; Alexander, D.L.; Garcia-Hernandez, M.; *et al.* The Arabidopsis information resource (TAIR): Improved gene annotation and new tools. *Nucleic Acids Res.* **2012**, *40*, D1202–D1210.

141. Geisler-Lee, J.; O'Toole, N.; Ammar, R.; Provart, N.J.; Millar, A.H.; Geisler, M. A predicted interactome for Arabidopsis. *Plant Physiol.* **2007**, *145*, 317–329.

142. Stark, C.; Breitkreutz, B.-J.; Chatr-Aryamontri, A.; Boucher, L.; Oughtred, R.; Livstone, M.S.; Nixon, J.; van Auken, K.; Wang, X.; Shi, X. The biogrid interaction database: 2011 update. *Nucleic Acids Res.* **2011**, *39*, D698–D704.

143. Shannon, P.; Markiel, A.; Ozier, O.; Baliga, N.S.; Wang, J.T.; Ramage, D.; Amin, N.; Schwikowski, B.; Ideker, T. Cytoscape: A software environment for integrated models of biomolecular interaction networks. *Genome Res.* **2003**, *13*, 2498–2504.

144. Eden, E.; Navon, R.; Steinfeld, I.; Lipson, D.; Yakhini, Z. Gorilla: A tool for discovery and visualization of enriched go terms in ranked gene lists. *BMC Bioinform.* **2009**, *10*, 48.

145. Huang, D.W.; Sherman, B.T.; Lempicki, R.A. Bioinformatics enrichment tools: Paths toward the comprehensive functional analysis of large gene lists. *Nucleic Acids Res.* **2009**, *37*, 1–13.

146. Consortium, G.O. The gene ontology (GO) database and informatics resource. *Nucleic Acids Res.* **2004**, *32*, D258–D261.

147. Mulder, N.J.; Apweiler, R. The InterPro Database and Tools for Protein Domain Analysis. In *Current Protocols in Bioinformatics*; John Wiley & Sons, Inc.: Berlin, Germany, 2002.

148. Kanehisa, M.; Goto, S.; Sato, Y.; Furumichi, M.; Tanabe, M. KEGG for integration and interpretation of large-scale molecular data sets. *Nucleic Acids Res.* **2011**, *40*, D109–D114.

149. Van de Peer, Y. Calculate and draw custom Venn diagrams. Available online: http://bioinformatics.psb.ugent.be/webtools/Venn/ (accessed on 30 April 2014).

150. Kasprzyk, A. BioMart: Driving a paradigm change in biological data management. *Database* **2011**, *2011*, doi:10.1093/database/bar049.

151. Consortium, T.G.O. *Gene Ontology* 2014. Available online: http://geneontology.org/ (accessed on 30 April 2014).

152. Consortium, T.G.O. *AmiGO 2*. Available online: http://amigo2.berkeleybop.org/amigo (accessed on 30 April 2014).

153. Consortium, T.G.O. Gene ontology: Tool for the unification of biology. *Nat. Genet.* **2000**, *25*, 25–29.

154. Carbon, S.; Ireland, A.; Mungall, C.J.; Shu, S.; Marshall, B.; Lewis, S. AmiGO: Online access to ontology and annotation data. *Bioinformatics* **2009**, *25*, 288–289.

155. Hu, J. The gene ontology normal usage tracking system. Available online: http://gowiki.tamu.edu/wiki/ (accessed on 30 March 2014).

156. Yadav, S.R.; Yan, D.; Sevilem, I.; Helariutta, Y. Plasmodesmata-mediated intercellular signaling during plant growth and development. *Front. Plant Sci.* **2014**, *5*, 44.

157. Maule, A.J.; Gaudioso-Pedraza, R.; Benitez-Alfonso, Y. Callose deposition and symplastic connectivity are regulated prior to lateral root emergence. *Commun. Integr. Biol.* **2013**, *6*, e26531.

158. Vilaine, F.; Kerchev, P.; Clément, G.; Batailler, B.; Cayla, T.; Bill, L.; Gissot, L.; Dinant, S. Increased expression of a phloem membrane protein encoded by NHL26 alters phloem export and sugar partitioning in Arabidopsis. *Plant Cell* **2013**, *25*, 1689–1708.

159. Lee, D.K.; Sieburth, L.E. Plasmodesmata formation: Poking holes in walls with ise. *Curr. Biol.* **2010**, *20*, R488–R490.

160. Burch-Smith, T.M.; Cui, Y.; Zambryski, P.C. Reduced levels of class 1 reversibly glycosylated polypeptide increase intercellular transport via plasmodesmata. *Plant Signal. Behav.* **2012**, *7*, 62–67.

161. Zavaliev, R.; Levy, A.; Gera, A.; Epel, B.L. Subcellular dynamics and role of arabidopsis β-1,3-glucanases in cell-to-cell movement of tobamoviruses. *Mol. Plant Microbe Interact.* **2013**, *26*, 1016–1030.

162. Zalepa-King, L.; Citovsky, V. A plasmodesmal glycosyltransferase-like protein. *PLoS ONE* **2013**, *8*, e58025.

163. De Storme, N.; de Schrijver, J.; van Criekinge, W.; Wewer, V.; Dörmann, P.; Geelen, D. Glucan synthase-like8 and sterol methyltransferase2 are required for ploidy consistency of the sexual reproduction system in arabidopsis. *Plant Cell* **2013**, *25*, 387–403.

164. Xu, M.; Cho, E.; Burch-Smith, T.M.; Zambryski, P.C. Plasmodesmata formation and cell-to-cell transport are reduced in decreased size exclusion limit 1 during embryogenesis in arabidopsis. *Proc. Natl. Acad. Sci. USA* **2012**, *109*, 5098–5103.

165. Burch-Smith, T.M.; Brunkard, J.O.; Choi, Y.G.; Zambryski, P.C. Organelle-nucleus cross-talk regulates plant intercellular communication via plasmodesmata. *Proc. Natl. Acad. Sci. USA* **2011**, *108*, E1451–E1460.

166. Burch-Smith, T.M.; Zambryski, P.C. Loss of increased size exclusion limit (ise)1 or ise2 increases the formation of secondary plasmodesmata. *Curr. Biol.* **2010**, *20*, 989–993.

167. Xie, B.; Wang, X.; Zhu, M.; Zhang, Z.; Hong, Z. CalS7 encodes a callose synthase responsible for callose deposition in the phloem. *Plant J. Cell Mol. Biol.* **2011**, *65*, 1–14.

168. Benitez-Alfonso, Y.; Cilia, M.; San Roman, A.; Thomas, C.; Maule, A.; Hearn, S.; Jackson, D. Control of arabidopsis meristem development by thioredoxin-dependent regulation of intercellular transport. *Proc. Natl. Acad. Sci. USA* **2009**, *106*, 3615–3620.

169. Golomb, L.; Abu-Abied, M.; Belausov, E.; Sadot, E. Different subcellular localizations and functions of Arabidopsis myosin VIII. *BMC Plant Biol.* **2008**, *8*, doi:10.1186/1471-2229-8-3.

170. Maule, A.J. Plasmodesmata: Structure, function and biogenesis. *Curr. Opin. Plant Biol.* **2008**, *11*, 680–686.

3

Chemokine-Releasing Nanoparticles for Manipulation of the Lymph Node Microenvironment

Taissia G. Popova [1], Allison Teunis [1], Ruben Magni [1], Alessandra Luchini [1], Virginia Espina [1], Lance A. Liotta [1] and Serguei G. Popov [2,*]

[1] Center for Applied Proteomics and Molecular Medicine, Department of Molecular Microbiology, School of Systems Biology, George Mason University, Manassas, VA 20110, USA; E-Mails: tpopova@gmu.edu (T.G.P.); ateunis@gmu.edu (A.T.); rmagni@gmu.edu (R.M.); aluchini@gmu.edu (A.L.); vespina@gmu.edu (V.E.); lliotta@gmu.edu (L.A.L.)

[2] National Center for Biodefense and Infectious Diseases, Department of Molecular Microbiology, School of Systems Biology, George Mason University, Manassas, VA 20110, USA

* Author to whom correspondence should be addressed; E-Mail: spopov@gmu.edu

Academic Editor: Subramanian Tamil Selvan

Abstract: Chemokines (CKs) secreted by the host cells into surrounding tissue establish concentration gradients directing the migration of leukocytes. We propose an *in vivo* CK gradient remodeling approach based on sustained release of CKs by the crosslinked poly(N-isopropylacrylamide) hydrogel open meshwork nano-particles (NPs) containing internal crosslinked dye affinity baits for a reversible CK binding and release. The sustained release is based on a new principle of affinity off-rate tuning. The NPs with Cibacron Blue F3G-A and Reactive Blue-4 baits demonstrated a low-micromolar affinity binding to IL-8, MIP-2, and MCP-1 with a half-life of several hours at 37 °C. The capacity of NPs loaded with IL-8 and MIP-1α to increase neutrophil recruitment to lymph nodes (LNs) was tested in mice after footpad injection. Fluorescently-labeled NPs used as tracers indicated the delivery into the sub-capsular compartment of draining LNs. The animals administered the CK-loaded NPs demonstrated a widening of the sub-capsular space and a strong LN influx of leukocytes, while mice injected with control NPs without CKs or bolus doses of soluble CKs alone showed only a marginal neutrophil response. This technology provides a new means to therapeutically direct or restore immune cell traffic, and can also be employed for simultaneous therapy delivery.

Keywords: nanoparticles; chemokines; neutrophils; lymph node

---

1. Introduction

Lymphoid organs and tissues of the host play key roles in the protection of the host from infections and spread of tumors. The immune response induced by microbial or tumor antigens involves a coordinated activity by a multitude of host cell populations. One of the important features of this response is a sophisticated process of direct leukocyte trafficking regulated in part by a chemotactic communication system based on interactions of chemokines (CKs) with cognate cellular receptors [1,2].

CKs secreted by the host cells into the surrounding tissue establish soluble and/or immobilized concentration gradients directing the migration of leukocytes toward areas of high CK concentration. Macrophages, dendritic cells (DCs) and neutrophils (PMNs) use chemotaxis for delivery and presentation of antigens to secondary lymphoid organs. Many pathogens have evolved sophisticated means to thwart the host defense system, including the blockade of leukocyte chemotaxis toward the infection [3,4]. Development of therapeutic means to restore and predictably manipulate the chemotaxis of immune cells is an attractive concept that can influence the design of new vaccine adjuvants, anti-tumor reagents, anti-inflammatory and anti-microbial treatments. A promising approach is to locally engineer CK gradients with the goal of promoting the accumulation of key immune players, such as DCs and neutrophils, or eliciting cascades of immune responses to eliminate pathogens or tumors [5–7].

Recent advances in nanotechnology now offer innovative ways to reach this goal using controlled-release materials to engineer CK gradients for basic studies and therapeutic applications [8–12]. It has been reported that biodegradable NPs providing a sustained release of various CKs are able to direct *in vitro* migration of dendritic cells (DCs), monocytes (DC precursors), and T cells [8,9]. In addition to immune cells, CK-releasing NPs have been used to recruit endogenous progenitor cells to local sites to promote wound healing, tissue regeneration [5,13], and restore perfusion in an ischemic tissue [14] with a potential to treat myocardial infarctions [15]. Importantly, sustained release of CK eliminated the need for repeated injections, a major advantage for clinical applications.

However, the loading of CKs into existing scaffolds suffers considerable limitations. Obtaining functional release of many CKs of interest is problematic, likely due to the harsh microenvironment within eroding particles, low capacity, and complex experimental procedures [12,16,17]. Thus, the need exists for an alternative system that could more efficiently entrap CKs with higher net bioactivity, and provide "point-source" release of a broader range of attractants to engineer CK gradients [17]. We propose to overcome these methodological limitations using a new multifunctional micro- and nano-particle (NP) platform technology recently invented by us [18–28].

The NPs we synthesized consist of the non-toxic polyacrylamide hydrogel scaffold demonstrating long-time persistence in tissues [29–31]. A variety of inner baits of different chemical nature can be quickly incorporated into the NPs to capture from the environment in one step a wide range of molecules including the CKs [19]. Among such baits the textile dyes of triazine, acidic, basic and disperse types attracted attention due to their affinity interactions with broad classes of protein ligands on the basis of specific molecular recognition processes [32,33]. The dye-ligand matrices used in affinity chromatography are thought to mimic the structural features of the corresponding natural substrates, cofactors, *etc.*, and are therefore considered pseudo-affinity matrices [34].

The NPs can be preloaded with the substances of interest which can be reversibly released from the baits at a controlled rate. The hydrogel structure protects the loaded cargo from degradation while the process of loading and release under mild physiological conditions assures preservation of its biological function. We have currently tested more than 20 different bait formulations ensuring capture and release of proteins [19,21]. Based on this platform, we propose a simple and versatile approach for creating CK-loaded NPs which could serve both as a research tool and a prototype of future immunotherapies for manipulating leukocyte trafficking. Our NPs are ideally suited as carriers for reversible loading and controlled release of CKs, which generally are cationic small proteins. The baits we use mimic the natural interactions of proteoglycans with CKs essential for *in vivo* activity. Our methodology can be applied beyond CKs to many host immunomodulatory proteins.

The main goal of this work was to demonstrate feasibility of the CK gradient remodeling approach using our NP technology to increase an influx of neutrophils into draining lymph nodes (LNs) of mice by the CK-releasing NPs. As a first step, we investigated the affinity of binding and dynamics of a sustained release of some CKs by NPs chosen to target an immune response. Secondly, we demonstrated quick NP delivery to draining lymph nodes (LNs) by subcutaneous injection of NP suspension into footpads of mice. Finally, the capacity of IL-8 and MIP-1α-loaded NP to increase neutrophil recruitment to LNs was tested after subcutaneous administration to mice. Our approach can be considered as a model of future therapeutic treatments targeting abnormalities in the recruitment of immune cells.

2. Results

2.1. NPs Containing Different Chemical Baits Can Be Loaded with CKs

The NPs we synthesized are based on poly(N-isopropylacrylamide) (pNIPAm) and methylenebisacrylamide as a cross-linker co-polymerized with allylamine or acrylic acid (AAc,) for incorporation of different chemical baits [19]. The NPs were characterized by their light scattering properties as described previously [19,21]. The average particle diameter in PBS at 25 °C for different batches was in the range of 600–700 nm with a standard deviation of size from 3 to 17 nm. The polydispersity index was found to be 0.2–0.4 indicating a low level of aggregation. These particles contain molecular pores of sufficient size to allow diffusion of the small proteins such as CKs (8–20 kDa) inside the particle core.

As the baits we used the triazine dyes Cibacron Blue F3G-A (Cibacron) and Reactive Blue 4 (Reactive Blue), which were chemically coupled through the amino group of allylamine-containing NPs [33]. The toluidine dye Trypan Blue was coupled to the carboxyl groups of the AAc-co-polymerized NPs activated by the water-soluble carbodiimide [27]. The chemical structures of the dyes containing multiple aromatic, condensed, and heterocyclic rings, as well as 2–4 negatively-charged sulfate groups per molecule, are shown in Figure A1. Overall, the dyes are capable of electrostatic, hydrophobic, and hydrogen-bonding interactions with the proteins [32–34].

The bait-coupled NPs were tested for their binding with IL-8. Since the AAc-containing NPs carry a polyelectrolytic negative charge in physiological conditions, these particles were also included in the test. Taking into account that the CK-binding capacity of the NPs can be influenced by pH [18], all experiments were carried out in PBS at the slightly alkaline pH 7.4 of lymph and blood. The 5% particle suspension was incubated with the CK in PBS at room temperature and the amount of bound CK was tested using sandwich ELISA. The CK experiments are often reported to be complicated by protein aggregation, which was minimized using freshly diluted stock solutions and low CK concentration (250 pg/mL) within the analytical interval of our ELISA test. The NPs demonstrated a range of affinities which allowed extraction of 66%, 68%, 76% and 84% of CK being present in solution by the AAc-, Trypan Blue-, Reactive Blue- and Cibacron-containing NPs, correspondingly (Figure A2). The Cibacron and Reactive Blue NPs were chosen for further experiments because of their superior binding.

To demonstrate the dynamic equilibrium nature of CK–bait interactions required for the release of CKs from NPs *in vivo*, we tested the binding of the CKs at different concentrations of NPs. In these experiments, we followed the procedure used later for the preparation of CK-loaded NPs for animal challenge. IL-8 and MCP-1 (at 250–500 pg/mL) were loaded onto Cibacron and Reactive Blue NPs at 4 °C overnight, quickly spun at room temperature to pellet the NPs after the loading step, and the amount of free CK in supernatants (Sups) was measured with ELISA. The degree of binding depended on the amount of particles interacting with CK, in general agreement with the mass law-driven equilibrium binding (Figure 1). The binding curves, however, somewhat deviated from the expected straight lines, indicating a certain degree of heterogeneity of the binding sites.

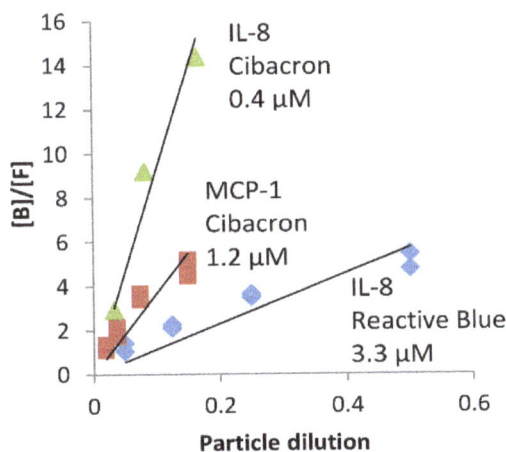

Figure 1. Binding isotherms of IL-8 (0.25 ng/mL) and MCP-1 (1 ng/mL) in PBS at 4 °C, 18 h with different concentrations of NPs (5% wet volume suspension diluted as indicated). Triangles and squares correspond to Cibacron, and diamonds to Reactive Blue. [F] and [B] stand for concentrations of free and particle-bound CKs, respectively.

The dissociation equilibrium constants (K_d) were calculated from the slopes of the straight lines as an approximate measure of affinity using the following equation, assuming independent CK binding to equal binding sites on the NPs:

$$K_d = [F][P]/[B]$$

where F and B stand for free and particle-bound CKs, correspondingly. P refers to the CK binding sites on the NPs.

In the case of the high-capacity NPs (estimated in our case to be close to 4×10^{-5} M based on the amount of dye used for coupling; see Materials and Methods), binding of CK at the concentrations used in our experiments (below 1 μg/mL or 10^{-7} M) does not change [P] to any considerable extent. Therefore, [P] = αP_0, where P_0 is the total concentration of binding sites, and α reflects a dilution of the stock NP suspension upon mixing with CK solution. The values of K_d calculated from the slopes of the plots in Figure 1 were found to be in the low-micromolar to high-nanomolar range favorable for the sustained CK release.

Although the above experiments demonstrated the binding affinity of CKs to bait-containing NPs in diluted solutions (<1 ng/mL), the experiments *in vivo* may require a delivery of small volumes of NPs containing high concentrations of CKs (up to 1 μg/mL [35,36]). Therefore, we tested the extent of binding using Western blot at CK levels of 0.2–2 μg/mL. To reach higher binding capacity, the CK loading onto NPs was carried out in the three-fold diluted PBS (1/3 PBS) which was expected to increase affinity due to reduced shielding of electrostatic interactions by the buffer ions. As a representative example, Figure A3 shows that the MIP-2 binding with Reactive Blue NPs at the decreased ionic strength of the buffer led to a virtually complete removal of the CK from solution. Similar results were obtained with Cibacron NPs (not shown). However, the amount of CK released from the NPs in Figure A3 remained lower than the input amount in control wells. It is likely that due to the CK self-aggregation, a small fraction of the CK was present in the dimeric or multimeric form and therefore was not accounted for. It is also possible that a portion of CK remained tightly bound to the NPs.

2.2. CK-Loaded NPs Provide a Sustained Release of Their Cargo

To test the CK release rates the NPs were loaded with CKs in 1/3 PBS at 4 °C overnight to ensure maximum binding and then quickly pelleted to remove supernatants (Sups). The NPs were then re-suspended in a much larger volume of buffer (compared to the loading step) to initiate the CK dissociation, and the suspension was incubated at the indicated temperature (22 °C or 37 °C) with slight agitation. The NPs from the aliquots of this suspension corresponding to a certain fraction of the total volume were pelleted and the amount of bound CK was determined by Western blot with a CK-specific antibody. An equal fraction of the total CK amount used for loading served as a control representing the ideal case of 100% binding and release. The intensities of the bands at different time points (Figure A4) were compared with the control band intensity. Typical time courses of the IL-8 and MIP-2 release by Cibacron and Reactive Blue NPs are shown in Figure 2. It was found that after loading at 4 °C almost all amounts of CKs were extracted from solution by the NPs. However, a re-suspension of the NP pellet in the fresh buffer released a portion of the CK which appeared to be loosely bound. It was evident from the comparison between the amount of CK in the control and on the NPs immediately after the re-suspension. However, the rest of the CK displayed an expected gradual release rate. Another caveat in these experiments was an observed increase of the MIP-2 amount in the NP pellet after the 10-h incubation at 37 °C. This effect which might reflect the aggregation of the dissociated CK and its co-precipitation with NPs during centrifugation was not studied further. An approximation of the release kinetic with the first-order rate equation was used to estimate the dissociation constants (k_d) and half-lives ($\ln2/k_d$) of the bound CKs (Table 1). According to these data, at the physiological temperature of 37 °C the release of both CKs is expected to take place for more than 20 h (corresponding to ~5× half-life times). The activation energies of the dissociation process calculated from the Table 1 data for IL-8 and MIP-2 were found to be close to each other (70 and 68 kJ/mol, respectively).

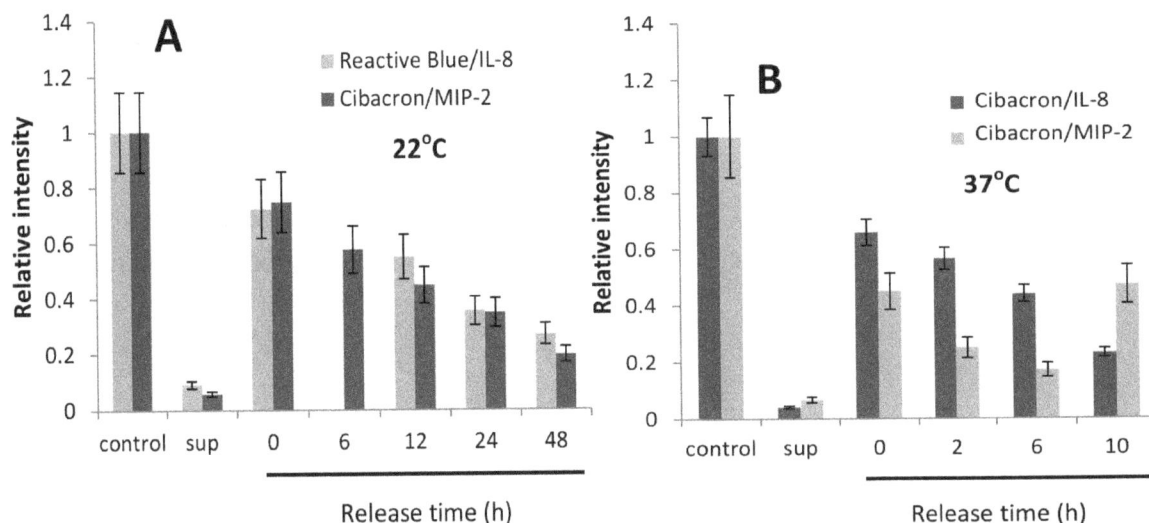

Figure 2. Release of IL-8 and MIP-2 from reactive Blue and Cibacron NPs. CKs (2 μg/mL) were mixed with indicated NPs (10% v/v suspension) in 100 μL of three-fold diluted PBS (1/3 PBS) and incubated at 4 °C overnight. After incubation the NPs were pelleted for 5 min at 16,000 g and room temperature, re-suspended in 1 mL of 1/3 PBS, and incubated at 22 °C (**A**) or 37 °C (**B**). Portions of the suspension were withdrawn at indicated times, the NPs were pelleted for 5 min at 16,000 *g*, and supernatants removed. The remaining pellets were boiled for 5 min in the SDS loading buffer and the amount of CK in solution determined by Western blot. In (**B**) the 1/3 PBS buffer was supplemented with 1 mg/mL BSA. The experiments were run in duplicate. Error bars indicate SD of relative band intensities for pairwise measurements ($n = 6$).

Table 1. Kinetic dissociation constants and half-dissociation times in PBS of IL-8 and MIP-2 bound to Cibacron (CB) and Reactive Blue (RB) NPs.

| Bait | 22 °C | | 37 °C | |
	IL-8	MIP-2	IL-8	MIP-2
CB	k_d 0.016 ± 0.005 * h^{-1} $t_{1/2}$ 43.5 h	k_d 0.030 ± 0.002 * h^{-1} $t_{1/2}$ 23.7 h	k_d 0.094 ± 0.017 * h^{-1} $t_{1/2}$ 7.44 h (BSA 1 mg/mL) ***	k_d 0.17 ± 0.05 * h^{-1} $t_{1/2}$ 4.1 h (BSA 1 mg/mL) ***
RB	k_d 0.022 ± 0.004 * h^{-1} $t_{1/2}$ 31.3 h	ND **	ND **	ND **

Notes: * Standard deviations of k_d calculated from the linear approximations of the kinetic curves; ** Not determined; *** BSA was included in the dissociation buffer.

2.3. BSA Does Not Interfere with CK Loading and Release

In the above experiments, at 37 °C, the binding buffer was supplemented with 1 mg/mL of BSA which is commonly used at concentration up to 10 mg/mL to decrease self-aggregation of proteins in solutions. However, BSA is potentially capable of interfering with the CK binding to bait dyes due to hydrophobic and electrostatic interactions. BSA is well known to bind a variety of proteins [37] and carries a negative charge of about 10 in neutral conditions [38]. On the other hand, the crosslinked structure of the particles is supposed to prevent BSA from entering the particle core [19], thus excluding a direct competition of BSA with the bait. To examine this, we tested the effect of BSA on the extent of CK binding and evaluated the CK dissociation rates in the presence of 1 and 10 mg/mL of BSA. Figure 3 shows that although BSA demonstrated a tendency to reduce the CK binding, its effect was not statistically significant. This suggests that the NP behavior in biological locations such as LNs would not be a subject of strong influence by the serum albumin as a major protein component of lymph.

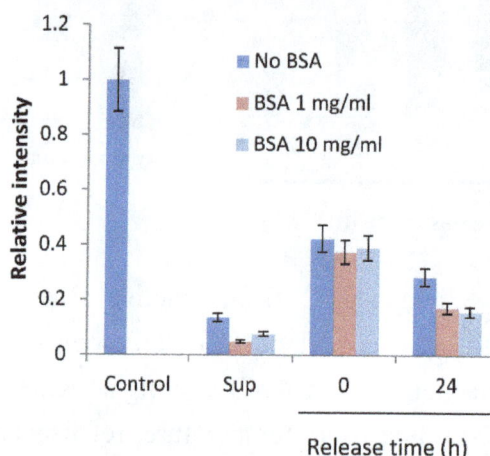

Figure 3. Effect of BSA on MIP-2 binding with and release from the Cibacron NPs. The CK (2 μg/mL) was loaded onto NPs (10% suspension) at 4 °C overnight in PBS buffer diluted 1:3 and supplemented with indicated concentrations of BSA in a total volume of 100 μL. After loading the particles were pelleted and Sups removed. The NP pellet was re-suspended in 1 mL of PBS with the indicated concentrations of BSA at room temperature. The amount of bound CK was determined by Western blot as described in Materials and Methods. The blot image was quantitated and relative intensities of the bands calculated. Error bars indicate SD calculated for three independent samples of control CK loaded on the same gel.

2.4. Subcutaneous Injection Quickly Delivers NPs to Regional LNs

The behavior *in vivo* of the NPs we synthesized has not been previously characterized. Therefore, to ensure effective targeting of the LNs with the NP-bound CKs, it was important to demonstrate that the subcutaneously injected NPs would be delivered by the lymphatic drainage into the regional LNs. For this purpose the pNIPAm NPs co-polymerized with allylamine were covalently labeled with the Alexa Fluor® 555 (Invitrogen, Waltham, MA, USA) fluorescent dye through a coupling reaction of the allylamine primary amino group with the succinimidyl ester-activated dye. The suspension of the fluorescent NPs in PBS was mixed with the equal volume of the 1% Evans Blue dye and injected into the hind footpads of mice. In preliminary experiments it was found that after 30 min the popliteal LNs became intensely stained blue and could be visually located for removal during surgery. The excised LNs were paraformaldehyde-fixed and paraffin-embedded for sectioning.

NPs from the periphery (such as the site of intradermal injection) are expected to quickly travel with the lymph flow to the LN *via* afferent lymphatic vessels entering at the convex side of the LN. Within the LN, the lymph drains through the subcapsular sinus, which is the space between the capsule and the inner part (cortex) of the LN. The lymph then flows inwards into trabecular sinuses, and finally into the medullary sinuses, before exiting through the efferent lymph vessels at the hilum on the concave side. Figure 4 shows a fluorescence microscopy image of the popliteal LN section after NP injection. The NPs were found to be well-dispersed and preferentially localized in the areas of subcapsular and medullar sinuses, while virtually absent from the trabecular sinuses in the cortical area. Similar results were observed in the more distant inguinal LNs (not shown). The subcapsular sinus localization favors the interaction of NPs with subcapsular macrophages and dendritic cells [39]. It is likely that the size of our NPs prevents them from flowing through the LN trabecular sinuses accessible to smaller NPs (<20 nm) [39].

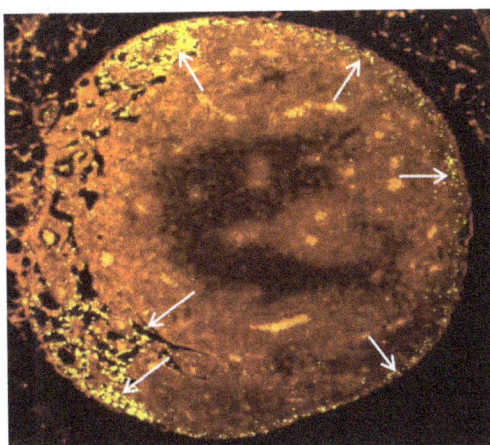

Figure 4. Fluorescent pNIPAm NPs labeled with Alexa Fluor 555 (yellow) quickly migrate to sub-capsular and medullar regions of popliteal LNs of mice (arrows). A suspension of NPs (20 μL) in PBS was injected into mouse hind foot pads for 30 min and the popliteal LNs surgically removed for histologic evaluation. The LNs were paraffin-embedded after fixation with paraformaldehyde, and the 8 μm tissue slices were mounted onto glass slide. The particles were observed at 555/570 nm using Olympus BX51 microscope with a TRITC filter set. Similar responses were detected in all three mice in the group challenged with CK-loaded NPs.

2.5. CK-Loaded NPs Mobilize Immune Cells to the LNs upon Administration to Mice

To demonstrate biological activity of CK-loaded NPs, we chose to test the combined activity of the neutrophil-attracting CKs (IL-8 and MIP-1α [40–42]). The neutrophils are major players during immune reactions demonstrating response to chemotactic stimuli within hours. These cells can be readily detected by immunohistochemistry due to the myeloperoxidase activity in their cytoplasmic granules. The NPs containing Reactive Blue bait were incubated with CKs to create concentrations of bound CKs of 1 and 0.1 μg/mL each. Small volumes (50 μL) of particle suspensions were injected into the hind footpads of mice. At certain times (30 min, 4 h, or 24 h) during post inoculation of the NPs, the mice received additional foot pad injections of the marker dye Evans Blue (1%). After 30 min the mice were euthanized, and the LNs (four per group) extracted for histological evaluation. Control mice received equal doses of the CKs as solutions without NPs or "empty" NPs without CKs. Representative images from these experiments are shown in Figures 5 and 6 along with the quantitation of results in Figure 7.

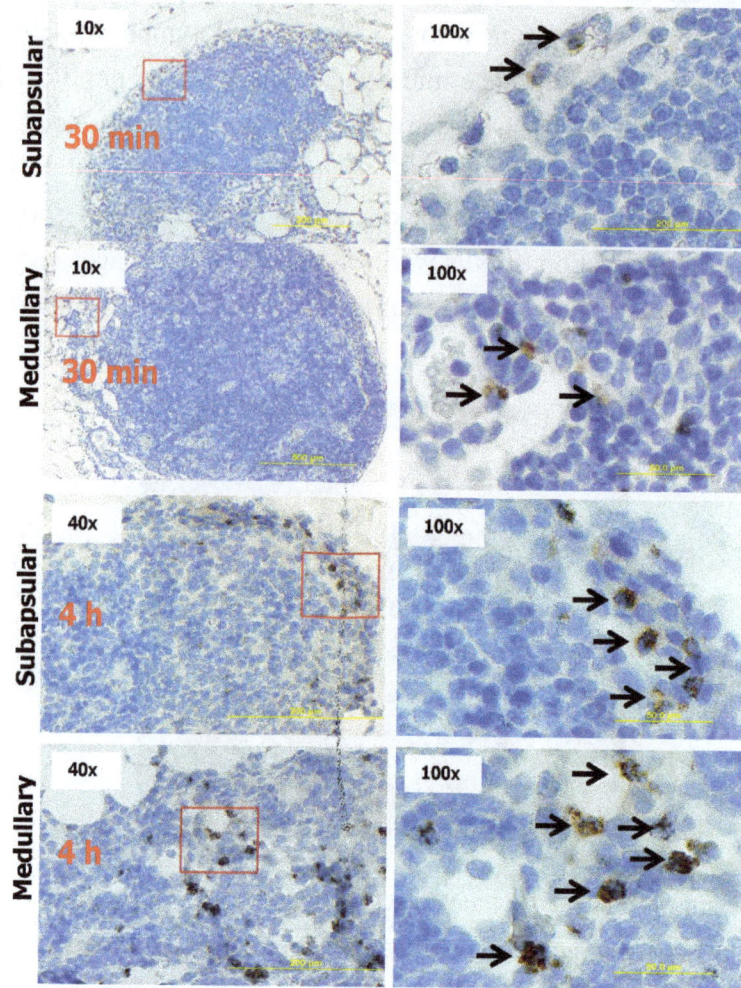

Figure 5. Representative images of the subcapsular and medullary regions of popliteal LNs after injection of Reactive Blue NP suspension (5% wet v/v, 50 μL) into each of the hind footpads of mice. H&E-stained sections after 30 min (**two top rows**) and 4 h (**two bottom rows**) post injection. Squared regions are shown on the right at higher magnification. Neutrophils (arrows) were immunostained brown for myeloperoxidase.

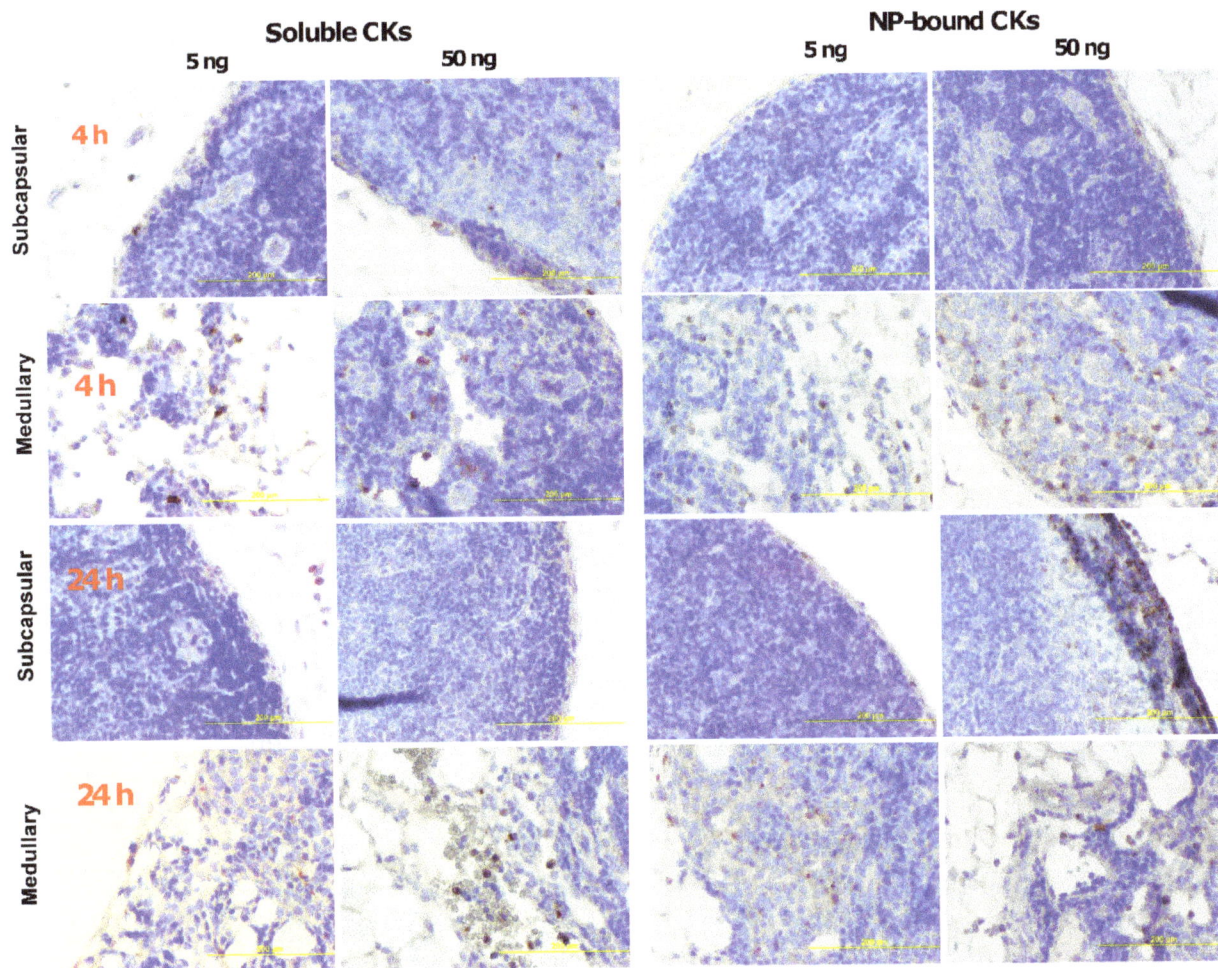

Figure 6. Representative images of the subcapsular and medullary regions of popliteal LNs after injection of the soluble (**two left columns**) and the Reactive Blue NP-loaded (**two right columns**) IL-8 and MIP-1α. H&E-stained sections after 4 h (**two top rows**) and 24 h (**two bottom rows**). The injected amount was 5 ng and 50 ng of each CK in the total volume of 50 μL. Neutrophils were immunostained brown for myeloperoxidase.

Analysis of the LNs after administration of control NPs at 30 min and 4 h post injection showed a low number of migrated neutrophils in subcapsular and medullar regions (Figure 5), in agreement with the distribution pattern seen in the experiments with the fluorescent NPs. These early time points reflect the response of the first wave of neutrophils from a pool immediately available in the circulation [43]. This wave typically subsides after 24 h, and the mobilization of the additional number of cells from the bone marrow during the second wave of neutrophils takes place after several days [44]. As expected, the neutrophil counts (per microscope field of view under × 100 magnification) after the injection of empty NPs in the subcapsular region became reduced from 5.9 ± 2.4 at 4 h post infection to 1.2 ± 1.3 at 24 h, close to the level of naïve mice (0.3 ± 0.01) (Figure 7A). In the medullary region, the counts at 4 h decreased from 10.9 ± 5.6 to 4.6 ± 2.6 at 24 h (Figure 7B).

The injections of soluble CKs revealed a low-grade, dose-dependent neutrophil infiltration which was detectable at 4 h as well as 24 h post injection (Figure 6). Due to the small number of migrated cells (similar to the above experiments with empty NPs) the subcapsular space remained narrow without a substantial increase in the total cell density. We suggested that in the tested conditions the soluble CKs were unable to trigger a substantial neutrophil response as a result of fast dissipation of the bolus CKs' doses. In support of this, at the highest soluble 50 ng dose the counts in the subcapsular region dropped from 13.8 ± 5.5 at 4 h to 1.5 ± 0.8 at 24 h (Figure 7A).

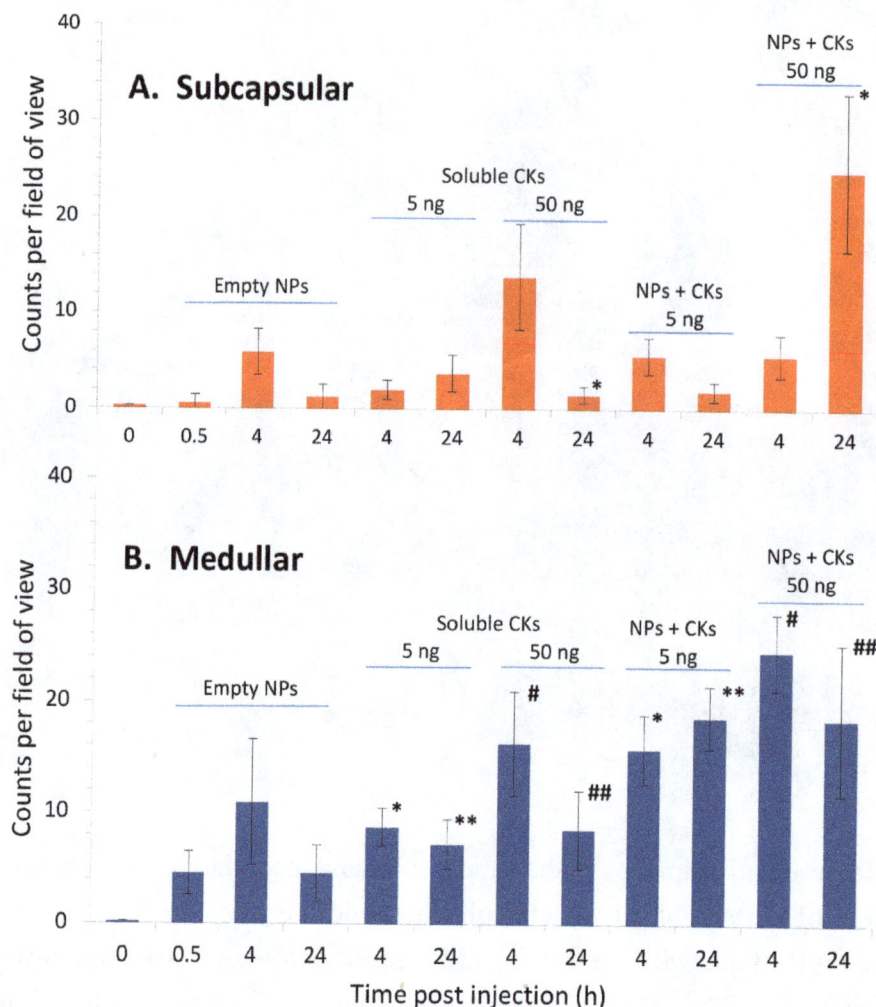

Figure 7. Enumeration of neutrophil counts in the subcapsular (**A**) and medullary (**B**) regions of popliteal LNs in the experiments corresponding to Figures 5 and 6. Mice were injected with the soluble and the Reactive Blue NP-loaded IL-8 and MIP-1α. The number of myeloperoxidase-positive neutrophils in the tissues slices was counted under microscope in five randomly selected fields of view (0.02 mm^2 each) for each of the indicated conditions (chemokine dosage, presence of NPs, and time post challenge). Error bars correspond to 95% confidence intervals. * and # indicate $p \leq 0.05$ between the corresponding counts with and without NPs.

According to the above consideration, the early time points the NP-bound CKs were expected to display even less activity than the equal dose of the soluble one because of the smaller CK amount released by the NPs. In contrast, at longer times post injection we expected a sustained release of the CKs by the NPs to attract more neutrophils than the soluble CKs. A side-by-side comparison of the effects of soluble and NP-bound CKs confirmed the above considerations (Figure 7A). In the subcapsular region, the early 4-h response to the NP-bound CKs at both 5 ng and 50 ng doses did not demonstrate an increase over the level of empty NPs (5.9 ± 2.4). On the other hand, the response to the

50 ng soluble dose (13.8 ± 5.5) was substantially stronger than to the 5 ng dose (2.0 ± 1.0). However, at 24 h post injection the NPs loaded with a 50 ng CK dose demonstrated a widening of the subcapsular space associated with a massive accumulation of the leukocytes (stained blue), including the neutrophils (stained brown for MP) at the level of 24.8 ± 8.2, in sharp contrast with the soluble CKs (1.5 ± 0.8).

In comparison with the subcapsular region, the medullar one demonstrated overall higher levels of infiltrating neutrophils and increased sensitivity to CK stimulation (Figure 7B), likely reflecting their appearance from blood vessels, but not the afferent lymphatics [44]. The contribution of empty NPs was slightly elevated but remained low at the level of occasional cells. The NP-loaded CKs relative to the soluble CKs displayed their stimulating effect at lower dose of 5 ng (15.7 ± 3.2 *vs.* 8.7 ± 1.7) and sustained this effect until 24 h (18.6 ± 2.8 *vs.* 7.2 ± 2.2 at 5 ng and 18.4 ± 6.8 *vs.* 8.5 ± 3.5 at 50 ng).

3. Discussion

The NP materials capable of controlled release of biologically active substances hold promise for the development of new therapeutic approaches for a variety of diseases and adverse conditions. In each particular application, such materials are required to display a range of specific properties relevant to the capacity, stability, release rates, toxicity and other potential side effects, as well as the ease of manufacture. From this standpoint, the pNIPAm NPs represent an attractive platform offering a possibility to generate non-toxic, sufficiently stable, controlled-porosity hydrogel materials amenable to different chemical modifications. The incorporation of more than 20 different dye baits allowed us to modify the affinity of NPs towards the substances of interest for a variety of applications [19,21].

In this study, we explored the possibility of loading the CKs onto the dye-coupled NPs with the aim of creating the nanomaterials which would release their cargo in a controlled manner based on the affinity of a particular CK-bait pair. In contrast to the commonly used biodegradable NPs, the bait affinity NPs allow a possibility of CK loading in mild physiological conditions. In addition, the hydrogel matrix protects the loaded CK from proteolytic damage. The release rate can be adjusted by using different baits. While the exact nature of the dye interactions with target proteins is often unclear, the triazine dyes we used in this study as baits, such as Cibacron F3G-A or Reactive Blue 4, contain sulfate groups and condensed aromatic and heterocyclic hydrophobic rings, which are expected to mimic the naturally occurring sulfated glycoseaminoglycans such as the heparan and dermatan sulfates interacting with CKs in animal tissues [35,45,46]. We found that the affinities of tested CKs (IL-8 and MCP-1) to the bait-containing NPs were in range with the low-micromolar values reported in the case of endothelial cells [47].

The NPs pre-loaded with CK passively release their cargo (that is not covalently bound) according to the affinity off-rate thereby generating a source of free CK. The release process was tested for IL-8 and MIP-2 loaded onto Cibacron and Reactive Blue baits during an overnight incubation at 4 °C. To determine the dissociation rate the NPs were quickly pelleted, the supernatant removed, and the pellet was resuspended in the fresh portion of the buffer (PBS with or without BSA). At a certain time point, the NP were pelleted and the amount of CK remaining in the pellet was analyzed by Western blot in comparison with a control sample representing the CK concentration used for loading. As expected, the initial drop corresponding to the release of the loosely bound CK was followed by a slow phase typical for ligand diffusion out of the cross-linked polymer hydrogels [48]. The half-life times estimated from the data assuming the first-order dissociation reactions were found to be in the range of several hours (at 37 °C), in favorable agreement with dynamics of the immune cells responses to administered CKs. For example, the infiltration of neutrophils in response to intradermal administration of IL-8 to rabbits and mice peaks after 30 min and remains significant until 8 h [49–51]. MIP-1α activity is detectable in mice at 2 h after CK injection and peaks at 10 h. Monocyte, lymphocyte, and, to a lesser degree, eosinophil infiltration was observed peaking at 10–24 h [52]. Calculations show that 1 μL of our NPs will release the amount of CK sufficient to create physiological levels of CKs (1–100 nM [36]) in the volume of 1–100 mL.

An important feature of the pNIPAm hydrogel is the exclusion of large proteins from the inner part of the NPs [19]. As a result, the bait molecules within the NPs are accessible to CKs (having the molecular masses <20 kD), but not to the much bigger BSA. This allows avoiding the competition of CK with BSA for the bait during the process of the CK loading in the presence of BSA, frequently used to prevent aggregation and inactivation of CKs in solution. The same consideration explains our observation that the presence of BSA does not appear to have a substantial effect on the CK dissociation rate.

The pNIPAm NPs are known to be thermoreactive. They shrink upon temperature increase above the lower critical solution temperature (LCST) of 32 °C [53]. The LCST can differ depending on the ratio of hydrophilic and hydrophobic segments of the polymer. Although this topic was not investigated in the current study, our results indicate that the structural changes in the NPs caused by the shift from the ambient temperature of 25 °C to the animal body temperature of 37 °C did not impose considerable constraints on the dissociation of the CKs. The activation energy estimate for this process based on the data from the Table averages around 70 kJ/mol for chemical reactions of this type.

To prove the feasibility of our approach *in vivo*, we carried out animal experiments with the mixture of IL-8 (human CXCL8) and MIP-1α (murine CCL3) loaded onto Reactive Blue NPs. The combined action of these CKs was expected to display a synergistic effect. IL-8 is one of the major pro-inflammatory human CKs which direct migration of neutrophils. It is also highly active upon administration to mice [40]. IL-8 activates the CXCR1 and CXCR2 CK receptors of neutrophils [54–56] implicating them in development and promotion of tumor progression and numerous inflammatory disorders [57]. Modulation of the function of CXCR2 is considered as a possible therapeutic strategy in the treatment of inflammatory conditions in humans [57]. MIP-1α is crucial for immune responses towards infection and inflammation [58]. It activates human granulocytes (neutrophils, eosinophils and basophils) which can lead to acute neutrophilic inflammation. It also participates in the inflammatory response through the induction of synthesis and release of pro-inflammatory cytokines such as IL-1, IL-6 and TNF-α from fibroblasts and macrophages.

In response to chemotactic factors, the neutrophils migrate to the LNs after infections or vaccination [43,59–61]. Our results demonstrate a profound effect of NP-bound CKs on the neutrophil numbers in LNs, in contrast to the control NPs or the bolus injections of CKs, which caused only a mild infiltration of neutrophils. The timing of responses in comparison with controls indicates a sustained release of CKs maintaining a stimulation of migration until 24 h post injection. The differences we detected in the neutrophil counts between the subcapsular and medullar regions of LNs may reflect their migration though distinct pathways. The increased presence of neutrophils in the cortical (subcapsular) sinuses is consistent with the NP entry by way of afferent lymphatics while the medullar neutrophils likely originate from blood in the process of extravasation across vascular portals termed high endothelial venules [43,44]. The latter pathway is used by neutrophils in response to pathogen-mediated inflammation [62], supporting the possibility of their targeting by CK-releasing NPs during infectious disease.

Overall, our results establish feasibility of manipulation with the biological effects of CKs *in vivo* using pNIPAm NPs containing the chemically-coupled dyes as delivery vehicles reversibly binding and releasing the CK molecules. In future studies, we plan to apply this approach to the design of controlled release formulations that deliver natural soluble factors targeting different immune cells for antimicrobial and anticancer applications.

4. Materials and Methods

4.1. Materials

N-Isopropylacrylamide (NIPAm), N-N′-methylenebis(acrylamide) (BIS), potassium persulfate (KPS) and allylamine (AA), were purchased from Sigma-Aldrich (St. Louis, MI, USA), Cibacron Blue F3G-A was purchased from Polysciences, Inc. (Warrington, PA, USA). Unless specified otherwise, all other reagents were from Sigma-Aldrich, and were used as received. Water for all reactions, solution preparation, and polymer washing was distilled, further purified using a Millipore Milli-Q system to a resistance of 18 MΩ, and passed through a 0.2 μm nylon filter. Recombinant mouse CCL3 (MIP-1α), CCL2 (MCP-1), CXCL2 (MIP-2), and human CXCL8 (IL-8) were carrier-free from BioLegend (San Diego, CA, USA).

4.2. Synthesis of NPs

The NP synthesis was carried out essentially as described in [19]. pNIPAm particles with ~7% molar content of acrylic acid (AAc) relative to the total monomer were prepared via precipitation polymerization. NIPAm (9.0 g) and BIS (0.28 g) were dissolved in 250 mL of water, and the solution was then partially degassed by vacuum filtration through a 0.45 μm nylon filter. The filtered solution was purged with nitrogen at room temperature and a medium rate of stirring for 15 min, before AAc (0.5 g) was added to the reaction. Following the addition of AAc, the solution was purged with nitrogen for another 15 min and then heated to 75 °C. Once the reaction mixture had attained a stable temperature of 75 °C, polymerization was initiated with the addition of KPS (0.1 g) in 1.0 mL of water. The reaction was maintained at a constant temperature of 75 °C with stirring under nitrogen for 3 h. After this time, the reaction was allowed to cool to room temperature overnight with stirring under nitrogen. For the preparation of NIPAm functionalized with allylamine, the AAc was replaced with allylamine (670 μL, 12 μmoles). The particles were then harvested and washed by centrifugation for 20 min at 23 °C and 16,000 g, with the supernatant subsequently discarded. The pelleted particles were then re-suspended in 300 mL of water, and the suspended particles pelleted by centrifugation. This centrifugation-dispersion process was repeated a total of five times. Particles were stored as a suspension in water with a few drops of chloroform as an antimicrobial.

Cibacron Blue F3G-A (Cibacron) and Reactive Blue 4 (Reactive Blue), the reactive triazine dyes, were immobilized via direct reaction with the amine group of the allylamine units within the particles, displacing the lone chlorine on the disubstituted triazine ring of the dye [33]. Reactive Blue NPs were a kind gift from Ceres Nanoscience, Inc. Trypan Blue was coupled to the NPs by condensation of the amino group of the dye to the carboxylic group of acrylic acid present in the pNIPAm-co-AAc NPs using activation with N-(3-dimethylaminopropyl)-N'-ethyl carbodiimide as described [27]. After the incorporation of the dyes, the particles were harvested and washed in water by five cycles of centrifugation-dispersion for 20 min at 23 °C and 16,000 g, with the supernatants discarded. Finally, NPs were re-suspended in water with a few drops of chloroform as antibacterial agent. To demonstrate the absence of bacterial contamination, 100 μL of particle suspension were plated on the Luria broth agar and incubated for up to 48 h at 37 °C. The N4 Plus PCS Submicron Particle Analyzer (Beckman Coulter, Brea, CA, USA) was used to determine the particle size and polydispersity index.

4.3. Analysis of CK Binding Using ELISA and Western Blot

The sandwich ELISA Ready-SET-Go!® IL-8 and MCP-1 kits (eBioscience, San Diego, CA, USA) were used according to the manufacturer's protocols to measure the IL-8 and MCP-1 binding to the Cibacron and Reactive Blue NPs. The CKs supplied with the kits were diluted with PBS to concentrations of 0.25–1.0 ng/mL and mixed with equal volumes of NPs washed 3× with PBS by pelleting the NPs at 16,000 g for 5–10 min and re-suspending the pellet in a fresh portion of PBS. Different dilutions of the stock suspension (5% wet pellet v/v) were used. The mixtures of NPs with CKs were incubated for indicated periods of time at 4 °C, the NPs were pelleted at 16,000 g for 5–10 min at room temperature, and the Sups were withdrawn for analyses. Triplicate wells in 96-well plates were used for Sups and control dilutions of the standard CKs.

For Western blot analysis, the samples of Sups and pelleted NPs were mixed with standard 2× SDS-PAGE loading buffer (Invitrogen), boiled for 5 min and loaded on 4%–20% polyacrylamide gel. Protein bands from the gel were transferred onto a nitrocellulose membrane using iBlot Gel Transfer Device (Invitrogen) and probed with the CK-specific polyclonal antibodies (from Biolegend for MIP-2 and LifeTechnologies for IL-8) followed by the appropriate secondary antibodies conjugated with horseradish peroxidase. SuperSignal West Femto Maximum Sensitivity Substrate (Pierce) was used to generate chemo-luminescence of the protein bands, which was measured with a Molecular Imager ChemiDoc XRS System (Bio-Rad, CA, USA). The relative intensities of bands were calculated after densitometry using the QuantityOne 4.6.5 software (Bio-Rad). According to control measurements with different amounts of CKs tested in triplicates within the same experiment, the standard deviations (SD) of relative band intensities were in the range of 7%–17% ($n = 3$).

4.4. Labeling of pNIPAm-co-AA Particles with Alexa Fluor 555

Alexa Fluor 555 (Invitrogen) is a bright orange dye widely used in fluorescent imaging. It is water-soluble and pH-insensitive from pH 4–10. The succinimidyl ester of Alexa Fluor® 555 was used for conjugating the dye to primary amines on pNIPAm-co-AA NPs. For this purpose, 100 μL of NPs were washed 2× with 1 mL of 50 mM bicarbonate buffer, pH 8.3, re-suspended in 500 μL of the buffer, and mixed with 50 μL of the dye solution (1 mg in 100 μL of DMF). After 1 h at room temperature, the particles were washed 3× with 1 mL of PBS (pH 7.4) and finally re-suspended in 500 μL of PBS. The particles were observed at 555/570 nm using Olympus BX51 microscope with a

TRITC filter set. The number of labeled NPs was counted after appropriate dilution and was found to be about 7×10^5 per 1 μL of original suspension. The particles suspended in water had an average size of 520 ± 54 nm with a dispersion index of 1.0 ± 0.3 indicating a slight tendency to aggregation. Overall, the particles were well-dispersed and migrated readily through the lymphatics. For injection into the hind leg footpads of mice, the NP suspension was mixed with equal volume of 2% tracer dye Evans Blue in PBS. This dye allowed us to locate the LNs during surgery and did not quench the fluorescence of Alexa Fluor 555. The LNs were prepared as described below and the particles were observed as described above.

4.5. Animal Challenge and LN Analysis

All animal procedures were approved by the George Mason University's Institutional Animal Care and Use Committee. Female 6–8-week-old DBA/2 mice (Jackson Labs) received food and water *ad libitum* and were challenged into both hind footpads with 50 μL of Reactive Blue NP suspensions or control CK solutions, three animals per challenge group. The NPs (5% wet *v/v*) suspension contained either 0.1 ng/μL or 1 ng/μL of each IL-8 and MIP-1α. Control solutions contained the same amounts of CKs without NPs, or the NPs without CKs. The mixtures were incubated for 18 h at 4 °C and then used for animal inoculations. Groups of mice were euthanized at 30 min, 4 h, and 24 h post inoculation. Thirty min before euthanasia, the animals were anesthetized with ketamine/xylazine and 20 μL of 1% tracer dye Evans Blue in PBS were injected into foot pads. In some experiments, the dye solution contained fluorescent nanoparticles. The LNs were surgically removed into 10% neutral buffered formalin solution for histological evaluation. After fixing in formalin, the tissues were embedded in paraffin, the paraffin blocks were sliced into 8 μm sections, and mounted onto glass slides for standard hematoxylin/eosine (H&E) staining and further microscopic evaluation. To detect the presence of neutrophils, sections after antigen retrieval were incubated in 3% hydrogen peroxide in methanol for 5 min to inhibit peroxidase activity, blocked with Dako Protein block (Dako) for 5 min, and then incubated with a primary anti-myeloperoxidase antibody (Ab9535 from AbCam, dilution 1:50) for 30 min, followed by Dako anti-rabbit EnVision+ HRP-Labeled Polymer (Dako). Colorimetric detection was completed with diaminobenzidine for 5 min, and slides were counterstained with hematoxylin.

5. Conclusions

We propose an *in vivo* CK gradient remodeling approach based on sustained release of CKs with a half-life of several hours at 37 °C by the crosslinked hydrogel open meshwork NPs containing internal covalently linked dye affinity baits for a reversible CK binding. The NPs loaded with IL-8 and MIP-1α demonstrated increased neutrophil recruitment to LNs of mice after footpad injection while mice injected with control NPs without CKs or bolus doses of soluble CKs alone showed only a marginal neutrophil response. This technology provides a new means to therapeutically direct or restore immune cell traffic, and can also be employed for simultaneous therapy delivery.

Acknowledgments

The authors thank Louis Sparace for help with carrying out experiments. This work was supported by the grant 5R21AI099851-2 from the National Institutes of Health, USA (S.G.P., V.E., and L.A.L.). The funders had no role in study design, data collection and analysis, decision to publish, or preparation of the manuscript. Publication of this article was funded in part by the George Mason University Libraries Open Access Publishing Fund.

Author Contributions

Serguei Popov, Lance Liotta, Virginia Espina, and Taissia Popova conceived and designed the experiments; Taissia Popova, Allison Teunis, and Ruben Magni performed the experiments; Alessandra Luchini contributed reagents/materials/analysis tools; Serguei Popov and Taissia Popova wrote the paper.

Appendix

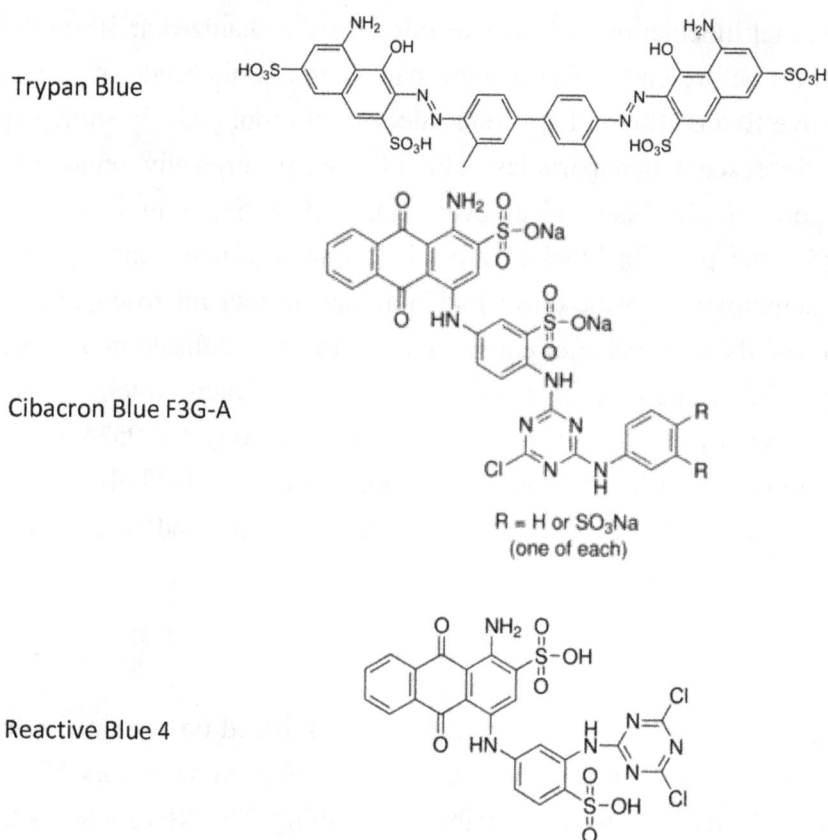

Figure A1. Chemical structures of the Trypan Blue, Cibacron Blue F3G-A (Cibacron), and Reactive Blue 4 (Reactive Blue) dyes.

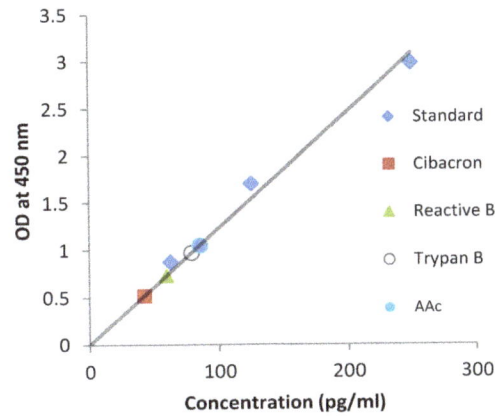

Figure A2. ELISA of IL-8 binding (250 pg/mL) to NPs containing different baits. The particle suspension (5% wet *v/v*) of the NPs was incubated with IL-8 for 30 min at room temperature. Diamonds correspond to the standard IL-8 concentrations used to draw a linear calibration plot.

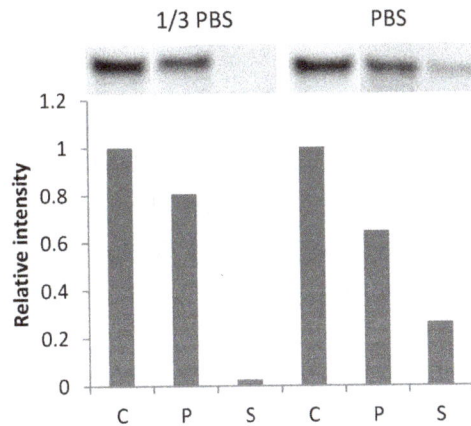

Figure A3. Binding of MIP-2 (200 ng/mL) with Reactive Blue NP suspension (10% wet *v/v*) in PBS and three-fold diluted PBS at 4 °C overnight. After incubation, the particles were quickly pelleted and the amount of CK determined by Western blot. Upper panels, the Western blot images; lower panels, band intensities corresponding to the control amount of CK before binding to the NPs (C), NP pellet (P), and Sups (S).

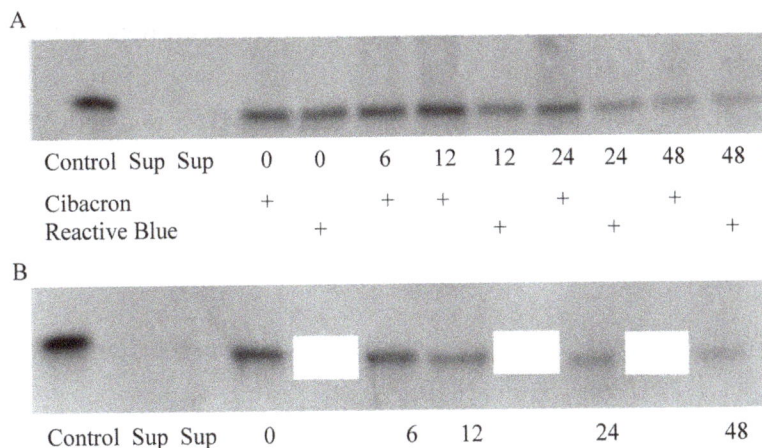

Figure A4. *Cont.*

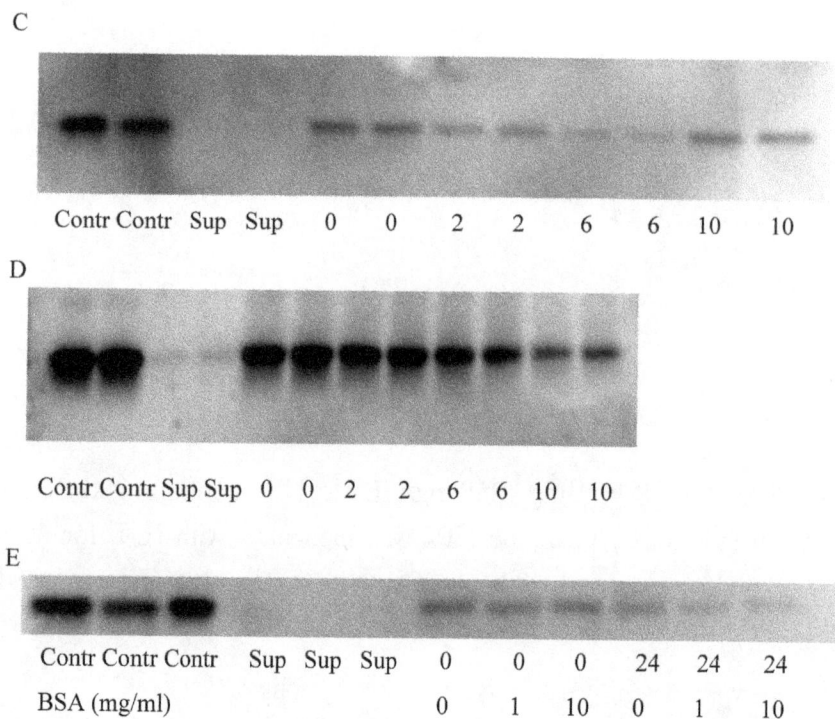

Figure A4. Raw data illustrating western blots in the experiments presented in Figures 2 and 3. (**A**) IL-8 release from Cibacron and Reactive Blue NPs, no BSA, 22 °C; (**B**) MIP-2, Cibacron NPs, no BSA, 22 °C; (**C**) MIP-2 release from Cibacron NPs, no BSA, 37 °C; (**D**) IL-8 release from Cibacron NPs, no BSA, 37 °C; (**E**) MIP-2 release from Cibacron NPs, with BSA, 22 °C. Numbers in the legends indicate time (h) after initiation of dissociation.

Conflicts of Interest

The authors declare no conflict of interest.

References

1. Sadik, C.D.; Luster, A.D. Lipid-cytokine-chemokine cascades orchestrate leukocyte recruitment in inflammation. *J. Leukoc. Biol.* **2012**, *91*, 207–215.
2. Kelly, M.; Hwang, J.M.; Kubes, P. Modulating leukocyte recruitment in inflammation. *J. Allergy Clin. Immunol.* **2007**, *120*, 3–10.
3. Godaly, G.; Bergsten, G.; Hang, L.; Fischer, H.; Frendéus, B.; Lundstedt, A.C.; Samuelsson, M.; Samuelsson, P.; Svanborg, C. Neutrophil recruitment, chemokine receptors, and resistance to mucosal infection. *J. Leukoc. Biol.* **2001**, *69*, 899–906.
4. Liston, A.; McColl, S. Subversion of the chemokine world by microbial pathogens. *BioEssays* **2003**, *25*, 478–488.
5. Hiroshi Kohara, Y.T. Review: Tissue engineering technology to enhance cell recruitment for regeneration therapy. *J. Med. Biol. Eng.* **2010**, *30*, 267–276.
6. Houshmand, P.; Zlotnik, A. Therapeutic applications in the chemokine superfamily. *Curr. Opin. Chem. Biol.* **2003**, *7*, 457–460.

7. Balmert, S.C.; Little, S.R. Biomimetic delivery with micro- and nanoparticles. *Adv. Mater.* **2012**, *24*, 3757–3778.

8. Wang, Y.; Irvine, D.J. Engineering chemoattractant gradients using chemokine-releasing polysaccharide microspheres. *Biomaterials* **2011**, *32*, 4903–4913.

9. Zhao, X.; Jain, S.; Benjamin Larman, H.; Gonzalez, S.; Irvine, D.J. Directed cell migration via chemoattractants released from degradable microspheres. *Biomaterials* **2005**, *26*, 5048–5063.

10. Kumamoto, T.; Huang, E.K.; Paek, H.J.; Morita, A.; Matsue, H.; Valentini, R.F.; Takashima, A. Induction of tumor-specific protective immunity by *in situ* Langerhans cell vaccine. *Nat. Biotechnol.* **2002**, *20*, 64–69.

11. Singh, A.; Suri, S.; Roy, K. *In-situ* crosslinking hydrogels for combinatorial delivery of chemokines and siRNA-DNA carrying microparticles to dendritic cells. *Biomaterials* **2009**, *30*, 5187–200.

12. Kress, H.; Park, J.-G.; Mejean, C.O.; Forster, J.D.; Park, J.; Walse, S.S.; Zhang, Y.; Wu, D.; Weiner, O.D.; Fahmy, T.M.; *et al.* Cell stimulation with optically manipulated microsources. *Nat. Methods* **2009**, *6*, 905–909.

13. Cross, D.P.; Wang, C. Stromal-derived factor-1 alpha-loaded PLGA microspheres for stem cell recruitment. *Pharm. Res.* **2011**, *28*, 2477–2489.

14. Kuraitis, D.; Zhang, P.; Zhang, Y.; Padavan, D.T.; McEwan, K.; Sofrenovic, T.; McKee, D.; Zhang, J.; Griffith, M.; Cao, X.; *et al.* A stromal cell-derived factor-1 releasing matrix enhances the progenitor cell response and blood vessel growth in ischaemic skeletal muscle. *Eur. Cell. Mater.* **2011**, *22*, 109–123.

15. Zhang, G.; Nakamura, Y.; Wang, X.; Hu, Q.; Suggs, L.J.; Zhang, J. Controlled release of stromal cell-derived factor-1 alpha *in situ* increases c-kit+ cell homing to the infarcted heart. *Tissue Eng.* **2007**, *13*, 2063–2071.

16. Van de Weert, M.; Hennink, W.E.; Jiskoot, W. Protein instability in poly(lactic-co-glycolic acid) microparticles. *Pharm. Res.* **2000**, *17*, 1159–1167.

17. Zhu, G.; Mallery, S.R.; Schwendeman, S.P. Stabilization of proteins encapsulated in injectable poly (lactide-co-glycolide). *Nat. Biotechnol.* **2000**, *18*, 52–57.

18. Luchini, A.; Geho, D.H.; Bishops, B.; Tran, D.; Xia, C.; Dufour, R.L.; Jones, C.D.; Espina, V.; Patanarut, A.; Zhou, W.; *et al.* Smart hydrogel particles: Biomarker harvesting: One-step affinity purification, size exclusion, and protection against degradation. *Nano Lett.* **2008**, *8*, 350–361.

19. Longo, C.; Patanarut, A.; George, T.; Bishop, B.; Zhou, W.; Fredolini, C.; Ross, M.M.; Espina, V.; Pellacani, G.; Petricoin, E.F.; *et al.* Core-shell hydrogel particles harvest, concentrate and preserve labile low abundance biomarkers. *PLoS One* **2009**, *4*, e4763.

20. Luchini, A.; Fredolini, C.; Espina, B.H.; Meani, F.; Reeder, A.; Rucker, S.; Petricoin, E.F.; Liotta, L.A. Nanoparticle technology: Addressing the fundamental roadblocks to protein biomarker discovery. *Curr. Mol. Med.* **2010**, *10*, 133–141.

21. Tamburro, D.; Fredolini, C.; Espina, V.; Douglas, T.A.; Ranganathan, A.; Ilag, L.; Zhou, W.; Russo, P.; Espina, B.H.; Muto, G.; *et al.* Multifunctional core-shell nanoparticles: Discovery of previously invisible biomarkers. *J. Am. Chem. Soc.* **2011**, *133*, 19178–19188.

22. Longo, C.; Gambara, G.; Espina, V.; Luchini, A.; Bishop, B.; Patanarut, A.S.; Petricoin, E.F.; Beretti, F.; Ferrari, B.; Garaci, E.; *et al.* A novel biomarker harvesting nanotechnology identifies Bak as a candidate melanoma biomarker in serum. *Exp. Dermatol.* **2011**, *20*, 29–34.

23. Douglas, T.A.; Tamburro, D.; Fredolini, C.; Espina, B.H.; Lepene, B.S.; Ilag, L.; Espina, V.; Petricoin, E.F.; Liotta, L.A.; Luchini, A. The use of hydrogel microparticles to sequester and concentrate bacterial antigens in a urine test for Lyme disease. *Biomaterials* **2011**, *32*, 1157–1166.

24. Luchini, A.; Tamburro, D.; Magni, R.; Fredolini, C.; Espina, V.; Bosch, J.; Garaci, E.; Petricoin, E.F.; Liotta, L. Application of analyte harvesting nanoparticle technology to the measurement of urinary HGH in healthy individuals. *J. Sports Med. Doping Stud.* **2012**, *2*, 2–4.

25. Bosch, J.; Luchini, A.; Pichini, S.; Tamburro, D.; Fredolini, C.; Liotta, L.; Petricoin, E.; Pacifici, R.; Facchiano, F.; Segura, J.; *et al.* Analysis of urinary human growth hormone (HGH) using hydrogel nanoparticles and isoform differential immunoassays after short recombinant HGH treatment: Preliminary results. *J. Pharm. Biomed. Anal.* **2013**, *85*, 194–197.

26. Magni, R.; Espina, B.H.; Liotta, L.A.; Luchini, A.; Espina, V. Hydrogel nanoparticle harvesting of plasma or urine for detecting low abundance proteins. *J. Vis. Exp.* **2014**, e51789.

27. Castro-Sesquen, Y.E.; Gilman, R.H.; Galdos-Cardenas, G.; Ferrufino, L.; Sánchez, G.; Valencia Ayala, E.; Liotta, L.; Bern, C.; Luchini, A. Use of a novel chagas urine nanoparticle test (Chunap) for diagnosis of congenital chagas disease. *PLoS Negl. Trop. Dis.* **2014**, *8*, e3211.

28. Shafagati, N.; Patanarut, A.; Luchini, A.; Lundberg, L.; Bailey, C.; Petricoin, E.; Liotta, L.; Narayanan, A.; Lepene, B.; Kehn-Hall, K. The use of nanotrap particles for biodefense and emerging infectious disease diagnostics. *Pathog. Dis.* **2014**, *71*, 164–176.

29. Wenger, Y.; Schneider, R.J.; Reddy, G.R.; Kopelman, R.; Jolliet, O.; Philbert, M.A. Tissue distribution and pharmacokinetics of stable polyacrylamide nanoparticles following intravenous injection in the rat. *Toxicol. Appl. Pharmacol.* **2011**, *251*, 181–190.

30. Lemperle, G.; Morhenn, V.; Charrier, U. Human histology and persistence of various injectable filler substances for soft tissue augmentation. *Aesthet. Plast. Surg.* **2003**, *27*, 354–366.

31. Gao, D.; Xu, H.; Philbert, M.A.; Kopelman, R. Bioeliminable nanohydrogels for drug delivery. *Nano Lett.* **2008**, *8*, 3320–3324.

32. Kumar, S.; Punekar, N.S. High-throughput screening of dye-ligands for chromatography. *Methods Mol. Biol.* **2014**, *1129*, 53–65.

33. Denizli, A.; Pişkin, E. Dye-ligand affinity systems. *J. Biochem. Biophys. Methods* **2001**, *49*, 391–416.

34. Sereikaite, J.; Bumelis, V.A. Examination of dye-protein interaction by gel-permeation chromatography. *Biomed. Chromatogr.* **2006**, *20*, 195–199.

35. Proudfoot, A.E.I.; Handel, T.M.; Johnson, Z.; Lau, E.K.; LiWang, P.; Clark-Lewis, I.; Borlat, F.; Wells, T.N.C.; Kosco-Vilbois, M.H. Glycosaminoglycan binding and oligomerization are essential for the *in vivo* activity of certain chemokines. *Proc. Natl. Acad. Sci. USA* **2003**, *100*, 1885–1890.

36. Al-Alwan, L.A.; Chang, Y.; Mogas, A.; Halayko, A.J.; Baglole, C.J.; Martin, J.G.; Rousseau, S.; Eidelman, D.H.; Hamid, Q. Differential roles of CXCL2 and CXCL3 and their receptors in regulating normal and asthmatic airway smooth muscle cell migration. *J. Immunol.* **2013**, *191*, 2731–2741.

37. Dennis, M.S.; Zhang, M.; Meng, Y.G.; Kadkhodayan, M.; Kirchhofer, D.; Combs, D.; Damico, L.A. Albumin binding as a general strategy for improving the pharmacokinetics of proteins. *J. Biol. Chem.* **2002**, *277*, 35035–35043.

38. Böhme, U.; Scheler, U. Effective charge of bovine serum albumin determined by electrophoresis NMR. *Chem. Phys. Lett.* **2007**, *435*, 342–345.

39. Manolova, V.; Flace, A.; Bauer, M.; Schwarz, K.; Saudan, P.; Bachmann, M.F. Nanoparticles target distinct dendritic cell populations according to their size. *Eur. J. Immunol.* **2008**, *38*, 1404–1413.

40. Das, S.T.; Rajagopalan, L.; Guerrero-Plata, A.; Sai, J.; Richmond, A.; Garofalo, R.P.; Rajarathnam, K. Monomeric and dimeric CXCL8 are both essential for *in vivo* neutrophil recruitment. *PLoS One* **2010**, *5*, e11754.

41. Baggiolini, M.; Walz, A.; Kunkel, S.L. Neutrophil-activating peptide-1/interleukin 8, a novel cytokine that activates neutrophils. *J. Clin. Investig.* **1989**, *84*, 1045–1049.

42. Ramos, C.D.L.; Canetti, C.; Souto, J.T.; Silva, J.S.; Hogaboam, C.M.; Ferreira, S.H.; Cunha, F.Q. MIP-1alpha[CCL3] acting on the CCR1 receptor mediates neutrophil migration in immune inflammation via sequential release of TNF-alpha and LTB4. *J. Leukoc. Biol.* **2005**, *78*, 167–177.

43. Yang, C.-W.; Strong, B.S.I.; Miller, M.J.; Unanue, E.R. Neutrophils influence the level of antigen presentation during the immune response to protein antigens in adjuvants. *J. Immunol.* **2010**, *185*, 2927–2934.

44. Yang, C.-W.; Unanue, E.R. Neutrophils control the magnitude and spread of the immune response in a thromboxane A2-mediated process. *J. Exp. Med.* **2013**, *210*, 375–387.

45. Massena, S.; Christoffersson, G.; Hjertström, E.; Zcharia, E.; Vlodavsky, I.; Ausmees, N.; Rolny, C.; Li, J.P.; Phillipson, M. A chemotactic gradient sequestered on endothelial heparan sulfate induces directional intraluminal crawling of neutrophils. *Blood* **2010**, *116*, 1924–1931.

46. Kuschert, G.S.; Coulin, F.; Power, C.A.; Proudfoot, A.E.; Hubbard, R.E.; Hoogewerf, A.J.; Wells, T.N. Glycosaminoglycans interact selectively with chemokines and modulate receptor binding and cellular responses. *Biochemistry* **1999**, *38*, 12959–12968.

47. Hoogewerf, A.J.; Kuschert, G.S.; Proudfoot, A.E.; Borlat, F.; Clark-Lewis, I.; Power, C.A.; Wells, T.N. Glycosaminoglycans mediate cell surface oligomerization of chemokines. *Biochemistry* **1997**, *36*, 13570–13578.

48. Fu, Y.; Kao, W.J. Drug release kinetics and transport mechanisms of non-degradable and degradable polymeric delivery systems. *Expert Opin. Drug Deliv.* **2010**, *7*, 429–444.

49. Colditz, I.; Zwahlen, R.; Dewald, B.; Baggiolini, M. *In vivo* inflammatory activity of neutrophil-activating factor, a novel chemotactic peptide derived from human monocytes. *Am. J. Pathol.* **1989**, *134*, 755–760.

50. Zwahlen, R.; Walz, A.; Rot, A. *In vitro* and *in vivo* activity and pathophysiology of human interleukin-8 and related peptides. *Int. Rev. Exp. Pathol.* **1993**, *34B*, 27–42.

51. Taub, D.D.; Anver, M.; Oppenheim, J.J.; Longo, D.L.; Murphy, W.J. T lymphocyte recruitment by interleukin-8 (IL-8). IL-8-induced degranulation of neutrophils releases potent chemoattractants for human T lymphocytes both *in vitro* and *in vivo*. *J. Clin. Investig.* **1996**, *97*, 1931–1941.

52. Menten, P.; Wuyts, A.; van Damme, J. Macrophage inflammatory protein-1. *Cytokine Growth Factor Rev.* **2002**, *13*, 455–481.

53. Qiu, Y.; Park, K. Environment-sensitive hydrogels for drug delivery. *Adv. Drug Deliv. Rev.* **2001**, *53*, 321–339.

54. Ahuja, S.K.; Lee, J.C.; Murphy, P.M. CXC chemokines bind to unique sets of selectivity determinants that can function independently and are broadly distributed on multiple domains of human interleukin-8 receptor B. Determinants of high affinity binding and receptor activation are distinct. *J. Biol. Chem.* **1996**, *271*, 225–232.

55. Holmes, W.E.; Lee, J.; Kuang, W.J.; Rice, G.C.; Wood, W.I. Structure and functional expression of a human interleukin-8 receptor. *Science* **1991**, 253, 1278–1280.

56. Murphy, P.M.; Tiffany, H.L. Cloning of complementary DNA encoding a functional human interleukin-8 receptor. *Science* **1991**, *253*, 1280–1283.

57. Konrad, F.M.; Reutershan, J. CXCR2 in acute lung injury. *Mediat. Inflamm.* **2012**, *2012*, 740987.

58. Ren, M.; Guo, Q.; Guo, L.; Lenz, M.; Qian, F.; Koenen, R.R.; Xu, H.; Schilling, A.B.; Weber, C.; Ye, R.D.; *et al.* Polymerization of MIP-1 chemokine (CCL3 and CCL4) and clearance of MIP-1 by insulin-degrading enzyme. *EMBO J.* **2010**, *29*, 3952–3966.

59. Abadie, V.; Badell, E.; Douillard, P.; Ensergueix, D.; Leenen, P.J.M.; Tanguy, M.; Fiette, L.; Saeland, S.; Gicquel, B.; Winter, N. Neutrophils rapidly migrate via lymphatics after *Mycobacterium bovis* BCG intradermal vaccination and shuttle live bacilli to the draining lymph nodes. *Blood* **2005**, *106*, 1843–1850.

60. Maletto, B.A.; Ropolo, A.S.; Alignani, D.O.; Liscovsky, M.V.; Ranocchia, R.P.; Moron, V.G.; Pistoresi-Palencia, M.C. Presence of neutrophil-bearing antigen in lymphoid organs of immune mice. *Blood* **2006**, *108*, 3094–3102.

61. Chtanova, T.; Schaeffer, M.; Han, S.J.; van Dooren, G.G.; Nollmann, M.; Herzmark, P.; Chan, S.W.; Satija, H.; Camfield, K.; Aaron, H.; *et al.* Dynamics of neutrophil migration in lymph nodes during infection. *Immunity* **2008**, *29*, 487–496.

62. Brackett, C.M.; Muhitch, J.B.; Evans, S.S.; Gollnick, S.O. IL-17 promotes neutrophil entry into tumor-draining lymph nodes following induction of sterile inflammation. *J. Immunol.* **2013**, *191*, 4348–4357.

Metal Organic Framework Micro/Nanopillars of Cu(BTC)·3H₂O and Zn(ADC)·DMSO

Arben Kojtari and Hai-Feng Ji *

Department of Chemistry, Drexel University, Philadelphia, PA 19104, USA;
E-Mail: ak865@drexel.edu

* Author to whom correspondence should be addressed; E-Mail: hj56@drexel.edu

Academic Editor: Jiye Fang

Abstract: In this work, we report the optical and thermal properties of Cu(BTC)·3H₂O (BTC = 1,3,5-benzenetricarboxylic acid) and Zn(ADC)·DMSO (ADC = 9,10-anthracenedicarboxylic acid, DMSO = dimethyl sulfoxide) metal-organic frameworks (MOFs) micro/nanopillars. The morphologies of MOFs on surfaces are most in the form of micro/nanopillars that were vertically oriented on the surface. The size and morphology of the pillars depend on the evaporation time, concentration, solvent, substrate, and starting volume of solutions. The crystal structures of the nanopillars and micropillars are the same, confirmed by powder XRD. Zn(ADC)·DMSO pillars have a strong blue fluorescence. Most of ADC in the pillars are in the form of monomers, which is different from ADC in the solid powder.

Keywords: nanopillars; micropillars; nanowire; metal-organic frameworks (MOFs)

1. Introduction

Metal-organic frameworks (MOFs), a porous, three-dimensionally linked coordination network materials [1,2] have been used for storage, purification, and separations of gases [3,4] as well as for drug release/delivery [5] heterogeneous catalysis [6,7] and sensing [8]. Due to their unique inorganic–organic hybrid nature and nanometer porous structures, MOFs are rich in fundamental properties which promise revolutionary new device concepts. Most of the studies in the past focused on the rational design, synthesis, characterization, and applications of MOFs in their micro-sized cubic crystals obtained from traditional MOF synthesis reaction. Recently, 2D coatings of MOFs and 1D micro/nanostructures, and MOF-based devices are gaining interests because they can be processed into integrated devices. The 2D coatings include polycrystalline thin films [9,10], SURMOF crystalline nanofilms [11], MOF-coated silicon nanowires [12], patterned growth [13], and single crystal arrays [14]. 1D nanowire structures of MOFs have also emerged [15], but a key barrier to wide-scale integration of functional 1D nanostructures into devices is the difficulty of reproducibly forming programmed contacts between nanowires and substrates.

Recently, we developed a surface-assisted approach for mass-production of 1D MOF micro/nanowire arrays that are vertically oriented on substrates [16]. Research in vertically oriented micro/nanowires, which are also called micro/nanopillar [17], is a rapidly growing area in the last decade because of the unique orientation of pillars and their easy-of-use for wide applications, such as photonic devices, cell growth and imaging [18], antireflection [19], light trap [20], battery [21], laser [22], photodetector [23], photovoltaics [24], light-emitting diodes [25], surface-enhance Raman spectroscopy (SERS) signal enhancing [26], drug delivery [27], sensors [28], and enhanced selective catalysis [29]. The uniqueness and advantage of the micro/nanopillar is that one end of each micro/nanopillar is mechanically connected to the surface when the micro/nanopillars are fabricated. In our preliminary work [30], we reported the fabrication procedure and single crystal structures of the first examples of the MOF micro/nanopillars on surfaces. In this work, we report the characterizations of these pillars with powder crystal diffraction pattern, thermal analysis, and fluorescence of the pillars.

2. Results and Discussion

2.1. SEM Images of the Nanopillars

Figure 1 shows the Scanning electron micrographs (SEM) images of $Cu(BTC) \cdot 3H_2O$ and $Zn(ADC) \cdot DMSO$ micro/nanopillar arrays on gold surfaces of substrates. The fabrication procedure has been reported in our previous work [30].

(A)   (B)

Figure 1. Scanning electron micrographs (SEM) images of (A) dense $Cu(BTC) \cdot 3H_2O$ nanopillar array and (B) dense $Zn(ADC) \cdot DMSO$ nanopillars grown on gold substrates. Samples were prepared by mixing a saturated BTC in water or ADC in DMSO solutions along with 30 mM $CuSO_4 \cdot 5H_2O$ or 30 mM $Zn(NO_3)_2 \cdot 6H_2O$, respectively, in equal volume quantities.

In general, the arrays were prepared by immersing gold coated silicon substrates in solutions that are used for synthesizing MOFs. MOFs form in both solutions and on the surfaces. The morphologies of MOFs on surfaces are most in the form of micro/nanopillars that were vertically oriented on the surface. Typical lengths ranged from 10 to 40 μm and diameters in the nanometer range. As evaporation time is extended to weeks both the diameter and the length of the micropillars increase. A similar result can be achieved by increasing the starting volume of both metal ion and ligand solutions used.

Both MOF systems efficiently produced epitaxial growth of micro/nanopillars on gold substrates. Most Cu(BTC)·3H2O pillars were oriented perpendicular to the plane of the substrate, but most of the Zn(ADC)·DMSO pillars have variable tilt angles.

2.2. Effect of Solvent and Substrate on the Formation of Nanopillars

Choices of solvent and substrates for both MOF systems were crucial in controlling pillared growth on the substrates. For Cu(BTC)·3H2O, methanol was initially used as a solvent to dissolve $CuSO_4 \cdot 5H_2O$ and BTC ligand in millimolar to micromolar concentrations to facilitate crystallization. However, mixing of equivalent amounts of both metal ion and ligand at millimolar concentrations yielded mostly square sheets on substrate surfaces (Figure 2A) while micromolar concentrations resulted in amorphous or mesh structures on gold and silicon substrates (Figure 2B,C). Distilled water supplanted methanol as a solvent choice for CuBTC systems, which yielded cubic structures at 10 μM concentrations on glass or silicon (Figure 2D) and nanopillared systems on gold (Figure 1).

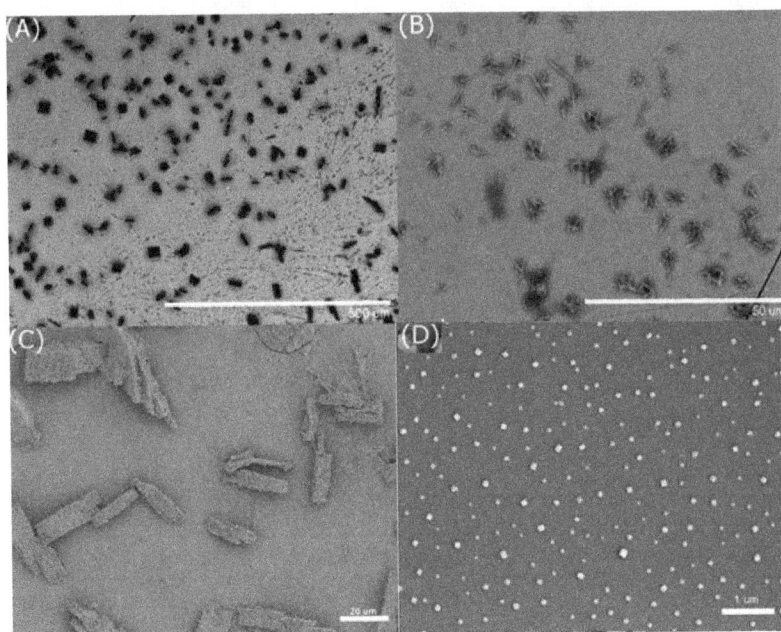

Figure 2. Formation of metal-organic frameworks (MOF) structures of Cu(BTC)·3H2O. (**A**) 100 μL of both 30 mM $CuSO_4 \cdot 5H_2O$ and 30 mM BTC in methanolic solutions mixed and completely dried onto gold substrate; (**B**) 0.3 mM of $CuSO_4 \cdot 5H_2O$ and 0.3 mM BTC in methanolic solutions mixed and completely dried onto gold substrate; (**C**) 100 μL of both 30 mM $CuSO_4 \cdot 5H_2O$ and 30 mM BTC in methanolic solutions mixed and completely dried onto silicon substrate; (**D**) 100 μL of both 10 μM $CuSO_4 \cdot 5H_2O$ and 10 μM BTC in aqueous solutions mixed and completely dried onto glass substrate.

For the ZnADC system, it was first thought to deprotonate both carboxylic acids of the ADC ligand using 1% NaOH to increase the its solubility in distilled water and mixing equimolar and equivalent amounts of both $Zn(NO_3)_2 \cdot 6H_2O$ and ADC solutions on gold, glass, and silicon substrates. However, precipitation immediately occurs after mixing and non-uniform particles are visible on the surface using SEM with both high and low concentration mixing, which indicates that nucleation did not initiate on the surface but rather in solution (Figure 3A,B). This method was abandoned and a co-solvent system was utilized by dissolving the ADC ligand in DMSO and $Zn(NO_3)_2 \cdot 6H_2O$ in distilled water. Results showed the growth of nanopillars on gold substrates (Figure 1B), but not on silicon surfaces. Control experiments showed that ADC alone did not form ordered structures in the mixed solvents.

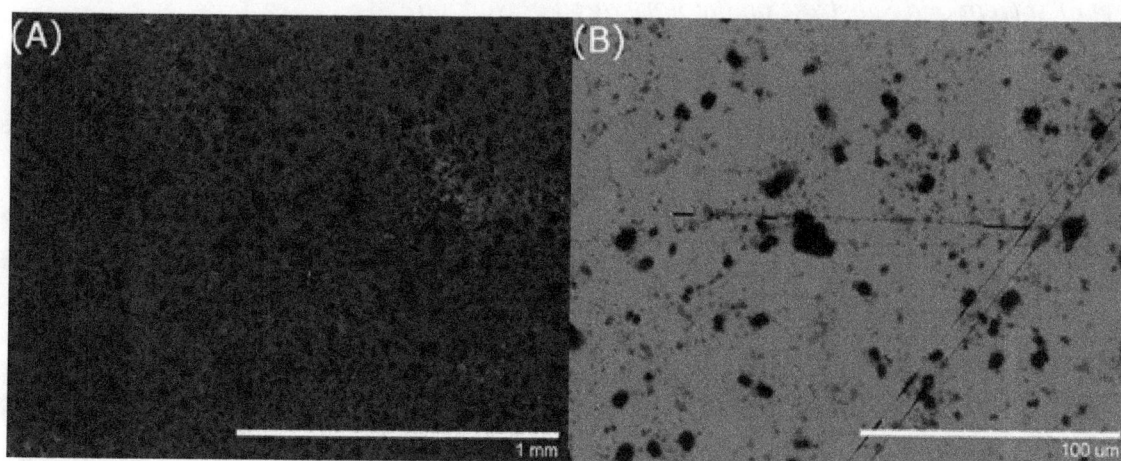

Figure 3. (**A**) 200 µL of 30 mM ADC aqueous solution with 1% NaOH and 30 mM $Zn(NO_3)_2 \cdot 6H_2O$ aqueous solution mixed and dried completely on gold substrate; (**B**) 200 µL of 0.3 mM ADC aqueous solution with 1% NaOH and 0.3 mM $Zn(NO_3)_2 \cdot 6H_2O$ aqueous solution mixed and dried completely on gold substrate.

2.3. XRD of the Micropillars and Nanopillars

Single-crystal data and refinement parameters for both $Cu(BTC) \cdot 3H_2O$ and $Zn(ADC) \cdot DMSO$ systems are summarized in Table 1.

Table 1. Crystal data and structure refinement for the two metal coordination polymers.

Condition and Parameter	Cu(BTC)·3H₂O	Zn(ADC)·DMSO
Empirical formula	$C_9H_{10}O_9Cu$	$C_{19}H_{17}SO_5Zn$
Formula weight	325.71	422.76
Temperature (K)	143(1)	143(1)
Wavelength (Å)	0.71073	0.71073
Crystal system	monoclinic	orthorhombic
Space group	$P2_1/n$	Pnma
a (Å)	6.7778(7)	7.3053(7)
b (Å)	18.8206(18)	17.5056(14)
c (Å)	8.5384(8)	12.6731(12)
β (°)	92.471(4)	-

Table 1. *Cont.*

Condition and Parameter	Cu(BTC)·3H₂O	Zn(ADC)·DMSO
Volume (Å^3)	1088.16(18)	1620.7(3)
Z	4	4
Density (calculated, Mg/m³)	1.988	1.733
μ (mm^{-1})	2.052	1.674
$F(0\ 0\ 0)$	660	868
Crystal size	$0.30 \times 0.22 \times 0.12$ mm³	$0.25 \times 0.02 \times 0.01$ mm³
θ range (°)	2.16 to 27.53	1.98 to 27.53
Index ranges	$-8 \le h \le 8, -24 \le k \le 24, -11 \le l \le 11$	$-9 \le h \le 9, -22 \le k \le 22, -15 \le l \le 16$
Reflections collected	37602	22496
Independent reflections	2488 (R(int) = 0.0170)	1929 (R(int) = 0.0557)
Completeness to θ = 27.53°	99.7%	99.9%
Absorption correction	Semi-empirical from equivalents	Semi-empirical from equivalents
Max. and min. transmission	0.7456 and 0.6410	0.7456 and 0.6511
Refinement method	Full-matrix least-squares on F^2	Full-matrix least-squares on F^2
Data/restraints/parameters	2488/0/173	1929/108/128
Goodness-of-fit on F^2	1.112	1.090
Final R indices (I > 2sigma(I))	$R1 = 0.0194$, w$R2 = 0.0526$	$R1 = 0.0366$, w$R2 = 0.0894$
R indices (all data)	$R1 = 0.0196$, w$R2 = 0.0526$	$R1 = 0.0578$, w$R2 = 0.0996$
Largest diff. peak and hole	0.442 and -0.383 e·Å^{-3}	0.970 and -0.660 e·Å^{-3}

The single crystal XRD study revealed the crystallographic structure of the Cu(BTC)·3H₂O micropillars. However, the crystals of nanopillars are too small to be determined by using single crystal XRD. In order to confirm the structure and crystallinity of the fabricated nanopillars, we performed powder XRD to compare the diffraction patterns of the micropillars and nanopillars, as well as to the calculated pattern of Cu(BTC)·3H₂O. For these experiments, both micro- and nanopillars were prepared in bulk and carefully removed from the gold surfaces for the analyses. Our powder XRD pattern results show strong similarities between the two experimental conditions described in this paper (Figure 4). In addition, the diffractograms for both crystal sizes are agreeable with the calculated pattern using the structure previously published [16].

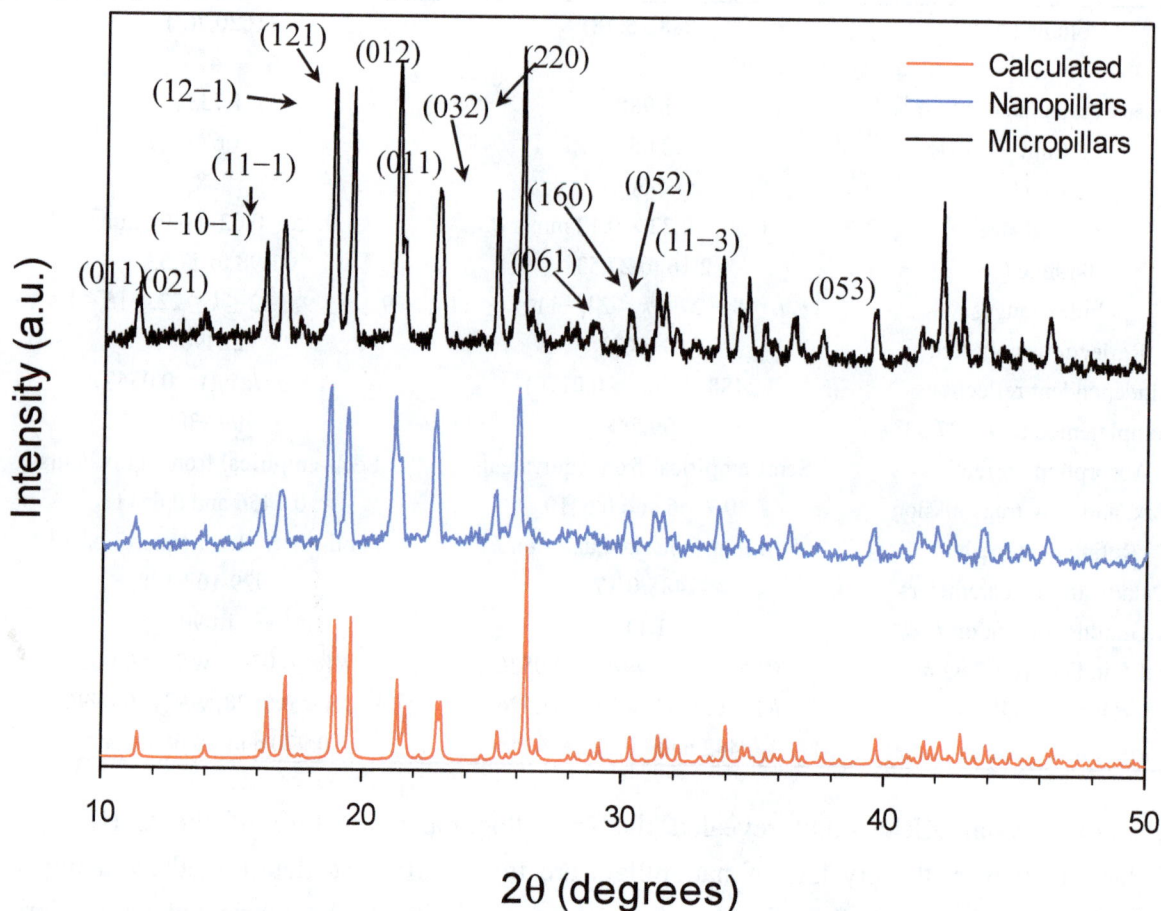

Figure 4. XRD diffraction patterns of Cu(BTC)·3H₂O micropillars and nanopillars, compared to the calculated pattern obtained via Cambridge Crystallographic Data Center (CCDC).

2.4. Thermal Analysis of the MOF Pillars

The thermal properties of the metal-coordination polymer nanopillars are shown in Figure 5. Three weight losses were observed in the thermalgravimetric analysis of Cu(BTC)·3H₂O crystals (Figure 5A). The onset of the first weight loss occurs at 75 °C, which can be attributed to the release of two out of the three Cu^{2+} coordinated water molecules. The loss of the third coordinated water molecule within the structure is removed with an observed weight loss at 205 °C. The observed total weight loss of all three coordinated water molecules is 17.2%, which is agreeable with the calculated percent of water weight loss of 16.5%. The 62.7% weight loss observed at the onset of 270 °C corresponds to the loss of the BTC ligand, which is close to the calculated value of 64.1%. The remaining mass can be attributed to copper metal (calculated 19.4%) and the degraded sample, with an observed percent mass of 20.1% at 800 °C. The weight losses observed correspond to a 3:1:1 ratio between the water molecules, copper metal, and the BTC ligand, respectively. These results are highly agreeable with the crystal structure of the network.

To understand thermal stability of the crystals for future gas absorption studies, thermal analysis was studied. For the Zn(ADC)·DMSO crystals, the thermalgravimetric analysis showed three weight losses (Figure 5B). The onset of the first mass loss was observed at 120 °C, which would correspond to the loss of DMSO solvent absorbed within the pores of the crystals. The mass loss associated with the loss of absorbed DMSO was found to be approximately 10.0%. The onset of the second loss in mass occurred shortly afterwards with a total mass loss of 14.9%. It is thought that is associated with the loss of DMSO coordinated to Zn^{2+} within the coordination network. The onset of the third mass loss occurs at 310 °C, which corresponds to the loss of the ADC ligand (50.4%). The remaining 24.7% mass is mostly due to Zinc metal and degraded carbon in the sample. The calculated ratio of 1:1 for DMSO and ADC, respectively, is agreeable with the crystal structure.

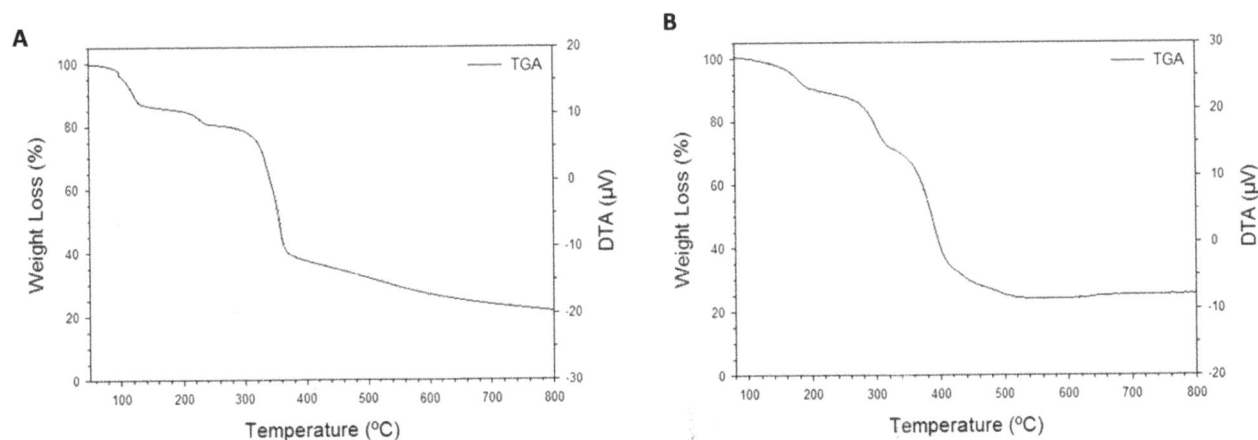

Figure 5. Thermal gravimetric (TG) analyses of (**A**) Cu(HBTC)·3H2O and (**B**) Zn(ADC)·DMSO crystals harvested from gold surfaces after pillared growth.

2.5. Fluorescence of Zn(ADC)·DMSO Pillars

Zn(ADC)·DMSO pillars have a strong blue fluorescence. Figure 6 shows fluorescence spectra (λ_{ex} = 320 nm) of Zn(ADC)·DMSO micro/nanopillars, ADC powder, and a 30 mM DMSO solution of ADC. The ADC solid exhibited a far red-shift with an emission maximum at 494 nm, compared to the maxima of the ADC DMSO solution at 448 nm. The peak in the solution is from the ADC monomer. The emission profiles observed in the powder are attributed to the excimer of ADC, which is common for solid state of polyaromatic hydrocarbons [30–32]. This 46 nm red-shift from solution to the powder state is attributed to J-aggregation of the anthracene molecules due to π–π stacking [31].

The emission peaks of Zn(ADC)·DMSO micro/nanowires are at 434 nm and 457 nm. These peaks also belong to the monomer of ADC [30]; however, with a slight blue-shift. The blue shift is more often due an increase in energy between S_0 and S_1 energy states, which likely arises from the change of the carbonyl resonance by the Zn^{2+} ion [33]. This result is consistent with the observation in XRD, i.e., no π–π interaction between ADC molecules in ZnADC nanopillars. The aforementioned luminescence results were expected for a MOF system using a fluorescent PAH-based ligand.

Figure 7 shows a fluorescence microscope image of Zn(ADC)·DMSO micro/nanopillars. Due to the light propagation along the long direction of nano/microwires, the fluorescence of the Zn/ADC nano/microwires is significantly brighter at the end of the wires, which is the outcoupling light resulting from highly oriented molecular arrangement of the crystal [34], suggesting the Zn/ADC nano/micropillars may be used as a waveguide material for optical applications. The MOF nanopillars prepared in this work can be made in large quantities at a low cost due to the facile self-assembling method.

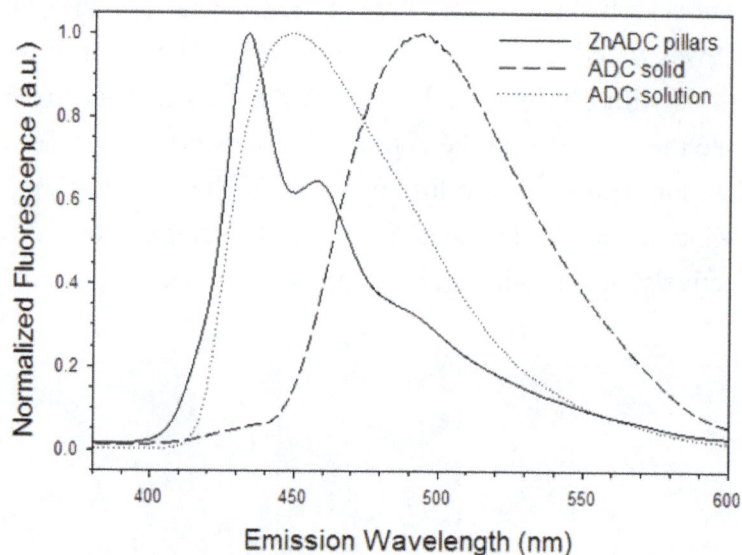

Figure 6. Fluorescence emission spectra of Zn(ADC)·DMSO and 9,10-anthracenedicarboxylic acid using solid-state and solution (λ_{ex} = 320 nm).

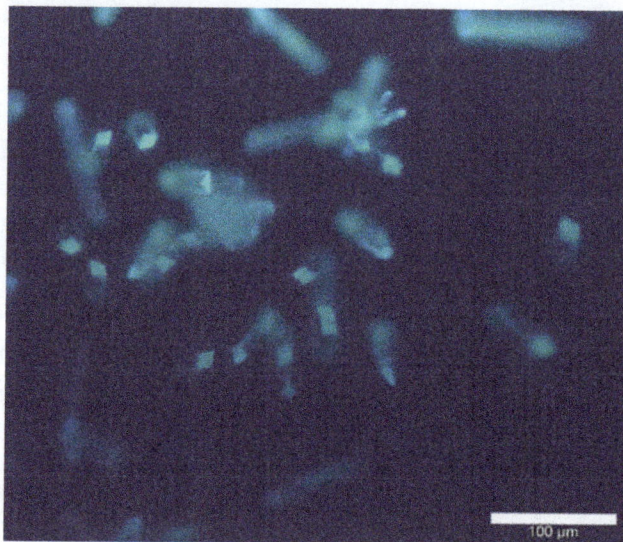

Figure 7. Fluorescence microscope image of Zn(ADC)·DMSO micro/nanopillars.

3. Experimental Section

CuSO$_4$·5H$_2$O (Fisher Scientific), Zn(NO$_3$)$_2$·6H$_2$O (Baker Chemical), 1,3,5-benzenetricarboxylic acid (BTC, Alfa Aesar, Ward Hill, MA, USA), 9,10-anthracenedicarboxylic acid (ADC, Sigma Aldrich, St. Louis, MO, USA), and solvents methanol and dimethyl sulfoxide (Reagent/ACS grade, Pharmco-AAPER, Brookfield, CT, USA) were used as received. Thin-film gold substrates were prepared by e-beam evaporation (BOC Edwards, Crawley, UK). 150 mm-diameter test silicon wafers (Wacker Silicone, Munich, Germany) were coated with a 3 nm thin film of chromium, followed by a

20 nm-thick deposition of gold. Prior to use, substrates were gently rinsed with deionized water and methanol, followed by treatment with UV/ozone (Bioforce ProCleaner, Ames, IA, USA) for 20 min.

Solid state and solution fluorescence spectra were obtained by using a Hitachi F-7000 fluorescence spectrophotometer using an excitation wavelength (λ_{ex}) of 320 nm with both the excitation and emission slit width at 5.0 nm. Fluorescence and optical microscopy images were taken using an Olympus BX51 microscope with an Olympus U-RFL-T Mercury burner. ATR infrared spectra were recorded using a Smiths IlluminatIR microscope. Thermal analyses were conducted using a thermogravimeter/differential thermal analyzer (TG/DTA 6300, SII Nanotechnology, Tokyo, Japan) using a heating rate of 10 °C·min^{-1} with a regulated nitrogen gas flow of 50 cm^3·min^{-1}. Scanning electron micrographs (SEM) were obtained using a Hitachi Tabletop Microscope (TM-1000) and Zeiss Supra 50VP with Oxford energy-dispersive X-ray detection (EDX). Micrographs obtained using the Zeiss Supra 50VP were obtained using the SE2 Inlens. Crystallographic data was collected using a Bruker Kappa APEX II Duo CCD diffractometer. Structure solution and refinement was performed using Bruker's SHELXTL package. Powder X-ray diffraction (XRD) patterns were obtained using a Rigaku Smartlab diffractometer using CuKα1 radiation in Bragg-Brentano focusing and Debye-Scherrer geometry. Calculated XRD patterns were obtained using the Mercury crystal visualization software package. Crystal structure files were obtained from the Cambridge Crystallographic Data Center (CCDC).

4. Conclusions

Crystalline micro/nanowires of Cu(BTC)·3H2O and Zn(ADC)·DMSO were characterized in this work. The facile approach for micro/nanopillar synthesis allows rapid pillared growth with high crystallinity of the MOF systems presented. The MOF micro/nanopillars may have controllable interconnection of the electronic and optoelectronic devices through a vertical integration process using no lithography. This process is low cost and can be high throughput, and offers significant advantages over other methods to prepare nanowire based devices.

Acknowledge

This work was supported from the National Science Foundation (No. DMR-1104835) and the Drexel-SARI Grant.

Author Contributions

Arben Kojtari did all the experiments and wrote the paper draft. Hai-Feng Ji advised the work and modified the MS.

Conflicts of Interest

The authors declare no conflict of interest.

References

1. Delgado-Friedrichs, O.; O'Keeffe, M.; Yaghi, O.M. Taxonomy of periodic nets and the design of materials. *Phys. Chem. Chem. Phys.* **2007**, *9*, 1035–1043.

2. Kuppler, R.J.; Timmons, D.J.; Fang, Q.-R.; Li, J.-R.; Makal, T.A.; Young, M.D.; Yuan, D.; Zhao, D.; Zhuang, W.; Zhou, H.-C. Potential applications of metal-organic frameworks. *Coord. Chem. Rev.* **2009**, *253*, 3042–3066.

3. Banerjee, R.; Phan, A.; Wang, B.; Knobler, C.; Furukawa, H.; O'Keeffe, M.; Yaghi, O.M. High-throughput synthesis of zeolitic imidazolate frameworks and application to CO_2 capture. *Science* **2008**, *319*, 939–943.

4. Sumida, K.; Horike, S.; Kaye, S.S.; Herm, Z.R.; Queen, W.L.; Brown, C.M.; Grandjean, F.; Long, G.J.; Dailly, A.; Long, J.R. Hydrogen storage and carbon dioxide capture in an iron-based sodalite-type metal–organic framework (Fe-BTT) discovered via high-throughput methods. *Chem. Sci.* **2010**, *1*, 184–191.

5. McKinlay, A.C.; Morris, R.E.; Horcajada, P.; Férey, G.; Gref, R.; Couvreur, P.; Serre, C. BioMOFs: Metal-organic frameworks for biological and medical applications. *Angew. Chem. Int. Ed. Engl.* **2010**, *49*, 6260–6266.

6. Lee, J.; Farha, O.K.; Roberts, J.; Scheidt, K.; Nguyen, S.T.; Hupp, J.T. Metal–organic framework materials as catalysts. *Chem. Soc. Rev.* **2009**, *38*, 1450–1459.

7. Robinson, D.M.; Go, Y.B.; Mui, M.; Gardner, G.; Zhang, Z.; Mastrogiovanni, D.; Garfunkel, E.; Li, J.; Greenblatt, M.; Dismukes, G.C. Photochemical water oxidation by crystalline polymorphs of manganese oxides: Structural requirements for catalysis. *J. Am. Chem. Soc.* **2013**, *135*, 3494–3501.

8. Sumida, K.; Rogow, D.L.; Mason, J.A.; McDonald, T.M.; Bloch, E.D.; Herm, Z.R.; Bae, T.-H.; Long, J.R. Carbon dioxide capture in metal–organic frameworks. *Chem. Rev.* **2012**, *112*, 724–781.

9. Bradshaw, D.; Garai, A.; Huo, J. Metal–organic framework growth at functional interfaces: Thin films and composites for diverse applications. *J. Chem. Soc. Rev.* **2012**, *41*, 2344–2381.

10. Huang, A.; Bux, H.; Steinbach, F.; Caro, J. Molecular-sieve membrane with hydrogen permselectivity: ZIF-22 in LTA topology prepared with 3-aminopropyltriethoxysilane as covalent linker. *Angew. Chem. Int. Ed. Engl.* **2010**, *49*, 4958–4961.

11. Xu, G.; Yamada, T.; Otsubo, K.; Sakaida, S.; Kitagawa, H. Facile "modular assembly" for fast construction of a highly oriented crystalline MOF nanofilm. *J. Am. Chem. Soc.* **2012**, *134*, 16524–16527.

12. Liu, N.; Yao, Y.; Cha, J.J.; McDowell, M.T.; Han, Y.; Cui, Y. Functionalization of silicon nanowire surfaces with metal–organic frameworks. *Nano Res.* **2011**, *5*, 109–116.

13. Zhuang, J.; Friedel, J.; Terfort, A. The oriented and patterned growth of fluorescent metal–organic frameworks onto functionalized surfaces. *Beilstein J. Nanotechnol.* **2012**, *3*, 570–578.

14. Carbonell, C.; Imaz, I.; Maspoch, D. Single-crystal metal–organic framework arrays. *J. Am. Chem. Soc.* **2011**, *133*, 2144–2147.

15. Hu, S.-M.; Niu, H.-L.; Qiu, L.-G.; Yuan, Y.-P.; Jiang, X.; Xie, A.-J.; Shen, Y.-H.; Zhu, J.-F. Design, synthesis and characterization of a modular bridging ligand platform for bio-inspired hydrogen production. *Inorg. Chem. Commun.* **2012**, *17*, 147–150.

16. Kojtari, A.; Carrol, P.; Ji, H.-F. Metal organic framework (MOF) micro/nanopillars. *Cryst. Eng. Comm.* **2014**, *16*, 2885–2888.

17. Chandra, D.; Yang, S. Stability of high-aspect-ratio micropillar arrays against adhesive and capillary forces. *Acc. Chem. Res.* **2010**, *43*, 1080–1091.

18. Xie, C.; Hanson, L.; Cui, Y.; Cui, B. Vertical nanopillars for highly localized fluorescence imaging. *Proc. Natl. Acad. Sci. USA* **2011**, *108*, 3894–3899.

19. Li, X.; Li, J.; Chen, T.; Tay, B.K.; Wang, J.; Yu, H. Periodically aligned Si nanopillar arrays as efficient antireflection layers for solar cell applications. *Nanoscale Res. Lett.* **2010**, *5*, 1721–1726.

20. Fan, Z.; Kapadia, R.; Leu, P.W.; Zhang, X.; Chueh, Y.-L.; Takei, K.; Yu, K.; Jamshidi, A.; Rathore, A.A.; Ruebusch, D.J.; *et al.* Ordered arrays of dual-diameter nanopillars for maximized optical absorption. *Nano Lett.* **2010**, *10*, 3823–3827.

21. Ji, L.; Tan, Z.; Kuykendall, T.; An, E.J.; Fu, Y.; Battaglia, V.; Zhang, Y. Multilayer nanoassembly of Sn-nanopillar arrays sandwiched between graphene layers for high-capacity lithium storage. *Energy Environ. Sci.* **2011**, *4*, 3611–3616.

22. Lo, M.-H.; Cheng, Y.-J.; Kuo, H.-C.; Wang, S.-C. Enhanced stimulated emission from optically pumped gallium nitride nanopillars. *Appl. Phys. Express* **2011**, *4*, doi:10.1143/APEX.4.022102.

23. Chen, K.J.; Hung, F.Y.; Chang, S.J.; Young, S.J. Optoelectronic characteristics of UV photodetector based on ZnO nanopillar thin films prepared by sol-gel method. *Mater. Trans.* **2009**, *50*, 922–925.

24. Santos, A.; Formentín, P.; Pallarés, J.; Ferré-Borrull, J.; Marsal, L.F. Fabrication and characterization of high-density arrays of P3HT nanopillars on ITO/glass substrates. *Sol. Energy Mater. Sol. Cells* **2010**, *94*, 1247–1253.

25. Jeon, D.-W.; Choi, W.M.; Shin, H.-J.; Yoon, S.-M.; Choi, J.-Y.; Jang, L.-W.; Lee, I.-H. Nanopillar InGaN/GaN light emitting diodes integrated with homogeneous multilayer graphene electrodes. *J. Mater. Chem.* **2011**, *21*, 17688–17692.

26. Caldwell, J.D.; Glembocki, O.; Bezares, F.J.; Bassim, N.D.; Rendell, R.W.; Feygelson, M.; Ukaegbu, M.; Kasica, R.; Shirey, L.; Hosten, C.; *et al.* Plasmonic nanopillar arrays for large-area, high-enhancement surface-enhanced raman scattering sensors. *ACS Nano* **2011**, 4046–4055.

27. Shalek, A.K.; Robinson, J.T.; Karp, E.S.; Lee, J.S.; Ahn, D.-R.; Yoon, M.-H.; Sutton, A.; Jorgolli, M.; Gertner, R.S.; Gujral, T.S.; *et al.* Vertical silicon nanowires as a universal platform for delivering biomolecules into living cells. *Proc. Natl. Acad. Sci. USA.* **2010**, *107*, 1870–1875.

28. Shin, C.; Shin, W.; Hong, H.-G. Electrochemical fabrication and electrocatalytic characteristics studies of gold nanopillar array electrode (AuNPE) for development of a novel electrochemical sensor. *Electrochim. Acta* **2007**, *53*, 720–728.

29. Zaera, F. Nanostructured materials for applications in heterogeneous catalysis. *Chem. Soc. Rev.* **2012**, *42*, 2746–2762.

30. Mizobe, Y.; Miyata, M.; Hisaki, I.; Hasegawa, Y.; Tohnai, N. Anomalous anthracene arrangement and rare excimer emission in the solid state: Transcription and translation of molecular information. *Org. Lett.* **2006**, *8*, 4295–4298.

31. Dong, Y.; Xu, B.; Zhang, J.; Tan, X.; Wang, L.; Chen, J.; Lv, H.; Wen, S.; Li, B.; Ye, L.; *et al.* Piezochromic luminescence based on the molecular aggregation of 9,10-bis((E)-2-(pyrid-2-yl)vinyl)anthracene. *Angew. Chem. Int. Ed. Engl.* **2012**, *51*, 10782–10785.

32. Abdel-Mottaleb, M.S.A.; Galal, H.R.; Dessouky, A.F.M.; El-Naggar, M.; Mekkawi, D.; Ali, S.S.; Attya, G.M. Fluorescence and photostability studies of anthracene-9-carboxylic acid in different media. *Int. J. Photoenergy* **2000**, *2*, 47–53.

33. Kim, H.S.; Bae, C.O.; Kwon, J.Y.; Kim, S.K.; Choi, M.W.; Yoon, J.Y. Study on the fluorescence of 1,8-anthracenedicarboxylic acid and 1,8-anthracenediformamide. *Bull. Korean Chem. Soc.* **2001**, *22*, 929–931.

34. Balzer, F.; Bordo, V.G.; Simonsen, A.C.; Rubahn, H.-G. Isolated hexaphenyl nanofibers as optical waveguides. *Appl. Phys. Lett.* **2003**, *82*, 10–12.

5

Magneto-Plasmonics and Resonant Interaction of Light with Dynamic Magnetisation in Metallic and All-Magneto-Dielectric Nanostructures

Ivan S. Maksymov

School of Physics, University of Western Australia, Crawley, WA 6009, Australia;
E-Mail: ivan.maksymov@uwa.edu.au

Academic Editor: Lorenzo Rosa

Abstract: A significant interest in combining plasmonics and magnetism at the nanoscale gains momentum in both photonics and magnetism sectors that are concerned with the resonant enhancement of light-magnetic-matter interaction in nanostructures. These efforts result in a considerable amount of literature, which is difficult to collect and digest in limited time. Furthermore, there is insufficient exchange of results between the two research sectors. Consequently, the goal of this review paper is to bridge this gap by presenting an overview of recent progress in the field of magneto-plasmonics from two different points of view: magneto-plasmonics, and magnonics and magnetisation dynamics. It is expected that this presentation style will make this review paper of particular interest to both general physical audience and specialists conducting research on photonics, plasmonics, Brillouin light scattering spectroscopy of magnetic nanostructures and magneto-optical Kerr effect magnetometry, as well as ultrafast all-optical and THz-wave excitation of spin waves. Moreover, readers interested in a new, rapidly emerging field of all-dielectric nanophotonics will find a section about all-magneto-dielectric nanostructures.

Keywords: magneto-plasmonics; magnonics; light-magnetic-matter interaction; ferromagnetic nanostructures; all-dielectric nanostructures

1. Introduction

Light technology has revolutionised society through its applications in medicine, communications, entertainment and culture. Nowadays, industries based on light are major techno-economic drivers, enabling cutting-edge technologies such as solar panels (see, e.g., the introduction to a focus issue on photovolatics [1]), LED lighting [2], computer monitors and mobile phone screens, digital cameras and projectors [3], as well as optical tomography and intravascular photoacoustic imaging [4]. These and many other devices give us better access to information, more reliable health care, better ways of saving energy and new forms of entertainment. To highlight the important role of light-based technologies in our society, the year 2015 was proclaimed by UNESCO to be the International Year of Light and Light-Based Technologies.

It is noteworthy that the research and commercial potentials of light-based technologies are much larger than the state-of-the-art of science and technology and the current market, respectively. For example, microelectronic chips made from silicon are cheap and easy to mass-fabricate, and can be densely integrated. Consequently, researchers in the photonics sector try to create silicon-based optical devices that can exploit the benefit of silicon while also being fully compatible with electronics (see the introduction to a focus issue on silicon photonics [5] and, e.g., [6,7]).

The development of new photonic technologies is very important because semiconductor microelectronics is expected to reach the fundamental limits of its functionality in the observable future. These limits are often discussed in the context of the Moore's law [8] and its limitations [9]. In particular, it is foreseen that integrated photonic circuits will allow optical systems to be more compact than discrete optical components and to have higher performance as compared with microelectronic circuits. Integrated photonic circuits are also expected to be compatible with nanoelectronic components [10] and to be better protected from electromagnetic interference. Finally, it is envisioned that integrated photonic circuits will be capable of exploiting the fundamental aspects of the quantum nature of light and matter, and to develop applications relating to quantum information and quantum computing, as well as measurements at and beyond the standard quantum limits [11,12].

However, it is unknown how fast integrated photonic circuits or at least their individual components (e.g., optical interconnects [13,14]) can complement and/or substitute semiconductor microelectronic chips. In practice, an integrated photonic circuit has to bring together multiple photonic functions [15] on a single chip. In order to achieve this goal one has to have reliable mechanisms of ultrafast all-optical signal processing at the micro- and nanoscale, as well as suitable materials and fabrication technologies. However, whereas many existing photonic technologies are excellent for, e.g., ultrafast all-optical signal processing [15] and connection to the outside of the integrated device [16], no one photonic technology can be used to realise all functions of electronics, microwave and magnetic devices as well as radio frequency antennas.

Consequently, experts agree that integrated devices of the future, which are expected to have better performance and offer previously unavailable nonclassical (quantum) functionality, will most probably rely on employing different systems (see, e.g., [7,11,12]). In turn, each subsystem of these systems will be specialised on fulfilling distinct tasks, and the physics of operation of each subsystem will be different.

Microwave magnetic nanodevices [17–24] are often viewed as the most probable candidates for the integration with photonic devices. This is because many processes involved in physics of magneto-optical phenomena are well understood (see, e.g., [25]), and many magneto-optical devices [25,26] (e.g., magneto-optical drives) had huge commercial success in the past and remain in use nowadays. Moreover, in magnetic data storage media, magnetic nanostructures have already been combined with nanoelectronics (e.g., in read heads and magnetic random access memories).

The aim of this review paper is to critically evaluate the recent progress in the integration of nanophotonic and nanomagnetic technologies. Research on this topic is being currently conducted by both photonics and magnetism sectors, which results in huge amount of literature but also leads to insufficient exchange of results between the two research communities. Consequently, this review paper also aims to bridge this gap. Of course, this review cannot cover the entire field. Several highly relevant areas have been selected without attempting to provide a full survey.

2. Magneto-Plasmonics

Surface plasmons are waves that propagate along a metal-dielectric interface [27–29]. A localised surface plasmon is the result of the confinement of a surface plasmon in a metallic nanoparticle of size comparable to or smaller than the wavelength of light used to excite the plasmon. Metallic nanoparticles at or near their plasmonic resonance generate highly localised electric field intensities. Varieties of nanoparticles and their constellations (e.g., nanoantennas [30]) were shown to enhance the local field and thus improve light-matter interaction.

Since all metals absorb light in the visible and infrared spectral ranges, maximum efficiency of a plasmonic device can be achieved by using metals with the lowest absorption cross-section. Therefore, most of the plasmonic devices are made of gold or silver because these two metals exhibit the lowest absorption losses at optical frequencies.

However, in many practical cases, these metals must be combined with optically active materials in order to provide active control of plasmons [28,29]. In particular, plasmons can be controlled by a magnetic field applied to a hybrid device consisting of a plasmonic metal nanostructure combined with a ferromagnetic layer. While the former supports propagating or localised plasmon modes with the lowest possible absorption losses, the latter exhibits a large magneto-optical activity that opens up routes for ultrafast control of light such as, e.g., magneto-plasmonic switching or high-sensitivity biosensing.

The research field that combines magnetic and photonic functionalities is called magneto-plasmonics. Controlling the optical properties of surface plasmons using magnetic effects was probably first suggested by Chiu and Quinn [31], who investigated the effect of the external static magnetic field on the dispersion relation of surface plasmons in a metal. The review papers by Armelles *et al.* [32] and Temnov [33], a special issue on magneto-plasmonics [34] as well as the book [35] laid out the basics of modern magneto-plasmonics. The paper by Armelles *et al.* [32] also provides an excellent introduction into the magneto-optical Faraday and Kerr effects. Consequently, in this section we focus on developments that have emerged since those works appeared. We also analyse results presented in earlier works that were not discussed in detail in the previous review papers, but are of immediate relevance to this review paper.

2.1. Plasmon-Enhanced Transverse Magneto-Optical Kerr Effect

As shown in [32,33], many works on magneto-plasmonic nanostructures demonstrate the possibility to enhance the magneto-optical effects by exploiting resonance properties of surface plasmons in gratings, which combine a nanopatterned thin layer of a noble metal (gold) with magneto-insulating thin film (bismuth iron garnet, BIG) [36]. Such structures offer a combination of a large Faraday rotation (owing to the BIG film) and small optical losses for wavelengths longer than 650 nm (owing to the nanostructured noble metal). The cross-polarised transmission and polar Kerr rotation in similar structures were measured in separate works as a function of external static magnetic field [37]. However, enhancement of magneto-optical effects by means of surface plasmons was not demonstrated. Previous theoretical works have also predicted that the Faraday and Kerr effects can be resonantly increased in hybrid gold-BIG structures, in particular near the Wood's anomalies [38,39]. These predictions have been confirmed in [36] by observing significant enhancement of the transverse magneto-optical Kerr effect (TMOKE) in transmission.

The TMOKE is defined as a change of reflected intensity of p-polarised light when the direction of the external static magnetic field is changed from the saturated state $+M_s$ to $-M_s$, being M_s the saturation magnetisation [25]. However, as already mentioned above, the TMOKE signal can also be detected in transmission by replacing the reflection coefficient by the transmission coefficient. The attainment of a large TMOKE response is important for many practical applications that include, but not limited to, 3D imaging [3], magnonics (see Section 3), magnetically tuneable optical metamaterials, and magneto-optical data storage [25].

Plasmon-enhanced TMOKE in a planar waveguiding configuration may also be useful for miniaturised photonic circuits and switches, which can be controlled by external magnetic fields [40]. The cited paper experimentally demonstrates a new concept for the magnetic modulation of the light transmission through at a high level of optical transparency. This concept incorporates magneto-insulating thin films with waveguide-plasmon polaritons [41], which stem from the coupling between localised surface plasmons and guided optical modes. The TMOKE signal is quantified as the relative change in the intensity of the transmitted light when the magnetisation is reversed. A large TMOKE signal of and a high experimental transmittance of 45% were demonstrated. However, as will be shown below, this value must be increased in order to make the observed effect more suitable for practical applications.

2.2. Plasmon-Enhanced Faraday Effect

Relative change in the intensity of the transmitted light due to the TMOKE is also known as the "transverse Faraday effect" [42]. However, more often under the Faraday effect (or Faraday rotation) one understands a rotation of the plane of polarisation which is proportional to the component of the magnetic field in the direction of propagation [32]. Apart from a few applications in measuring instruments, sensing, and spintronics, the Faraday rotation can be used for amplitude modulation of light, and are the basis of optical isolators and optical circulators, which are required in optical telecommunications and other laser applications.

It is noteworthy that dimensions of miniaturised commercial optical isolators are about the size of the tip of a pencil. These miniaturised devices offer opportunities for laser diode manufacturers requiring an in-package ultra-high performance isolator. However, their dimensions are not suitable for the application in integrated photonic circuits.

Consequently, one has to utilise arrays of metallic nanoparticles or metallic gratings combined with magneto-insulating films made from, e.g., BIG. The first approach have been used in [43], in which enhanced optical Faraday rotation has been reported in gold-coated maghemite nanoparticles. However, the enhancement of the Faraday rotation achieved using a single nanoparticle may be insufficient for practical applications.

The second approach relying on the combination of metallic gratings with BIG films have been used in [44] (Figure 1a–c). The BIG film was prepared by pulsed laser deposition (see, e.g., [45]). First, a thin buffering layer of 10-nm yttrium iron garnet (YIG) was deposited on a glass substrate and was annealed at 1000 °C. After that, another layer of 140 nm BIG was deposited on the YIG buffering layer. As the next step, gold nanowires were fabricated on the BIG film by electron beam lithography and a subsequent lift-off process.

The fabricated system allows one to simultaneously excite plasmonic and photonic modes of the hybrid structure. A localised plasmon resonance is excited by incoming light polarised perpendicular to the wires. Moreover, the nanowires introduce periodicity to the hybrid structure and enable light to couple into the thin BIG film. Consequently, the nanostructure acts as a planar photonic crystal waveguide for photonic modes [41]. The localised plasmon resonance and the waveguide resonance interact strongly, and this interaction can be controlled by applying external static magnetic field. As a result, Faraday rotation has been increased by up to ~ 9 times compared with the bare BIG film, while high transparency is maintained (Figure 1d,e).

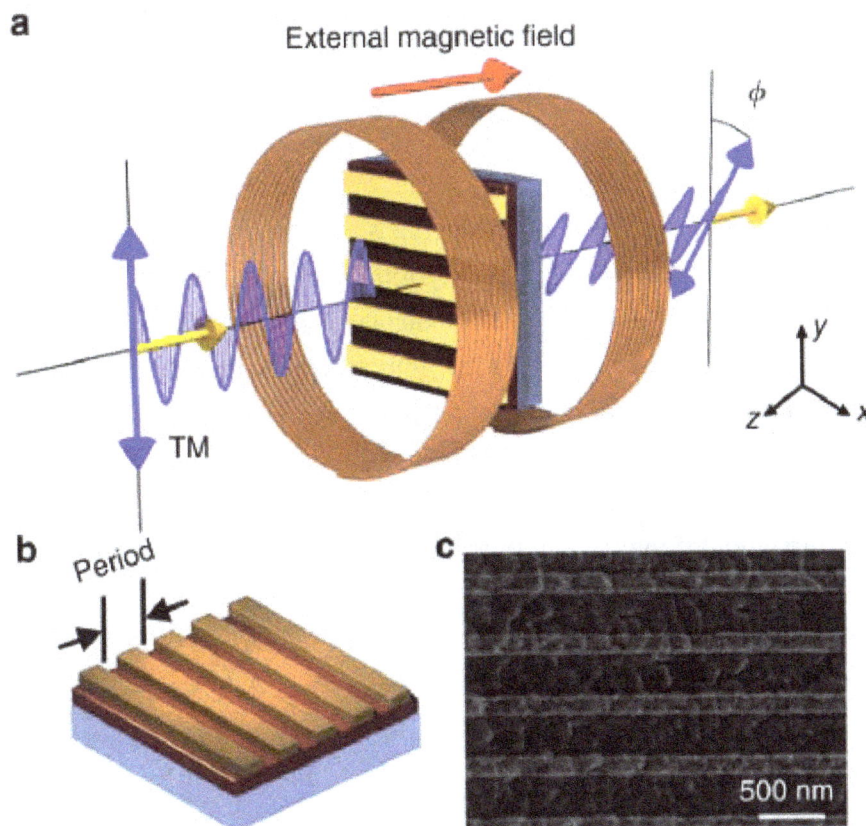

Figure 1. *Cont.*

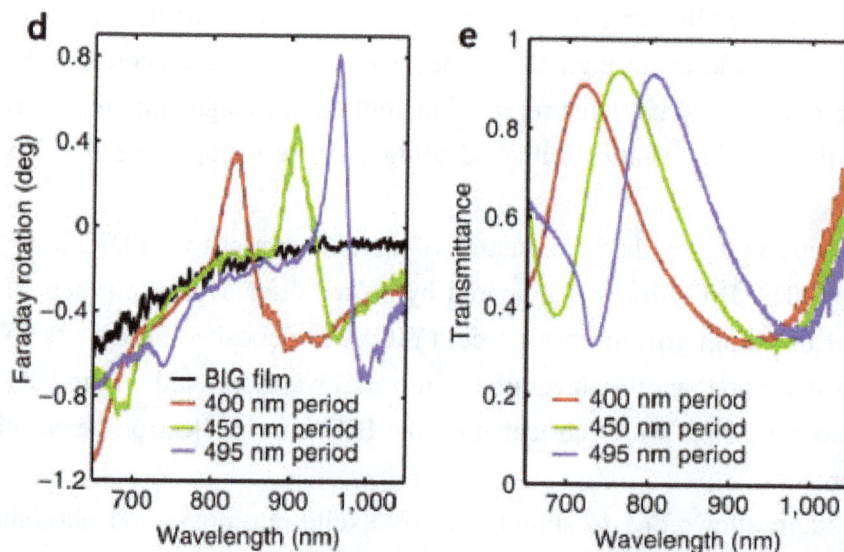

Figure 1. (a) Faraday rotation by a magneto-plasmonic photonic crystal for the TM-polarised incident light, where ϕ is the Faraday rotation angle. At normal incidence, TM-polarised light has the electric field perpendicular to the gold wires; (b) Schematic of the magneto-optical photonic crystal, where the BIG film (dark red) is deposited on a glass substrate (blue) and the periodic gold nanowire structure is sitting atop; (c) A scanning electron microscopy image of one of the investigated hybrid structures; (d) Measured Faraday rotation of the three samples at normal incidence (TM polarization), compared with measured Faraday rotation of the bare BIG film; (e) Measured transmittance of the three samples at normal incidence (TM polarization). Reprinted from [44] with permission by Macmillan Publishers Ltd, Copyright 2013.

We would like to discuss the problem of achieving simultaneously high Faraday effect (and also TMOKE, see [40]) and high transparency of the hybrid plasmonic-magneto-optic nanostructure. As shown in Figure 1d-e, the spectra of Faraday rotation display a resonant feature. As the period increases, the resonance shifts to longer wavelength and the maximum Faraday rotation increases. The sample with the 495 nm period produces a maximum Faraday rotation at 963 nm, which is an $\sim$ 9-fold enhancement as compared with the Faraday rotation in the BIG film without gold wires. However, the same sample shows just 36% transmittance at 963 nm. Despite a considerable enhancement of the Faraday rotation and impressive miniaturisation of the device, this value of transmission is far from >80%–90%, which is required for practical applications.

The same problem of low transmission that accompanies the plasmon-enhanced TMOKE response also exists in all-ferromagnetic nanostructures, which are discussed in Section 2.3. The explanation of the origin of this problem in the framework of the Fano resonance [46] and possible ways of its solution will also be summarised in Section 2.3.

2.3. Plasmon-Enhanced TMOKE in All-Ferromagnetic Nanostructures

As discussed above, microwave magnetic devices are often viewed as the most probable candidates for the integration with photonic devices. Magnonic crystals, which are an important research direction in a broader field of microwave magnetism [17–21,23], are one- or two-dimensional nanostructures consisting, respectively, of ferromagnetic metal nanostripes or dots, anti-dots (holes in ferromagnetic films) and their combinations. In addition, all-ferromagnetic nanostructures are potential candidates for ultra-high-density magnetic recording and they are controllable by light (see, e.g., [47]).

Magnonic crystal can also be made from magneto-insulating materials such yttrium iron garnet (YIG) [18,48]. Similar to BIG, YIG possesses good optical properties, which makes it possible to develop devices exploiting the physics of light-spin wave interaction [26]. The progress in this direction is supported by advances in the fabrication of magneto-optical components compatible with silicon platform [49]. Resonant optical properties of all-magneto-dielectric nanostructures will be discussed in Section 4.

Apart from their importance in magnonics, all-ferromagnetic nanostructures have attracted significant attention because they can be used to enhance magneto-optical effects (see, e.g., [50–58]). Indeed, although ferromagnetic metals exhibit a stronger optical damping as compared with gold or silver [55], resonance excitation of surface plasmons in ferromagnetic nanostructures is significant, which makes it possible to enhance the TMOKE response.

For example, pure nickel subwavelength one-dimensional gratings (Figure 2a) without any noble-metal film inclusions [50,59] support the excitation of surface plasmons strong enough to achieve the TMOKE enhancement comparable with the case of trilayer Au/Co/Au films [32] and ten times larger than in a reference continuous nickel film (Figure 2b). It is noteworthy that plasmonic enhancement of magneto-optical effects was also demonstrated in all-ferromagnetic two-dimensional nanostructures (see, e.g., [35,60–62]). However, similar to the hybrid nanostructure used to enhance the Faraday effect (Figure 1), the maximum of the TMOKE signal produced by the grating is accompanied by a significant decrease in reflection (the green line in Figure 2b).

We would like to note that the full width of plasmonic resonances in noble-metal gratings is mostly defined by the radiative damping of surface plasmons, which leads to the Fano-type shape of the resonances that appear in the spectral vicinity of the Wood's anomaly [63]. It is well-known that in plasmonic nanostructures the lineshape of the Fano resonance can be manipulated by using different approaches (see, e.g., [64–66]).

The same strategy works when applied to all-ferromagnetic magneto-plasmonic nanostructures. For example, in [57] a strong plasmon-assisted TMOKE response was achieved in all-Permalloy one-dimensional grating simultaneously with the maximum of reflectivity attainable in this nanostructure (Figure 3). Although the maximum attainable reflectivity is just 20%, which is because this grating is in fact a one-dimensional magnonic crystal that was designed without optimising the optical properties, the work [57] clearly demonstrates a correlation between the TMOKE enhancement and the Fano resonance [46]. Indeed, as shown by the vertical lines in Figure 3b–e, both experiment and theory demonstrate that the maximum of the plasmon-enhanced TMOKE response has been achieved at the frequency of the Fano resonance, which for the given geometry of the Permalloy grating also corresponds to nearly the maximum of the grating reflectivity.

Of course, in magneto-plasmonic experiments the Fano resonance frequency depends on the applied static magnetic field. However, in many cases experimental conditions do not allow one to measure the absolute values of reflectivity for the two opposite directions of the magnetisation vector with accuracy which is sufficient for reliable extraction of the value of the shift in the Fano resonance frequency upon reversal of magnetisation. In [57], using the simulation data this shift was estimated to be of at least 1.5 nm. Therefore, in [57] a model of general asymmetric Fano resonance profile with magnified characteristic features was used to demonstrate that the maximum of the TMOKE occurs at the frequency at which the slope of the Fano resonance peak is maximum. This finding is consistent with a previous theoretical model proposed by Belotelov *et al.* [36].

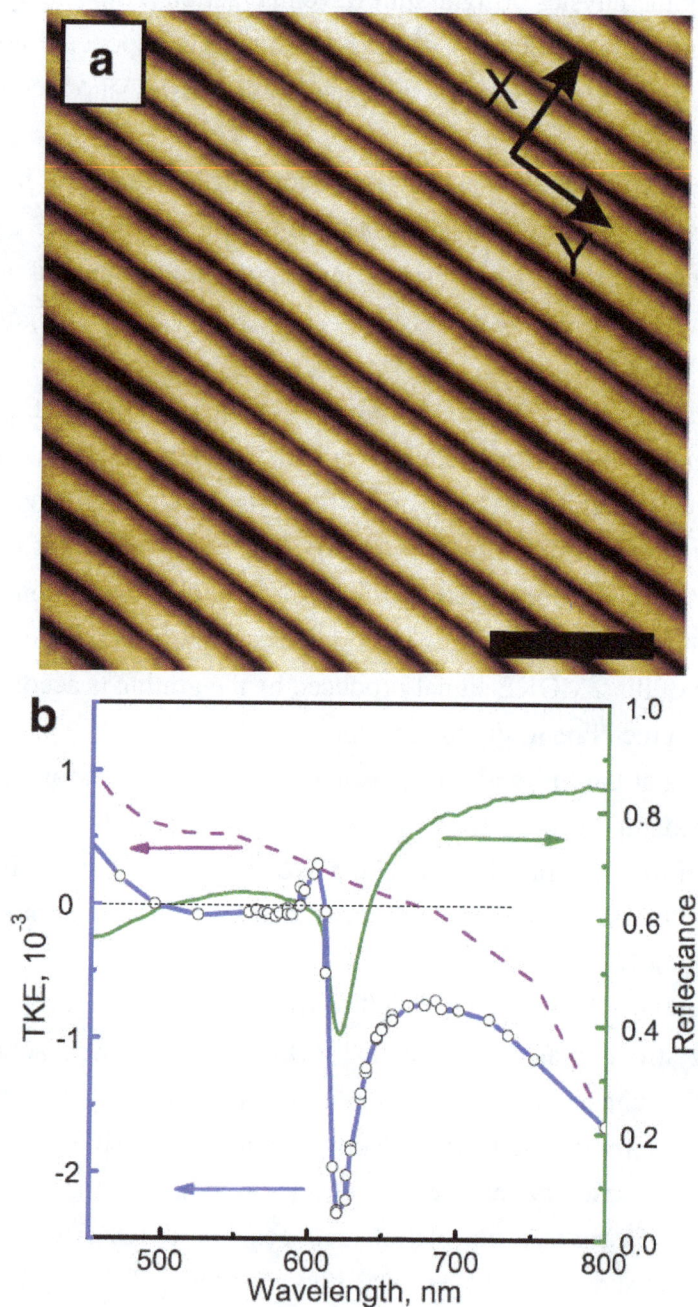

Figure 2. (a) The AFM image of the all-nickel magneto-plasmonic grating. Black bar is equal to 1 μm; (b) TMOKE spectra of the one-dimensional magneto-plasmonic grating (circles) and the reference plain nickel film (dashed curve). The solid line shows the reflection spectrum of the grating. Reproduced from [50] with permission by AIP Publishing LLC, Copyright 2010.

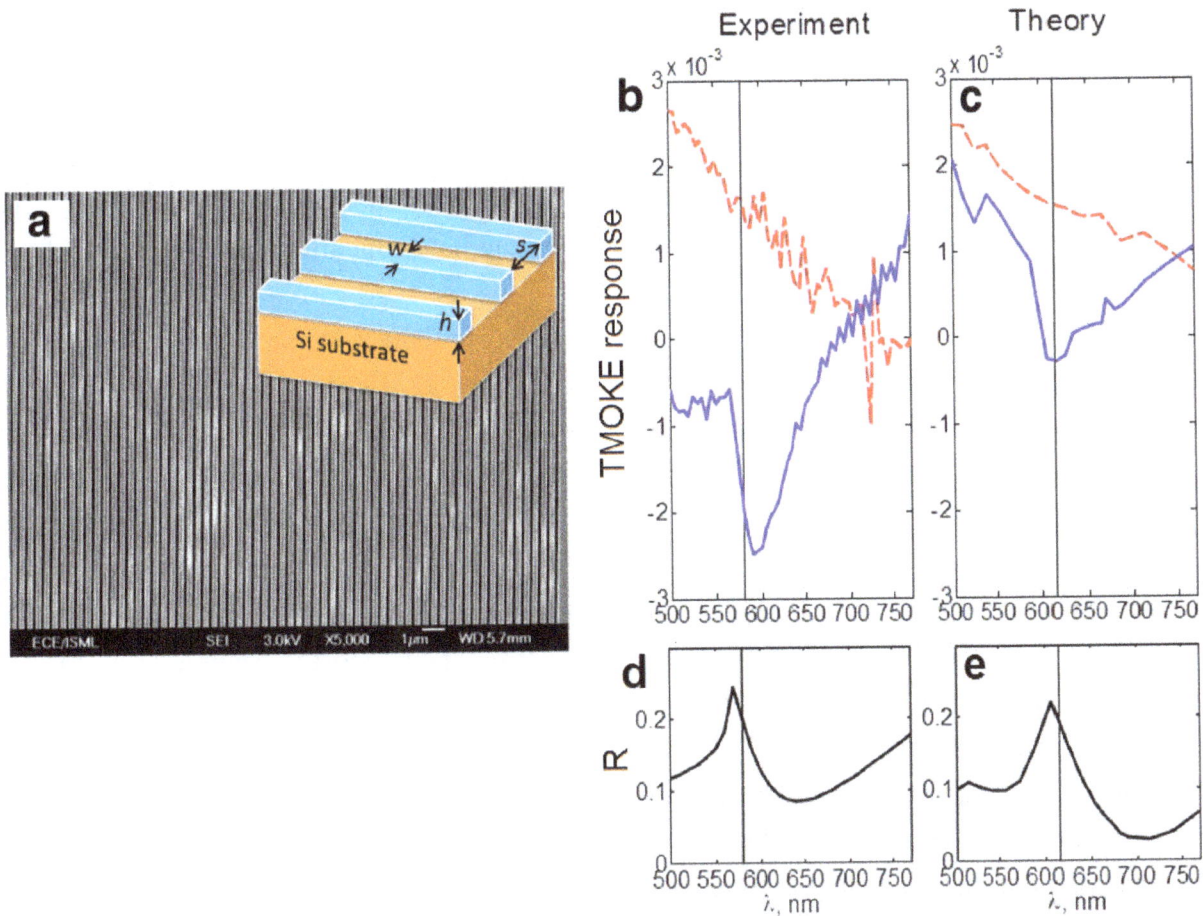

Figure 3. (a) Scanning electron micrograph and schematic of the Permalloy grating (one-dimensional magnonic crystal). The total area of the grating is 0.5×0.5 cm^2, $h = 100$ nm, $w = 264$ nm, and $s = 113$ nm. The thickness of the Si substrate is 0.8 mm; (b) Measured and (c) Simulated TMOKE response of the 100 nm-thick reference Permalloy film (dashed line) and Permalloy grating (solid line); Measured (d) and (e) Simulated reflectivity spectra of the Permalloy grating. The vertical straight solid line denotes the Fano resonance wavelengths. Reproduced from [57] with permission by AIP Publishing LLC, Copyright 2013.

To conclude the section on all-ferromagnetic nanostructures, we would like discuss the material of the grating (magnonic crystal) shown in Figure 3a - Permalloy (Ni$_{80}$Fe$_{20}$). As discussed in [17–24], Permalloy is paramount for all applications in microwave signal processing, magnetic memory, logics, and sensors. This is because of the optimum combination of microwave magnetic properties of Permalloy: the vanishing magnetic anisotropy and one of the smallest magnetic (Gilbert) damping among ferromagnetic metals. Similar to nickel that has been proven to support plasmonic resonances [50,55], Permalloy also possesses high magneto-optical properties as confirmed by experiments on thin Permalloy films [67–69]. However, thinking of application in spintronics and magnonics, nickel is characterised by large magnetic losses and is known as a material with very large magnetostriction. Furthermore, its saturation magnetisation ($6000/4\pi$ Oe) is small (almost two times smaller than for Permalloy and almost four times smaller than for iron). For all these reasons, use of nickel in spintronics and magnetisation dynamics is limited.

2.4. Longitudinal Magneto-Photonic Intensity Effect

The Faraday and Kerr effects have been known for more than 150 years and they, as well as their different configurations [25,32], have been exploited in various devices. However, it turns out that a combination of these effects with unique optical properties of high-quality nanostructures may lead to the discovery of new magneto-optical effects, which open up novel opportunities to control light with magnetic fields at the nanoscale.

Belotelov *et al.* [70] investigated a nanostructure consisting of three layers: a non-magnetic dielectric substrate, followed by a magneto-insulating layer, and a thin gold film sitting on top. The gold layer was periodically perforated by parallel slits with period $d = 661$ nm. The lower refractive index of the substrate relative to one of the magnetic layer ensured the existence of guided optical modes in the magnetic layer.

In absence of an external magnetic field, the magnetic layer is demagnetised and the investigated structure supports surface plasmons and waveguided transverse magnetic (TM) or transverse electric (TE) waves. Simulations demonstrated that the TM and TE modes are mostly localised in the magnetic layer. Moreover, it was shown that these modes are sensitive to the permittivity of the adjacent metal and therefore also have plasmonic character.

When an external static magnetic field is applied, the presence of non-diagonal terms of the dielectric permittivity tensor transforms the modes into "quasi-TM" and "quasi-TE" modes. Besides the TM components, the quasi-TM mode also contains TE components. Because all six field components are non-zero for both modes, those modes can be excited by the incident light of any polarisation.

The mode wavenumber of the quasi-TM and quasi-TE modes in the longitudinally magnetised investigated structure is proportional to g^2. As shown in Section 2.3, for typical values of g the Fano resonance shift induced by the applied magnetic field is small and therefore results only in small modifications of the transmittance/reflectance spectra [57]. However, a considerably larger effect is expected to originate from the magnetisation-induced changes of the field distributions of the modes. In this case, the key point is the appearance of the TM components in the quasi-TE mode, which makes it possible to excite this mode by light of TM polarisation. The excited quasi-TE mode takes a fraction of the incident optical energy, changing the overall absorbed energy by a factor proportional to g^2.

For zeroth order of diffraction, the considered magneto-optical intensity effect can be described by the relative difference between the transmittance coefficients T_M and T_0 of the magnetised and the demagnetised structure: $\delta = (T_M - T_0)/T_0$ observed at the TE mode frequencies. Because the longitudinal magnetisation of the investigated structure also modifies the field of the TM modes by inducing TE components, one should expect intensity variation at the TM-mode resonances as well.

This effect is called the "longitudinal magneto-photonic intensity effect" (LMPIE). The LMPIE should be viewed as a novel effect. For example, a similarly defined intensity-related effect, but of different origin, was studied in conventional magneto-optical ferromagnetic films, and was called orientational effect (see, e.g., [71]). However, for illumination polarised along the magnetisation, the orientational effect vanishes.

In [70], the LMPIE was observed in transmission. Intensities of the zeroth-diffraction-order transmitted light for demagnetised and longitudinally magnetised MPCs were compared to determine the parameter δ that characterises the LMPIE. Firstly, it was verified that no intensity modulation occurs for the bare magnetic film (Figure 4a, green curve). Secondly, the LMPIE for the nanostructure illuminated with normally incident TM-polarised light was measured and it was found that the longitudinally applied magnetic field resonantly increases the transparency of the nanostructure by 24% at 840 nm (Figure 4a). Numerical modelling results (blue curve in Figure 4a) reproduced the experimental data with good accuracy.

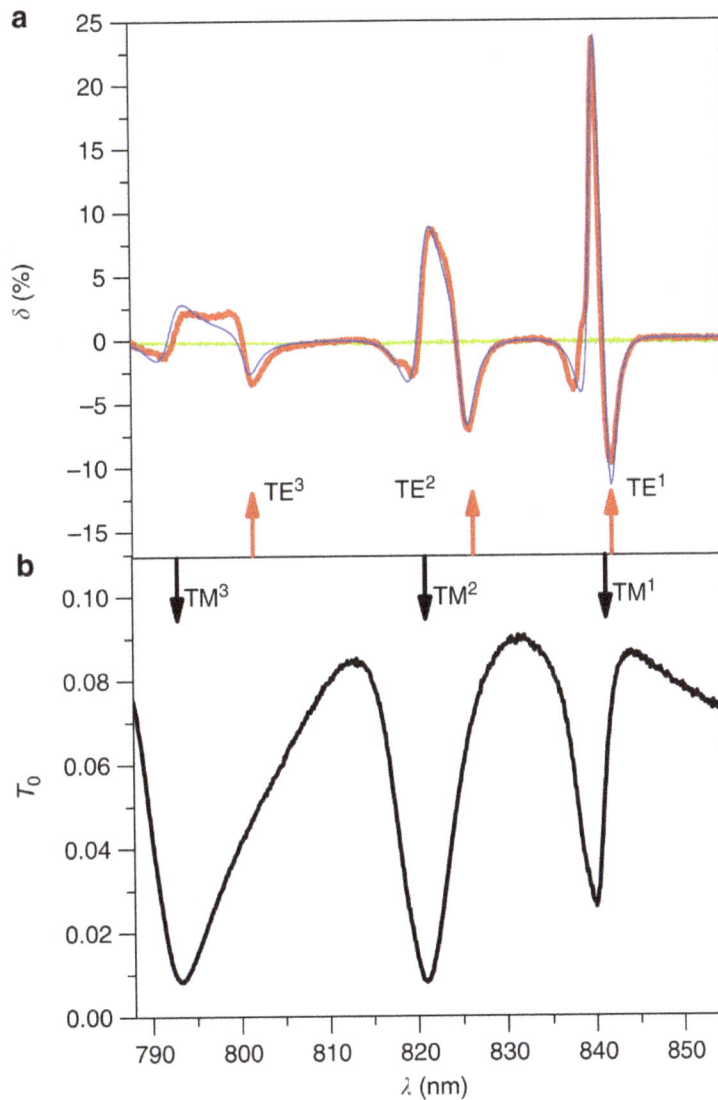

Figure 4. (a) Spectrum of the LMPIE when a static magnetic field reaching almost the saturation value. Blue curve shows calculated δ. There is no LMPIE for the bare magnetic film (green curve); (b) Spectrum of the optical transmittance for the demagnetised structure. Black and red arrows indicate calculated spectral positions of the quasi-TM and quasi-TE resonances, respectively. The modes are denoted by the number of their H_y or E_y field maxima along the z axis. The light is TM-polarized and hits the sample under normal incidence. Reprinted from [70] with permission by Macmillan Publishers Ltd, Copyright 2013.

3. Perspectives of Magneto-Plasmonics in Spin-Wave and Magnonic Applications

Let us now discuss how the plasmon-enhanced magneto-optical effects can be used in the emerging areas of spin wave technology and magnonics. Let us start with a simple example.

A personal computer (PC) consumes 60–250 W of power. There are more than one billion PCs in use around the world. The total power consumed and dissipated by PCs is alarming and they leave a considerable carbon footprint [72,73]. Therefore, the development of energy-efficient computing devices is recognised as a priority by Intel, Microsoft, and Google.

Significant reduction of power consumption in PCs can be achieved by using magnetic nanodevices–nonvolatile nanomagnetic logics [21], spin-wave logics [74–79], magnetoresistive random-access memory (MRAM) and spin-transfer torque magnetic random-access memory (STT-MRAM) [21], which store information magnetically and manipulate it using magnetisation dynamics. The fundamental motion of the vector of magnetisation is its precession about the equilibrium direction in the material (Figure 5a). Spin waves (Figure 5b) represent collective magnetisation precession, which is non-uniform in space [80,81]. In spin waves, the amplitude of dynamic magnetisation is small. This property is used in spin wave microwave devices and logics. Magnetic memory and logics based on the reversal of the magnetisation vector [Figure 1c] employ large-amplitude precessional dynamics to switch the direction of magnetisation.

High-quality nanopatterned ferromagnetic media - magnonic crystals [17–21,23] provide full control of dynamic magnetisation in a fashion similar to the control of light using photonic crystals [82]. Magnonic crystals are used as the "building blocks" of magnetic RAM and spin wave logic schemes integrated with spin transfer torque oscillators and switches [21,22]. These technologies are promoted actively by semiconductor giants [83].

The interaction of light with dynamic magnetisation is of utmost importance for the development of novel magnetic nanotechnologies. For example, the main challenge in the development of magnetic memory and logics is the real-time read-out of the logic state. The most probable technological solution is the use of magneto-optical Kerr effect (MOKE) spectroscopy, which allows probing dynamic magnetisation in real time [20,84]. MOKE spectroscopy manifests as a magnetisation-induced change in the intensity or polarisation of incident light upon reflection from the magnetised medium [25]. This effect has been exploited in commercial magneto-optical devices such as magneto-optical discs [25].

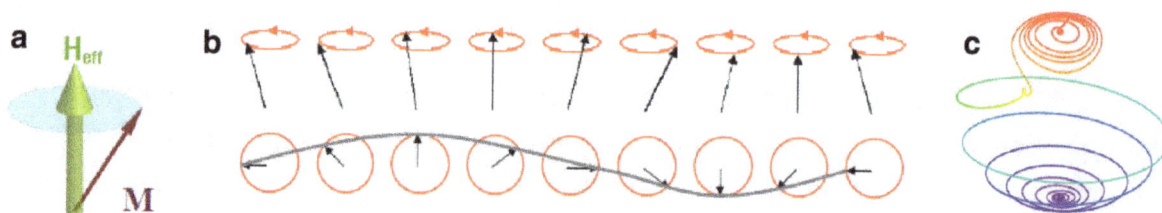

Figure 5. (a) Dynamic precession of the magnetisation vector **M**; (b) Spin wave (one complete precession distributed over a chain of spins); and (c) Trajectory of **M** in the magnetisation reversal process. Reproduced from [85] with permission by AIP Publishing LLC, Copyright 2009.

However, as shown schematically in Figure 6a, the natural interaction between light and dynamic magnetisation is weak. (For a more detailed discussion see, e.g., [32]). The strength of the interaction between light and dynamic magnetisation can be increased by exploiting magneto-plasmonic effects, which is possible because plasmon-enhanced magneto-optical response has been achieved in the same nanostructures which are used as magnonic crystals. The plasmon-enhanced light scattering from spin waves is schematically shown in Figure 6a,b. One sees that the excitation of surface plasmons is not possible in the case of a continuous Permalloy ferromagnetic thin film. Consequently, there is no plasmon resonance and the enhancement of the light-magnetic-matter interaction is impossible.

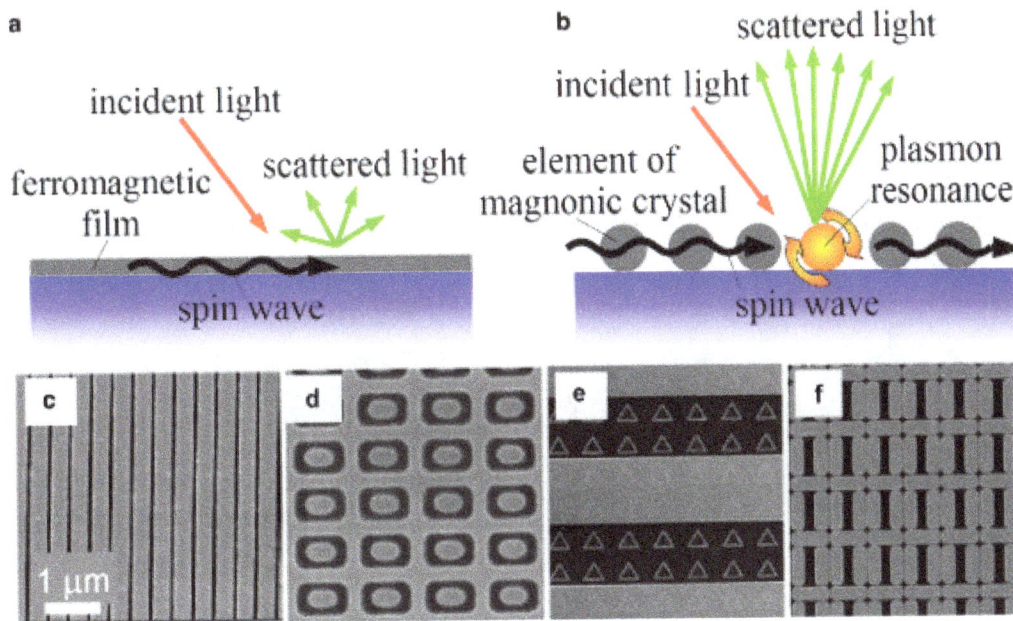

Figure 6. (a) Light interaction with a spin wave in a uniform ferromagnetic thin film sitting on top of a dielectric substrate; (b) Excitation of the plasmon resonance in a ferromagnetic magnonic crystal leads to enhancement of the intensity of light scattered from the spin wave; (c–f) Some examples of plasmon resonance-supporting Permalloy magnonic crystals fabricated at the National University of Singapore (see, e.g., [86]). Reproduced from [57,87–89] with permission by AIP Publishing LLC, Copyright 1012, 2013, and by American Physical Society, Copyright 2012.

Of course, one can cover the Permalloy film by a prism (see Figure 10 in [32]) in order to excite plasmons under the Attenuated Total Reflection (ATR) geometry [90]. Alternatively, in order to excite plasmons one can nanopattern the film as shown in Figure 6c–f. The structures shown in these figures are actually Permalloy magnonic crystals fabricated by deep UV lithography followed by lift-off [86,91]. We already demonstrated that a magnonic crystal (Figure 6a) combines good optical and magneto-plasmonic properties (Section 2.3). The magnonic crystals shown in Figure 6d–f also support plasmon resonances.

For the sake of illustration, let us consider a dolmen-like structure Figure 6f [89]. It is noteworthy that dolmen-like structures play important role in modern plasmonics thanks to their unique optical properties (see, e.g., [92]). To date, magnonics experiments have been conducted on relatively large arrays of identical dolmens, and the dimensions of these dolmens are comparable with the wavelength of incident light [89]. However, from the viewpoint of magnetism, the dimensions of the dolmens can be reduced readily to an optical subwavelength, which is the case of the dolmens in [92].

Figure 7 shows the simulated absolute values of the electric field in one unit cell of a periodic array of gold and Permalloy dolmens with the same dimensions as in [89]. The wavelength of normally incident light is 632.8 nm. One can see the excitation of plasmon modes in the 50 nm-wide gaps between the individual elements of the dolmens. Naturally, light absorption in Permalloy is larger than in gold. Consequently, amplitude of the electric field in the gaps of the Permalloy structure is smaller than that in the gold structure. Moreover, one can see that in the Permalloy structure the propagation length of plasmons localised in the gaps is significantly smaller. Nevertheless, simulations suggest that in dolmen-like magnonic crystals (and also in other types of magnonic crystals) plasmons may contribute to the magneto-optical response. Consequently, their contribution should be taken into account in order to facilitate understanding of the magnetisation reversal processes in magnetic nanostructures conducted by means of different modalities of the magneto-optic Kerr effect magnetometry and microscopy [93–96].

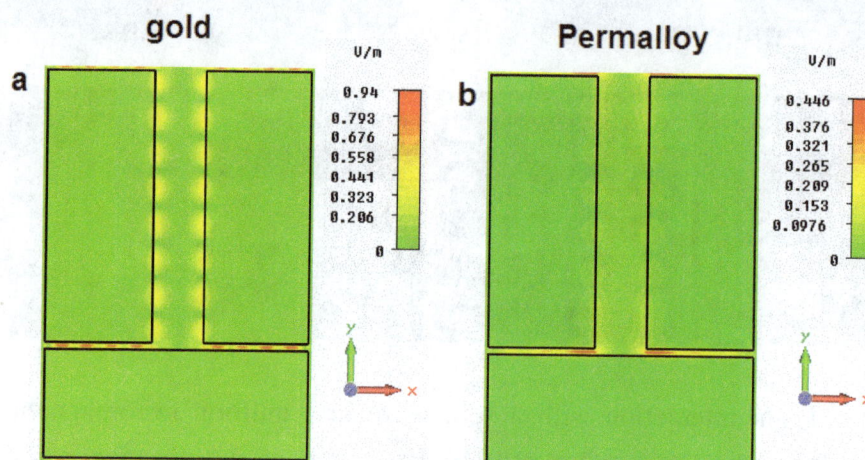

Figure 7. Simulated absolute values of the electric field in one unit cell of a periodic array of (**a**) Gold and (**b**) Permalloy dolmens with the same dimensions as in [89]. The wavelength of normally incident light is 632.8 nm. The width of the gaps between the individual elements of the dolmens is 50 nm. Simulations were conducted by using commercial CST Microwave Studio software.

Apart from the MOKE magnetometry and spectroscopy, one also needs a method with which to probe magnetisation dynamics simultaneously in the frequency and wave-vector domain as well as to map directly the intensity of magnetisation at the nanoscale. Brillouin light scattering (BLS) spectroscopy is the most suitable method [19,20,97–99]. BLS spectroscopy has been described as the reflection of incident light from a moving Bragg-diffraction grating produced in a magnetic medium due to propagating spin waves [97,100]. As a result, a portion of the scattered light is shifted by the frequency of the spin waves, *i.e.*, it is scattered in an inelastic manner.

In a typical BLS experiment (see, e.g., [19]), the dispersion relationship of spin waves is measured with p-polarised monochromatic light incident at angle θ (Figure 8a). (The angle θ is controlled in a broad range because it is linked to the wave vector k_{SW} of the probed spin wave by the relationship $k_{SW} = (4\pi/\lambda)sin\theta$, where λ is the wavelength of incident light.) Importantly, the same polarisation of incident light is required for the excitation of surface plasmons. Consequently, by analogy with the plasmon-enhanced TMOKE response, the conditions for the resonant enhancement of the BLS signal due to surface plasmons can be met in BLS measurements of metallic ferromagnetic magnonic crystals.

A typical BLS spectrum shows the measured intensity of the scattered light against the measured frequency shift (Figure 8b). The elastically scattered incident light is in the centre of the spectrum. The Stokes and anti-Stokes peaks correspond to inelastically scattered light. The frequency shift corresponding to the maxima of these peaks is denoted by dots in the dispersion diagram shown in Figure 8c. In this figure, the lines are the theoretical prediction and the shaded areas are the magnonic band gaps.

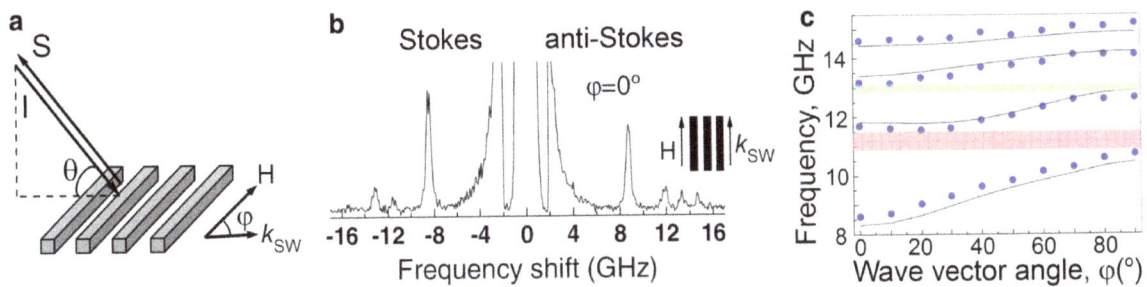

Figure 8. **(a)** Schematic of a typical BLS experiment; **(b)** Example of a measured BLS spectrum; **(c)** Experimental (dots) and theoretical (lines) dispersion curves of the Permalloy magnonic crystal shown in Figure 7a. Reproduced with permission from [19].

Technically, the central elastically scattered peak in the BLS spectra in Figure 8b is the reflectivity of light R detected in measurements of static transverse MOKE. Static MOKE magnetometry is very useful for probing the state of magnetic logic elements. The logic state changes as a result of magnetisation reversal, and can be "0" or "1", which correspond to opposite directions for the magnetisation vector. Time-domain MOKE spectroscopy allows probing ultrafast (sub-nanosecond scale) magnetisation dynamics with high temporal resolution. This makes time-domain MOKE spectroscopy very useful for investigating ultrafast magnetic switches and multiplexors driven by time-domain signals of different shape and duration [84,98].

It is noteworthy that several attempts have been made to increase the BLS from acoustic waves using plasmons [101–104]. Moreover, the dynamics of acoustic phonons generated by femtosecond impulsive optical excitation can be clearly resolved by a surface plasmon technique, with enhanced sensitivity orders of magnitude higher than regular optical probe measurements [33,105–107]. However, one needs to keep in mind that whereas the BLS from acoustic waves is photon-phonon scattering, the BLS from spin waves is photon-magnon scattering. Most significantly, to excite plasmons one needs to fabricate large-area metallic nanostructures on top of the acoustic wave waveguide. In contrast, magnonic crystals inherently support plasmons without the need for additional metallic nanostructures [57].

Nevertheless, in the remainder of this Section we discuss the results from [103] that demonstrates the surface-plasmon enhancement of BLS from acoustic waves in gold-discs arrays fabricated on top of glass (Figure 9a). Hereafter, the reader will see that this result is encouraging for further research on the plasmon-enhanced BLS from spin waves.

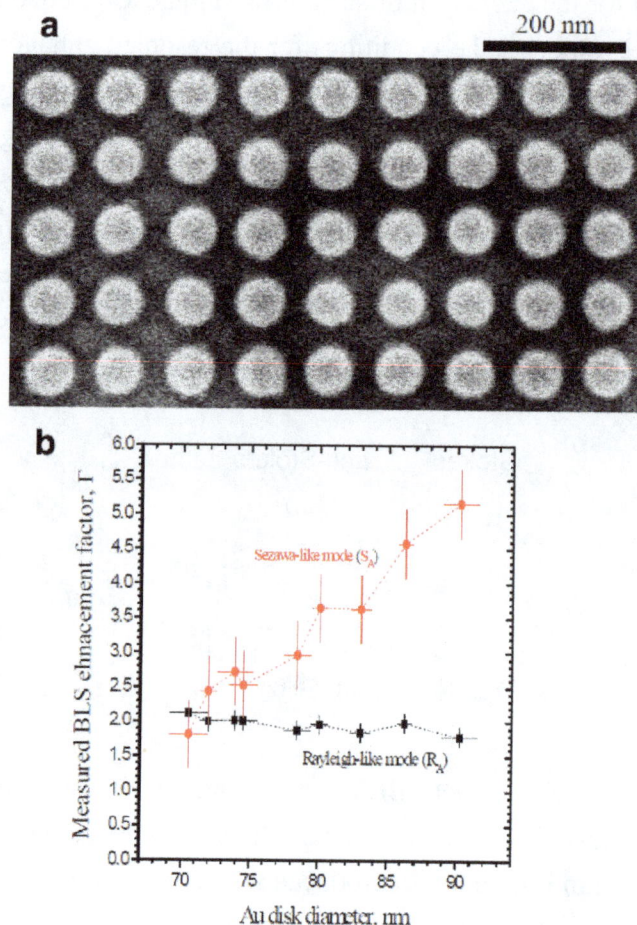

Figure 9. (a) Image of a section of a gold nanodisc array on glass; (b) Acoustic BLS enhancement factor Γ the Rayleigh-like and Sezawa-like modes as a function of the nanodisc diameter. Reproduced with permission from [103].

In [103], BLS anti-Stokes spectra were taken at $60°$ incident angle for the gold continuous film on glass, a bare glass substrate with the same composition, and a nanodisc array on glass. It is noteworthy that surface plasmon excitations do not occur when light is incident on a continuous gold film at the gold/air interface because the dispersion curves of light and surface plasmons do not overlap [90]. For the gold film on glass two peaks were detected, which correspond to the well known Rayleigh-like and Sezawa-like modes. In addition, additional peaks appeared in the spectrum from the gold nanodisc array. These peaks had no corresponding modes in the gold film.

Figure 9b shows the measured enhancement factors Γ for the Rayleigh-like and Sezawa-like modes as a function of the nanodisc diameter at an angle of $60°$. One can see that the maximum enhancement for the Rayleigh-like mode occurs for the smallest nanodisc size, but varies just $\sim 10\%$ over the 20 nm range of diameters. However, measurements demonstrate the opposite trend for the Sezawa-like mode enhancement, for which the largest values of the enhancement factor occurs with the largest nanodiscs. Interestingly, the demonstrated enhancement is consistent with the simulated values of electric-field intensity on the top and the bottom surfaces of the gold nanodiscs. (These values were normalised to the incident intensity.) It suggests that by improving the light-focusing properties of the discs one can improve the enhancement factor Γ. Most significantly, unlike in the case of the light-magnetic matter interaction, the interaction of light with acoustic waves does not involve magneto-optical effects. As discussed throughout this review, the strength of magneto-optical effects can be considerably enhanced by means of plasmon resonances. Consequently, in magnonic BLS experiments one may expect even larger values of Γ for spin wave modes.

4. Enhanced Magneto-Optical effects in All-Magneto-Dielectric Nanostructures

Since all metals absorb light in the visible and infrared spectral ranges, maximum efficiency of metallic plasmonic devices can be achieved only by using metals with the lowest absorption cross-section. Therefore, most of the plasmonic devices are made of gold or silver because these two metals exhibit the lowest absorption losses at optical frequencies. (However, losses remain considerable even if gold or silver are used as the model material.) For example, transmission spectra of a gold grating with exactly the same dimensions as the Permalloy grating in Figure 3a were investigated in [57]. (Of course, the gold grating does not exhibit magneto-optical properties; it was used as a reference to better understand the optical response of the Permalloy grating.) As shown in Figure 2 in [57], the reflectivity of the gold grating is more than 3 times larger than that of the Permalloy grating. A combination of such a high reflectivity with a strong plasmon-enhanced TMOKE response would be advantageous for practical applications.

It is noteworthy that inevitable absorption losses in metals at optical wavelengths has already motivated researchers in the photonics sector to investigate diffraction gratings (see, e.g., [108]) as well as metamaterials and nanoantennas based on alternative materials such as semiconductors [109] and dielectrics [30,110]. Of course, as discussed in the cited papers, all-dielectric nanostructures offer a significantly smaller local field enhancement as compared with metallic nanostructures. However, a combination of the enhancement offered by all-dielectric nanostructures with low losses often opens up novel opportunities as compared with metal nanostructures.

Naturally, the same strategy of substituting metals with dielectrics works in the field of magneto-optics. For example, Bai et al. [111] demonstrated theoretically strong polarisation conversion effects, including large Kerr rotation and ellipticity, in zero-order reflection of light from a BIG grating embedded in a dielectric material. The field localisation that gives rise to the enhanced magneto-optical effects was achieved due to the excitation of leaky guided waves, which propagate in (and around) the high-index BIG layer in directions perpendicular to the external static magnetic field. It was shown that whereas the resonant interaction with the incident field results in narrow-band reflection spectra, large magneto-optical polarisation conversion effects result from the splitting of the degeneracy of right- and left-circularly polarised eigenmode resonances with highly dispersive phase spectra near the resonance peak.

Similar results were obtained by Marinchio et al. [112] who demonstrated theoretically that subwavelength nanowire gratings made of arbitrary anisotropic materials can present a resonant response at specific geometric conditions. They showed that for a given polarisation of incident light a resonant magneto-optical response can be obtained by tuning the incidence angle and grating parameters to operate near the resonance condition for the opposite polarisation. These results are important for the understanding and optimisation of magneto-optical structures and devices based on resonant subwavelength gratings, which is used in sensing applications [113].

We would like to note that the gratings investigated in [111,112] are subwavelength, *i.e.*, their period is significantly smaller than the wavelength of incident light. In such gratings, at the resonance wavelength the zero-order reflected and transmitted light is present in the far field, with higher-order diffraction modes suppressed (see Figure 10). The high reflectivity exhibited by subwavelength gratings depends on the angle of incidence and it is attributed to the excitation of a waveguide mode through phase matching by the grating. The waveguide mode then recouples back (again through grating phase matching) to radiative modes and produces the observed high reflection [108]. The recoupling occurs at the frequencies of the resonant Wood's anomaly [63], which in turn is linked to the Rayleigh wavelength at the given angle of incidence. Furthermore, the lineshape of the resonance peaks in the reflectivity spectrum of the grating is asymmetric due to Fano interference [46,63].

The complex resonance behaviour of the proposed subwavelength dielectric grating is a mechanism that differs it from conventional magneto-optical diffraction gratings having a large period as compared with the wavelength of incident light (see, e.g., [114]). Moreover, conventional gratings were made of magneto-insulating wires on a silicon substrate. The application of these materials for the fabrication of nanostructures was a technological challenge until the recent progress in nanostructuring of BIG and YIG (see, e.g., [115]) and similar materials. New fabrication technologies enable the formation of complex (two- and three-dimensional) optical and magnetic nanostructures in YIG, which opens up opportunities to development novel devices for optical communication and photonic integration.

Consequently, in the theoretical work [116] it was proposed to employ BIG or YIG as one of the constituent materials of a subwavelength grating having exactly the same dimensions as the all-Permalloy grating in Figure 3a. In addition, in order to overcome the drawback of the all-Permalloy grating - a low reflectivity of just $\sim 20\%$ - it was suggested to use silicon as the second constituent material to fill the gaps between the BIG (or YIG) nanowires. It was shown by means of rigorous simulations that, for the same angle of incidence as in [57], one can achieve a one order of magnitude enhancement of the TMOKE response as compared with a continuous magnetic film. Furthermore, as shown in Figure 1 of [116], the silicon nanowires play the crucial role in achieving a high reflectivity of $\sim 80\%$. (Recall that for the all-Permalloy grating the reflectivity is just $\sim 20\%$.) Most significantly, as shown in Figure 10, both the TMOKE response of the all-magneto-dielectric grating and its reflectivity increase as the angle of incidence is decreased. This increase is not readily attainable using magneto-plasmonic gratings made of magneto-insulating films and gold nanowires.

To conclude this Section we would like to note that unique advantages of non-magnetic dielectric photonic nanostructures over non-magnetic metallic plasmonic structures are their low absorption losses. These advantages can boost performance of many photonic devices in the visible and near-IR frequency spectra [30,110]. By analogy, the application of all-magneto-dielectric nanostructures instead of metallic ones can is expected to boost performance of many magneto-optical devices.

However, the application of similar ideas in the field of magnonics is not straightforward. As shown in Section 2.3, Permalloy and other magnetic metal alloys are very important for magnonics and spintronics. Moreover, despite the recent progress [115], the fabrication of high-quality YIG nanostructures remains a challenge and also YIG is relatively expensive. On the other hand, as discussed in [18], until certain technical obstacles are overcome in modern magnetic alloy materials, YIG remains a valuable source of insight and the interest in YIG-based magnonic systems is particularly strong. Consequently, we anticipate the appearance of hybrid structures combining magnetic and non-magnetic conductive and insulating materials. Indeed, such structures have been proposed in the recent works [113,117].

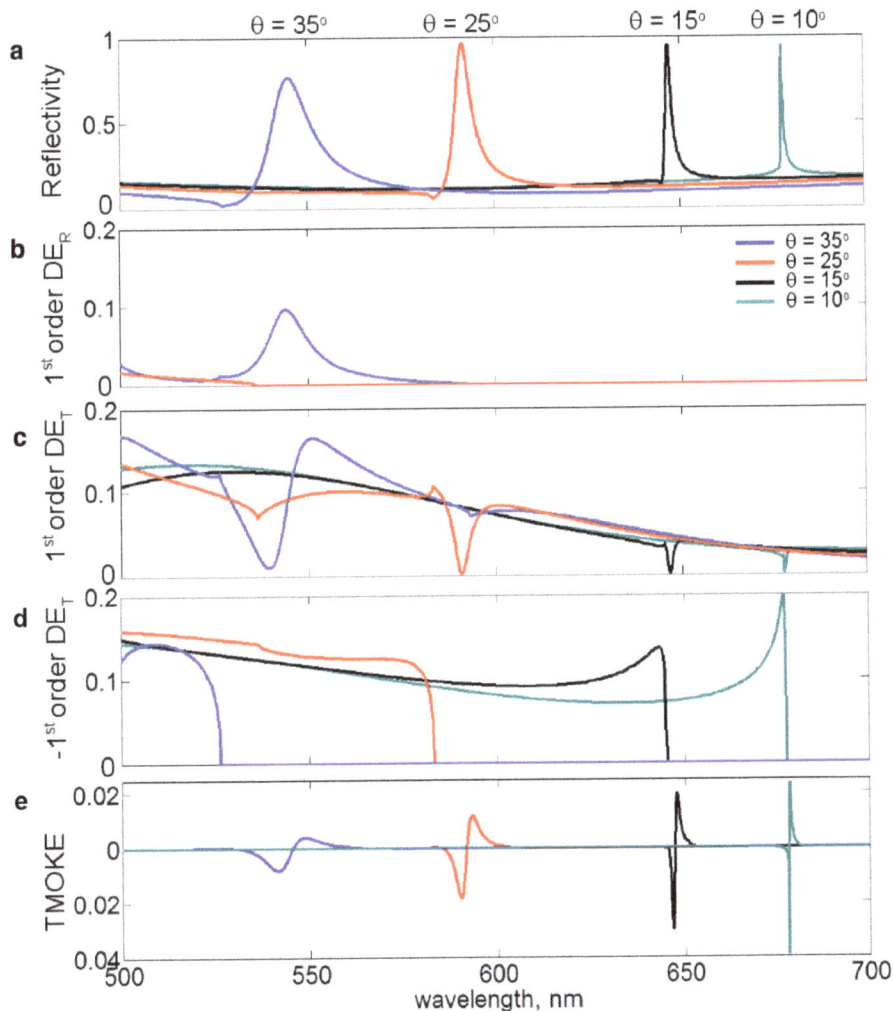

Figure 10. (a) Reflectivity and (b–d) Reflected (DE_R) and transmitted (DE_T) diffraction efficiency for the +1st and 1st orders of the Bi:YIG-Si grating at different angles of incidence θ; (e) TMOKE response of the grating at different θ. The minor secondary peaks in the spectra occur at the Rayleigh wavelengths. Reproduced with permission from [116].

5. Plasmon-Enhanced Inverse Magneto-Optical Effects and All-Optical Magnetisation Switching

5.1. Plasmon-Enhanced Inverse Faraday Effect

Ultrafast all-optical control of a medium magnetisation [47,84,118] and the achievement of time-dependent magneto-optical effects [119,120] at the subpicosecond time scale are of utmost importance for the development of novel magnetic data storage systems and generation of spin waves via light [121]. The optical way to control the magnetisation is enabled by the phenomenon in which a circularly polarised light induces static magnetisation in a gyrotropic medium. This phenomenon is called the inverse Faraday effect, which was predicted theoretically in the 1960s [122,123]. (Note that not only the Faraday effect has its inverse counterpart, but there is also the inverse Cotton-Mouton effect [124].) This effect is nonlinear [25,125], which should be taken into account when spin dynamics is manipulated directly and coherently by using multiple laser pulses [126].

The Faraday effect [25,84] is observed as a rotation of the polarisation plane of light transmitted through a magnetic medium and it can be expressed as $\alpha_F = \frac{\chi}{n}\mathbf{M}\cdot\mathbf{k}$, where α_F is the specific Faraday rotation, $\mathbf{M}$ is the magnetisation, n is the refractive index, $\mathbf{k}$ is the wave vector of light, and χ is the magneto-optical susceptibility. The inverse Faraday effect, which is observed as the induction of a static magnetisation $\mathbf{M}_0$ by a high-intensity laser radiation, is determined by the same magneto-optical susceptibility χ as $\mathbf{M}_0 = \frac{\chi}{16\pi}[\mathbf{E}(\omega)\times \mathbf{E}^*(\omega)]$, where $\mathbf{E}(\omega)$ and $\mathbf{E}^*(\omega)$ are the electric field of the light wave and its complex conjugate, respectively [84,107,118,127].

Importantly, in the case of the inverse Faraday effect χ is a ratio between the induced magnetisation and the laser intensity. Consequently, as discussed in [84,118], optical control of magnetisation is the most efficient in materials with high values of the Faraday rotation per unit magnetisation. Furthermore, χ is allowed in all media regardless their crystallographic and magnetic structures. Most significantly, the inverse Faraday effect does not require light absorption. This implies that the effect of light on magnetisation is non-thermal and can be considered as instantaneous because it takes place on a femtosecond time scale. The first experimental demonstration of such non-thermal ultrafast optical control of magnetisation was done by Kimel *et al.* [118], who used the inverse Faraday effect to excite magnetisation dynamics in 60 μm thick DyFeO$_3$ magnetic samples, in the hundreds of GHz frequency range. They demonstrated that when a circularly polarised light pulse of high intensity (pump pulse) is focused on a magnetic material, a spin precessional motion starts within the light spot via the inverse Faraday effect, whereby an effective magnetic field is generated along the beam direction.

Building up on the results in [118], Satoh *et al.* [121] investigated YIG samples. (Recall that YIG is often employed in magnonic and spintronic devices because of intrinsically low magnetic damping in YIG [18]). Initially, the magnetisation laid in the plane parallel to the sample surface. When a circularly polarised pump beam with a pulse width of $\sim$ 100 fs was focused into a circular spot on the sample surface, spin precessional motion occurred within the spot via the inverse Faraday effect. The magnetisation deviated from the plane and the out-of-plane component became nonzero. This out-of-plane magnetisation component was detected by the Faraday polarisation rotation of a low-intensity linearly polarised pulse (probe pulse). A time-resolved pump-probe experiment was carried out by measuring the Faraday rotation angle of the probe pulse, which is time-delayed with respect to the pump pulse. The precessional motion propagated in two dimensions out of the pump light spot in the form of a spin wave (Figure 11a). By scanning the relative position of the probe light spot on the sample with respect to the pump light spot, it was possible to observed time- and space-resolved spin wave propagation (Figure 11b). It was found that the wavelength of the spin wave was 200 $-$ 300 μm, and the group velocity was about 100 km/s [128]. Numerical simulations reproduced the experimental result with good accuracy (Figure 11c).

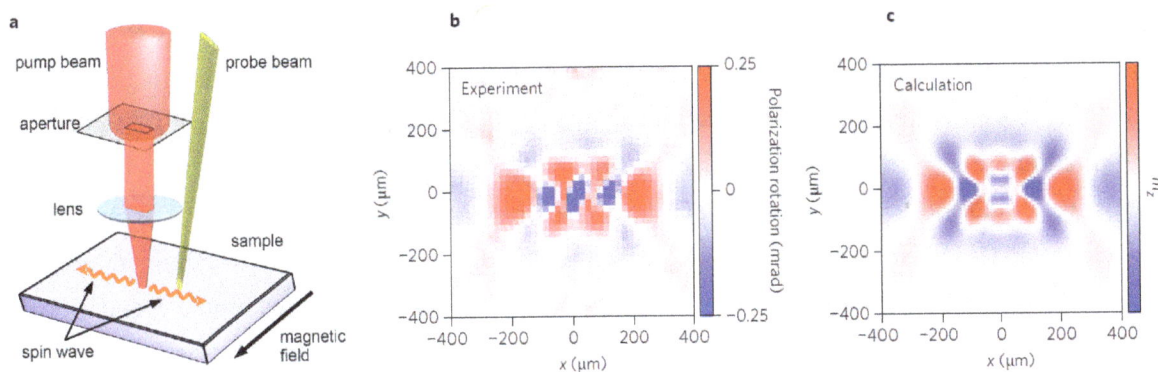

Figure 11. (a) Experimental configuration for directional control of spin wave emission; (b) Two-dimensional spin waveforms obtained by (b) Experiment and (c) Simulation. Reproduced with permission from Macmillan Publishers Ltd: [121]. Copyright 2012, and reproduced with permission from [128].

As discussed above, in the experiment from [121] spin precessional motion occurred within the spot via the inverse Faraday effect. Therefore, one may hypothesise that, similar to [129] in which magnetic recording using an integrated plasmonic antenna was demonstrated, the efficiency of all-optical spin wave excitation may be improved by enhancing the inverse Faraday effect by means of plasmons. This can be done, for example, by fabricating a plasmonic nanoantenna (for a review see, e.g., [30]) on top of the magnetic sample.

However, as discussed in [130], one should keep in mind that the coupling of a laser pulse to nanoantennas with a small footprint is challenging. Hence, one needs to employ arrays of nanoantennas or single nanoantennas that have a large footprint comparable with the spot of the laser beam. In both cases, one of the most probable candidates are the so-called tapered Yagi-Uda-type arrayed nanoantennas, which combine a large footprint with high efficiency in a broad spectral range [131,132]. Finally, we would like to note that arrayed nanoantennas can be made by using dielectric nanoparticles (for a review see, e.g., [30]). The potential advantage of such all-dielectric nanoantennas is their high coupling efficiency achievable due to low absorption losses in the constituent dielectric material.

5.2. Plasmon-Enhanced Inverse Transverse Magneto-Optical Effect

As an alternative strategy for the enhancement of light coupling to the magnetic medium one can employ all-ferromagnetic nanostructures (*i.e.*, magnonics crystal) and combine them with nanostructures made of plasmonic metals (*i.e.*, gold or silver). This combination was investigated by Belotelov and Zvezdin [133], who revealed the existence of the inverse TMOKE in continuous magnetic films (Figure 12) and showed that the strength of this effect can be increased by exploiting plasmonic resonances in magnetic nanostructures (Figure 13).

It is shown in Section 5.1 that the cross product $[\mathbf{E}(\omega) \times \mathbf{E}^*(\omega)]$ is non-zero for elliptically polarised light. In contrast, in the TMOKE configuration [32] the incident light is linearly polarised. However, as discussed in [133], due to the boundary conditions between a dielectric medium and a medium with absorption or with negative permittivity linearly polarised light is converted into elliptically polarised light. This implies that the cross product $[\mathbf{E}(\omega) \times \mathbf{E}^*(\omega)]$ is non-zero. It is noteworthy that this picture is for the *p*-polarised incident light (see Figure 12) and it does not hold for the case of the *s*-polarised light [133]. We also note that, by analogy with the inverse Faraday effect, the inverse TMOKE is also a nonlinear phenomenon.

A necessary condition for the appearance of an effective magnetic field in a ferromagnetic medium due to the inverse TMOKE is a decay of the electromagnetic field inside this medium due to optical absorption losses and/or a negative real part of the dielectric permittivity of plasmonic metals. However, for thin ferromagnetic films in which light can be reflected from the bottom film surface, the effective magnetic field does not vanish even for media with a purely real refractive index.

Consequently, one can see an important difference between the inverse TMOKE and the inverse Faraday effect. In the inverse Faraday effect, the effective magnetic field is induced by circularly polarised light and is detected along the wave vector. In contrast, the effective magnetic field induced due to the inverse TMOKE is orientated along the cross product of the incident wave vector and the normal to the surface of the ferromagnetic film. Due to this difference, the application of the inverse TMOKE in optically induced femtosecond magnetism is potentially more promising.

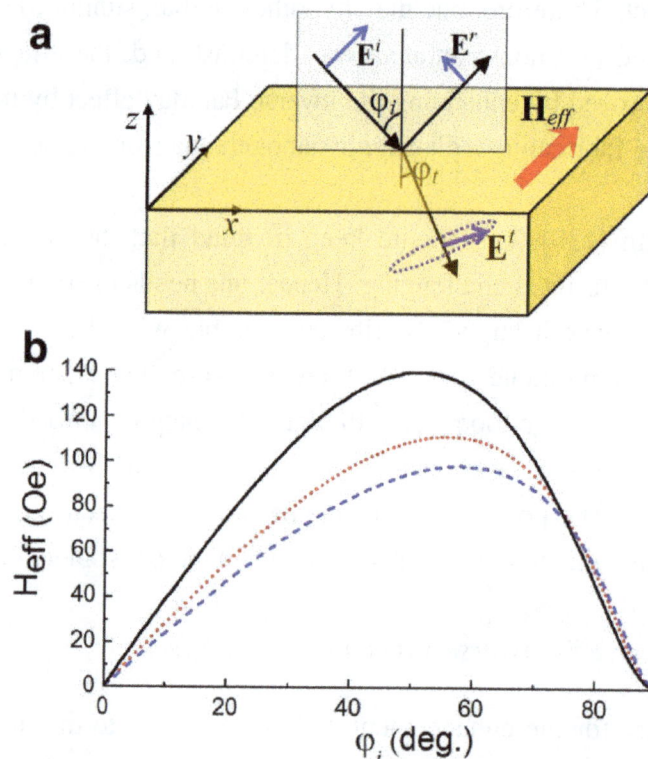

Figure 12. (a) Configuration for the inverse TMOKE. *p*-polarised light is obliquely incident on the ferromagnet film and generates an effective magnetic field H_{eff} inside the film; (b) The inverse TMOKE near the surface (at a point 5 nm in depth) of the magnetic films of iron (black solid line) at $\lambda = 630$ nm, nickel (red dotted line) at $\lambda = 630$ nm, and bismuth iron garnet (BIG) (blue dashed line) at $\lambda = 400$ nm as a function of angle of incidence of light. It is assumed that the samples are illuminated with a 40-fs duration laser pulse with a peak intensity of 500 W/μm^2. Reproduced from [133] with permission by American Physical Society, Copyright 2012.

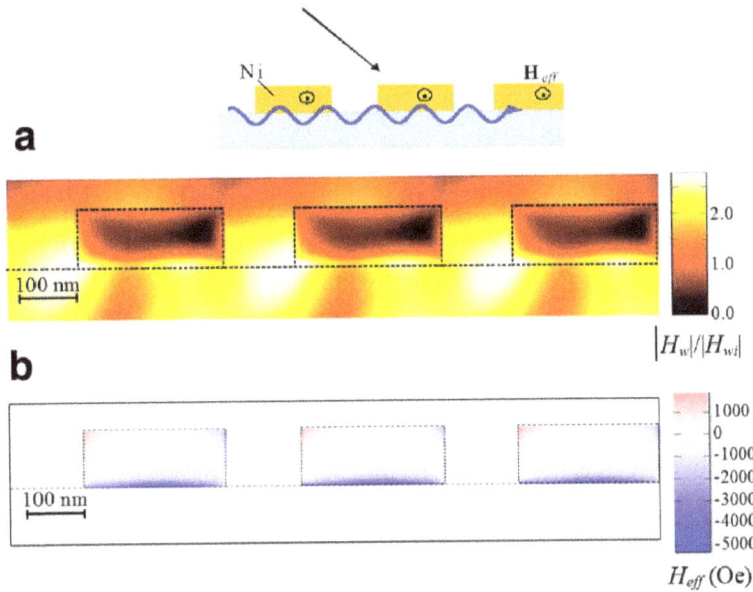

Figure 13. (a) Near-field distribution of the absolute value of the electromagnetic wave magnetic field $|H_w|$ in the plasmonic crystal (shown in the inset above) normalised to the magnetic field of the incident light $|H_{wi}|$; (b) Distribution of the effective magnetic field H_{eff} generated in the plasmonic crystal by the laser pulse. Three periods of the structure are shown. A laser pulse with a peak intensity of 500 W/μm^2 is p-polarised and is obliquely incident at $10°$. Period $d = 400$ nm, slit width is 120 nm, and nickel film thickness is 100 nm. Reproduced from [133] with permission by American Physical Society, Copyright 2012.

Before we discuss the enhancement of the inverse TMOKE in ferromagnetic nanostructures, we would like to note that, as shown in [133], the inverse TMOKE is not related to the magnetic dichroism. It is also different from the inverse Cotton-Mouton effect [124].

The simplest way to excite surface plasmons and enhance the strength of the inverse TMOKE is to use a continuous nickel film and a prism. This configuration were used, e.g., in [32] to enhance the "direct" TMOKE. (The term "direct" is used to avoid possible confusion with "inverse".) In the case of the inverse TMOKE this approach allowed to increase the induced effective magnetic field by ~ 4 times [133]. Moreover, a periodic composite of alternating gold and nickel nanostripes makes it possible to increase the effective magnetic field by ~ 6 times with respect to the plasmon-assisted enhancement in the continuous film. This is achievable because for the incident light wavelength of 630 nm the permittivities of gold and nickel have close real parts. (But their imaginary parts greatly differ because of significantly larger absorption losses in nickel.) Consequently, the wavelength of the surface plasmons waves at the gold and nickel surfaces are almost the same. (However, the corresponding absorption coefficients are very different.)

Belotelov and Zvezdin also considered the case of an all-nickel one-dimensional grating that is also often called plasmonic crystal. The dimensions of this structure are very similar to those of the one-dimensional Permalloy magnonic crystal in Figure 3a. As discussed in Section 2.3, microwave magnetic properties of Permalloy are suitable for applications in magnonics, but the application of nickel is limited. However, magneto-plasmonic properties of the nickel structure are very similar to the properties of the Permalloy structure because Permalloy is a nickel-iron magnetic alloy, with about

80% nickel content. Consequently, the result for the nickel grating presented below is also applicable to a broader class of devices – magnonic crystals.

The nickel nanostructure concentrates electromagnetic energy density at the surface plasmon resonance conditions, which leads to a local energy enhancement in the near-field region (Figure 13a). As a result, the effective magnetic field induced due to the inverse TMOKE is increased and it exceeds 5000 Oe near the nickel interface. A prominent feature of the inverse TMOKE in the nickel structure is that the effective magnetic field has opposite directions in different parts of the nickel nanowire cross-sections (Figure 13b), which opens up additional opportunities for the magnetisation control in ferromagnetic nanostructures.

To conclude this section, we would like to note that similar to the inverse Faraday effect the enhancement of the inverse TMOKE has also been predicted in magneto-insulating structures combined with gold nanowires [134].

6. Perspectives of Terahertz Magneto-Plasmonics in Ultrafast Control of Dynamic Magnetisation

Terahertz (THz) radiation is used in telecommunications, non-invasive imaging, and other areas of science and technology (see, e.g., [135]). The THz region (~ 0.1 THz to ~ 10 THz) is often described as the last frontier of the electromagnetic spectrum because of the relatively low maturity level of components and systems that operate in this region. As a result, the THz community continues to be active and growing. THz magneto-plasmonics is one of the research directions of this common effort.

The recent advances in THz magneto-plasmonics may be split into three groups. The first group includes non-reciprocal surface magneto-plasmon waveguides, gratings, resonators and metamaterials for THz radiation (see, e.g., [136–142]). This group also includes magnetically tuneable THz isolators [143,144]. The second group is concerned with tuneable magneto-plasmons to achieve efficient control of THz radiation by graphene monolayers. Some papers from this group are cited in the following Section in the context of graphene plasmonics for THz to infrared applications [145–148].

The works from the third group are concerned with probing and controlling the magnetisation dynamics in magnetic structures. It is known that in antiferromagnets the spin waves typically occur at 1 THz [149]. However, in contrast to ultrafast visible-light-based techniques enabling the dynamic magnetisation to be controlled with a temporal resolution of picoseconds to femtoseconds (see above), THz techniques are not yet well developed. At present, the progress in this direction is motivated by numerous potential advantages of using THz radiation instead of light.

The first advantage is that THz pulses can trigger spin precession on the timescale of subpicoseconds without inducing unwanted electronic excitation or thermal effects [150–152]. For example, the authors of [151] explored the interplay of the NiO antiferromagnet with highly intense THz pulses, with a time resolution of 8 fs. A highly intense THz pulse was focused onto the sample (for more details see Figure 1 in [151]). The THz pulse had a broad spectrum from 0.1 to 3 THz, fully overlapping with the magnon resonance at 1 THz. An 8 fs pulse probed the THz-driven magnetisation dynamics at a variable delay time t. In the experiment, the dynamic magnetisation $\mathbf{M}(t)$ induced by the pulse manifested itself by the Faraday effect, where the projection of $\mathbf{M}(t)$ onto the propagation direction of the probe pulse caused a transient circular birefringence that rotates the linear probe polarisation by an angle, which value is exclusively caused by the THz pulse [151]. The Fourier transformation of the resulting time trace of the measured Faraday rotation produced a narrow peak at 1 THz, clearly linking the signal to the high-frequency spin eigenmode in NiO.

An even more interesting effect would be the ultrafast all-optical magnetisation switching in antiferromagnets. This effect was predicted in the theoretical work [153]. The authors of this work suggested that THz laser pulses can in principle be used to switch the sublattice magnetisation of antiferromagnets on a picosecond scale.

As discussed in [154], another advantage is that the THz energy range is comparable with the low-energy terms of the magnetic Hamiltonian such as magnetic anisotropy and the Dzyaloshinsky-Moriya interaction. This property of THz radiation provides an opportunity to measure these effects directly. This is important in the development of novel materials for spintronic applications, but often is challenging for other experimental techniques such as intelatic neutron scattering (see [154] and references therein).

Although the possibility to use magneto-plasmons for the improvement of THz-bases control and spectroscopy of spin waves was not the topic in the cited papers, we point out that it should be possible to combine THz plasmonic properties with the capability of THz radiation to control and probe spin waves. Our suggestion is based on recent advanced in an adjacent research direction of THz metamaterials. For instance, Kurihara *et al.* [155] demonstrated enhancement of the spin precession of orthoferrite using the magnetic near-field produced by a split-ring resonator (SRR) using the THz spin-pump Faraday probe measurement technique. The precession amplitude was enhanced by ~ 8 times when the resonance frequency of spin precession was close to the magnetic resonance of SRR. It suggested that the local enhancement of the magnetic near-field in the THz range by metallic structures (see, e.g., [135]) may be used for further increase in the enhancement factor, leading to the nonlinear control of spin dynamics using THz radiation.

7. Relevant Topics

In this Section we would like to highlight several topics that go beyond the scope of this review paper, but remain of immediate relevance to magneto-plasmonics and also to a broader research area of non-reciprocal surface waves and one-way optical waveguides [156–163]. For instance, surface plasmons play a very prominent role in the optical response of different nanoscale materials systems that include not only metals and semiconductors but also graphene [109]. It is known that graphene supports surface plasmons at infrared and THz frequencies [145–148]. Hence, magneto-plasmonic interaction has been predicted theoretically and demonstrated experimentally in graphene structures (see, e.g., [164–171]).

The next research direction that we would like to briefly discuss is the possibility to control the extraordinary optical transmission [172] through subwavelength hole arrays or similar structures by applying a magnetic field. As shown, e.g., in [61,173–182], the transmission of light through a thin metal film with a periodic subwavelength hole array can be influenced by the presence of the externally applied magnetic field. For instance, the spectral locations of the transmission peak resonances can be shifted by varying the magnitude and direction of the applied field.

This effect may be of immediate relevance to light-magnetic-matter interaction in anti-dot magnonic crystals, *i.e.*, ferromagnetic metal films with a periodic array of air holes. In many cases the radius of the holes is considerably smaller than the wavelength of the incident light.

Interestingly, plasmon-enhanced magneto-optical effects can be exploited in nanosources of coherent light such as spaser [183]. A magneto-optical spaser representing a novel ultra-small source of coherent circularly polarised light was proposed in [184]. Unlike a non-magnetic spaser, this light source has two spasing modes with left and right cirlcular polarisation of the dipole moments. Each of the modes has different pumping thresholds so that for a certain range of values of pumping only a mode with a specific circular polarisation can be realised. This is especially important for prospective applications for quantum information, where manipulations of qubits require illumination with a circularly polarised light.

Plasmon-enhanced magneto-optical effects are also important for the development of Spatial Light Modulators capable of enhancing two-dimensional (2D) projection systems and 3D holographic displays for smartphones, projectors and 3D television [3]. We would like to discuss this important topic in more detail.

In the stereoscopic-based 3D displays, viewers can only artificially perceive 3D and they do not see actual 3D images, which may result in discomfort or even illness caused by eye fatigue. Holographic display, on the other hand, is a true 3D display technology, which can provide all depth cues such as binocular parallax, motion parallax, accommodation and convergence, without having to wear any special glasses. It promises no eye fatigue and has big market potential for various applications for entertainment, advertisement, medical imaging, training and education, defence and military, and scientific visualisation.

A spatial light modulator (SLM) is an optical device that can spatially modulate the amplitude (or intensity), phase or polarisation of an optical wave. SLM is a key device for digital 3D holographic display [185]. The pixel pitch and pixel count are two key SLM parameters that affect both the size and viewing angle of reconstructed 3D images. For practical applications of the 3D holographic display technology, advanced SLM devices with submicron resolution and ultrafast response time need to be developed and manufactured with high yield. However, the current commercial technologies do not meet these requirements.

Consequently, Aoshima *et al.* [3] proposed and demonstrated a spin-transfer-switching-based [21] SLM device that exploits the magneto-optical spatial light-modulation effect and is driven by spin-polarised current flow. The proposed SLM device has a spatial resolution as small as several hundred nanometers and possesses the potential for being driven at ultra-high speed of several tens of nanoseconds. However, even these characteristics just marginally satisfy both the size and speed requirements of SLMs for use in displaying holographic 3D moving images. We anticipate that the plasmon-enhanced magneto-optical interaction together with the possibility to focus plasmons into sub-100 nm spots may be used to further improve the performance of this SLM device.

Finally, we would like to mention that plasmon-enhanced magneto-optical effects may help to improve the efficiency of portable medical devices for clinical and in-field tests. For instance, malaria is a deadly infection caused by species of the Plasmodium parasite, passed on via the bite of an infected mosquito. Diagnosis is made by a blood test. However, sometimes it is necessary to repeat the test a number of times, as the parasites can be difficult to detect. A new method proposed in [186] exploits magneto-optical effects and it allows high-sensitivity detection of malaria pigment (hemozoin crystals) in blood via the magnetically induced rotational motion of the hemozoin crystals. We believe that the plasmon-enhanced magneto-optical interaction may further improve the sensitivity of the proposed method.

8. Conclusions

We reviewed several important directions in the study of magneto-plasmonics and resonant light-magnetic-matter interaction at the nanoscale. We endeavoured to present the material from two different points of view: magneto-plasmonics, and magnonics and magnetisation dynamics. Our main aim was to try to combine research outcomes produced by both research sectors, and we also tried to focus on novel results and/or results that were not highlighted in the previous review papers. It is clear that this research area is yet to reach its peak and to provide a valuable overall impact on the fields of photonics and nanomagnetism.

Acknowledgments

The author has been supported by the UPRF scheme of the University of Western Australia. He is very grateful to Mikhail Kostylev, Nikita Kostylev, Sergey Samarin, Jim Williams, Adekunle Adeyeye, Peter Metaxas, Lisa Willig and Jessica Hutomo for useful discussions and suggestions. He also would like to thank anonymous reviewers and colleagues from the magneto-plasmonics sector for comments that helped to improve the manuscript.

Conflicts of Interest

The author declares no conflict of interest.

References

1. Editorial. A sunny outlook. *Nat. Photon.* **2012**, *6*, 129.
2. The Nobel Prize in Physics 2014 - Press release 2014.
3. Aoshima, K.; Funabashi, N.; Machida, K.; Miyamoto, Y.; Kuga, K.; Ishibashi, T.; Shimidzu, N.; Sato, F. Submicron magneto-optical spatial light modulation device for holographic displays driven by spin-polarized electrons. *J. Disp. Technol.* **2010**, *6*, 374–380.
4. Jansen, K.; van Soest, G.; van der Steen, A.F. Intravascular Photoacoustic Imaging: A New Tool for Vulnerable Plaque Identification. *Ultrasound Med. Biol.* **2014**, *40*, 1037–1048.
5. Editorial. Simply silicon. *Nat. Photon.* **2010**, *4*, 491.
6. Almeida, V.R.; Barrios, C.A.; Panepucci, R.R.; Lipson, M. All-optical control of light on a silicon chip. *Nature* **2004**, *431*, 1081–1084.
7. Vivien, L.; Pavesi, L. *Handbook of Silicon Photonics*; CRC Press: Boca Raton, FL, USA, 2013.
8. Schaller, R. Moore's law: Past, present and future. *IEEE Spectr.* **1997**, *34*, 52–59.
9. Zhirnov, V.; Cavin, R.; Hutchby, J.; Bourianoff, G. Limits to binary logic switch scaling—A gedanken model. *Proc. IEEE* **2003**, *91*, 1934–1939.

10. Lu, W.; Lieber, C.M. Nanoelectronics from the bottom up. *Nat. Mater.* **2007**, *6*, 841–850.

11. Wallquist, M.; Hammerer, K.; Rabl, P.; Lukin, M.; Zoller, P. Hybrid quantum devices and quantum engineering. *Phys. Scripta* **2009**, *2009*, 014001.

12. Xiang, Z.L.; Ashhab, S.; You, J.Q.; Nori, F. Hybrid quantum circuits: Superconducting circuits interacting with other quantum systems. *Rev. Mod. Phys.* **2013**, *85*, 623–653.

13. Savage, N. Linking with light. *IEEE Spectr.* **2002**, *39*, 32–36.

14. Kinsey, N.; Ferrera, M.; Shalaev, V.M.; Boltasseva, A. Examining nanophotonics for integrated hybrid systems: A review of plasmonic interconnects and modulators using traditional and alternative materials. *J. Opt. Soc. Am. B* **2015**, *32*, 121–142.

15. Willner, A.E.; Khaleghi, S.; Chitgarha, M.R.; Yilmaz, O.F. All-Optical Signal Processing. *J. Lightwave Technol.* **2014**, *32*, 660–680.

16. Andryieuski, A.; Lavrinenko, A.V. Nanocouplers for Infrared and Visible Light. *Adv. Optoelectron.* **2012**, *2012*, 839747.

17. Kruglyak, V.V.; Demokritov, S.O.; Grundler, D. Magnonics. *J. Phys. D Appl. Phys.* **2010**, *43*, 264001.

18. Serga, A.A.; Chumak, A.V.; Hillebrands, B. YIG magnonics. *J. Phys. D Appl. Phys.* **2010**, *43*, 264002.

19. Gubbiotti, G.; Tacchi, S.; Madami, M.; Carlotti, G.; Adeyeye, A.O.; Kostylev, M. Brillouin light scattering studies of planar metallic magnonic crystals. *J. Phys. D Appl. Phys.* **2010**, *43*, 264003.

20. Lenk, B.; Ulrichs, H.; Garbs, F.; Münzenberg, M. The building blocks of magnonics. *Phys. Rep.* **2011**, *507*, 107–136.

21. Stamps, R.L.; Breitkreutz, S.; Åkerman, J.; Chumak, A.V.; Otani, Y.; Bauer, G.E.W.; Thiele, J.U.; Bowen, M.; Majetich, S.A.; Kläui, M.; *et al.* The 2014 Magnetism Roadmap. *J. Phys. D* **2014**, *47*, 333001.

22. Hirohata, A.; Takanashi, K. Future perspectives for spintronic devices. *J. Phys. D* **2014**, *47*, 193001.

23. Krawczyk, M.; Grundler, D. Review and prospects of magnonic crystals and devices with reprogrammable band structure. *J. Phys. Cond. Matter* **2014**, *26*, 123202.

24. Demokritov, S.O.; Slavin, A.N. *Magnonics: From Fundamentals to Applications*; Springer: Berlin, Germany, 2013.

25. Zvezdin, A.K.; Kotov, V.A. *Modern Magnetooptics and Magnetooptical Materials*; IOP Publishing: Bristol, UK, 1997.

26. Stancil, D.D.; Prabhakar, A. *Spin Waves: Theory and Applications*; Springer: Berlin, Germany, 2009.

27. Barnes, W.L.; Dereux, A.; Ebbesen, T.W. Surface plasmon subwavelength optics. *Nature* **2003**, *424*, 824–830.

28. Stockman, M.I. Nanoplasmonics: past, present, and glimpse into future. *Opt. Express* **2011**, *19*, 22029–22106.

29. Zayats, A.V.; Maier, S.A. *Active Plasmonics and Tuneable Plasmonic Metamaterials*; John Wiley and Sons: New Jersey, NJ, USA, 2013.

30. Krasnok, A.E.; Maksymov, I.S.; Denisyuk, A.I.; Belov, P.A.; Miroshnichenko, A.E.; Simovski, C.R.; Kivshar, Y.S. Optical nanoantennas. *Phys. Usp.* **2013**, *56*, 539–564.

31. Chiu, K.; Quinn, J. Magneto-plasma surface waves in solids. *Il Nuovo Cimento B* **1972**, *10*, 1–20.

32. Armelles, G.; Cebollada, A.; García-Martín, A.; González, M.U. Magnetoplasmonics: Combining Magnetic and Plasmonic Functionalities. *Adv. Opt. Mater.* **2013**, *1*, 10–35.

33. Temnov, V.V. Ultrafast acousto-magneto-plasmonics. *Nat. Photon.* **2012**, *6*, 728–736.

34. Armelles, G.; Dmitriev, A. Focus on magnetoplasmonics. *New J. Phys.* **2014**, *16*, 045012.

35. Inoue, M.; Levy, M.; Baryshev, A.V. *Magnetophotonics: From Theory to Applications*; Springer: Berlin, Germany, 2013.

36. Belotelov, V.I.; Akimov, I.A.; Pohl, M.; Kotov, V.A.; Kasture, S.; Vengurlekar, A.S.; Gopal, A.V.; Yakovlev, D.R.; Zvezdin, A.K.; Bayer, M. Enhanced magneto-optical effects in magnetoplasmonic crystals. *Nat. Nanotech.* **2011**, *6*, 370–376.

37. Wurtz, G.A.; Hendren, W.; Pollard, R.; Atkinson, R.; Guyader, L.L.; Kirilyuk, A.; Rasing, T.; Smolyaninov, I.I.; Zayats, A.V. Controlling optical transmission through magneto-plasmonic crystals with an external magnetic field. *New J. Phys.* **2008**, *10*, 105012.

38. Belotelov, V.I.; Doskolovich, L.L.; Zvezdin, A.K. Extraordinary Magneto-Optical Effects and Transmission through Metal-Dielectric Plasmonic Systems. *Phys. Rev. Lett.* **2007**, *98*, 077401.

39. Belotelov, V.I.; Bykov, D.A.; Doskolovich, L.L.; Kalish, A.N.; Zvezdin, A.K. Extraordinary transmission and giant magneto-optical transverse Kerr effect in plasmonic nanostructured films. *J. Opt. Soc. Am. B* **2009**, *26*, 1594–1598.

40. Kreilkamp, L.E.; Belotelov, V.I.; Chin, J.Y.; Neutzner, S.; Dregely, D.; Wehlus, T.; Akimov, I.A.; Bayer, M.; Stritzker, B.; Giessen, H. Waveguide-Plasmon Polaritons Enhance Transverse Magneto-Optical Kerr Effect. *Phys. Rev. X* **2013**, *3*, 041019.

41. Christ, A.; Zentgraf, T.; Kuhl, J.; Tikhodeev, S.G.; Gippius, N.A.; Giessen, H. Optical properties of planar metallic photonic crystal structures: Experiment and theory. *Phys. Rev. B* **2004**, *70*, 125113.

42. Mansuripur, M. The Faraday effect. *Opt. Photon. News* **1999**, *10*, 32–36.

43. Jain, P.K.; Xiao, Y.; Walsworth, R.; Cohen, A.E. Surface Plasmon Resonance Enhanced Magneto-Optics (SuPREMO): Faraday Rotation Enhancement in Gold-Coated Iron Oxide Nanocrystals. *Nano Lett.* **2009**, *9*, 1644–1650.

44. Chin, J.Y.; Steinle, T.; Wehlus, T.; Dregely, D.; Weiss, T.; Belotelov, V.I.; Stritzker, B.; Giessen, H. Nonreciprocal plasmonics enables giant enhancement of thin-film Faraday rotation. *Nat. Commun.* **2013**, *4*, 1599.

45. Wehlus, T.; Körner, T.; Leitenmeier, S.; Heinrich, A.; Stritzker, B. Magneto-optical garnets for integrated optoelectronic devices. *Phys. Status Solidi A* **2011**, *208*, 252–263.

46. Miroshnichenko, A.E.; Flach, S.; Kivshar, Y.S. Fano resonances in nanoscale structures. *Rev. Mod. Phys.* **2010**, *82*, 2257–2298.

47. Lambert, C.H.; Mangin, S.; Varaprasad, B.S.D.C.S.; Takahashi, Y.K.; Hehn, M.; Cinchetti, M.; Malinowski, G.; Hono, K.; Fainman, Y.; Aeschlimann, M.; *et al.* All-optical control of ferromagnetic thin films and nanostructures. *Science* **2014**, *345*, 1337–1340.

48. Mruczkiewicz, M.; Pavlov, E.; Vysotskii, S.; Krawczyk, M.; Filimonov, Y.; Nikitov, S. Magnonic Bandgaps in Metalized 1-D YIG Magnonic Crystals. *IEEE Trans. Magnet.* **2014**, *50*, 1–3.

49. Bi, L.; Hu, J.; Kimerling, L.; Ross, C.A. Fabrication and characterization of $As_2S_3/Y_3Fe_5O_{12}$ and $Y_3Fe_5O_{12}$/SOI strip-loaded waveguides for integrated optical isolator applications. *Proc. SPIE* **2010**, *7604*, 760406–760406–10.

50. Grunin, A.A.; Zhdanov, A.G.; Ezhov, A.A.; Ganshina, E.A.; Fedyanin, A.A. Surface-plasmon-induced enhancement of magneto-optical Kerr effect in all-nickel subwavelength nanogratings. *Appl. Phys. Lett.* **2010**, *97*, 261908.

51. Newman, D.M.; Wears, M.L.; Matelon, R.J.; Hooper, I.R. Magneto-optic behaviour in the presence of surface plasmons. *J. Phys.: Cond. Matter* **2008**, *20*, 345230.

52. Ctistis, G.; Papaioannou, E.; Patoka, P.; Gutek, J.; Fumagalli, P.; Giersig, M. Optical and Magnetic Properties of Hexagonal Arrays of Subwavelength Holes in Optically Thin Cobalt Films. *Nano Lett.* **2009**, *9*, 1–6.

53. Sapozhnikov, M.V.; Gusev, S.A.; Troitskii, B.B.; Khokhlova, L.V. Optical and magneto-optical resonances in nanocorrugated ferromagnetic films. *Opt. Lett.* **2011**, *36*, 4197–4199.

54. Bonanni, V.; Bonetti, S.; Pakizeh, T.; Pirzadeh, Z.; Chen, J.; Nogués, J.; Vavassori, P.; Hillenbrand, R.; Åkerman, J.; Dmitriev, A. Designer Magnetoplasmonics with Nickel Nanoferromagnets. *Nano Lett.* **2011**, *11*, 5333–5338.

55. Chen, J.; Albella, P.; Pirzadeh, Z.; Alonso-González, P.; Huth, F.; Bonetti, S.; Bonanni, V.; Åkerman, J.; Nogués, J.; Vavassori, P.; *et al.* Plasmonic Nickel Nanoantennas. *Small* **2011**, *7*, 2341–2347.

56. Chetvertukhin, A.V.; Baryshev, A.V.; Uchida, H.; Inoue, M.; Fedyanin, A.A. Resonant surface magnetoplasmons in two-dimensional magnetoplasmonic crystals excited in Faraday configuration. *J. Appl. Phys.* **2012**, *111*, 07A946.

57. Kostylev, N.; Maksymov, I.S.; Adeyeye, A.O.; Samarin, S.; Kostylev, M.; Williams, J.F. Plasmon-assisted high reflectivity and strong magneto-optical Kerr effect in permalloy gratings. *Appl. Phys. Lett.* **2013**, *102*, 121907.

58. Maccaferri, N.; Berger, A.; Bonetti, S.; Bonanni, V.; Kataja, M.; Qin, Q.H.; van Dijken, S.; Pirzadeh, Z.; Dmitriev, A.; Nogués, J.; *et al.* Tuning the Magneto-Optical Response of Nanosize Ferromagnetic Ni Disks Using the Phase of Localized Plasmons. *Phys. Rev. Lett.* **2013**, *111*, 167401.

59. Newman, D.M.; Wears, M.L.; Matelon, R.J. Plasmon transport phenomena on a continuous ferromagnetic surface. *Europhys. Lett.* **2004**, *68*, 692.

60. Krutyanskiy, V.L.; Kolmychek, I.A.; Gan'shina, E.A.; Murzina, T.V.; Evans, P.; Pollard, R.; Stashkevich, A.A.; Wurtz, G.A.; Zayats, A.V. Plasmonic enhancement of nonlinear magneto-optical response in nickel nanorod metamaterials. *Phys. Rev. B* **2013**, *87*, 035116.

61. Melander, E.; Östman, E.; Keller, J.; Schmidt, J.; Papaioannou, E.T.; Kapaklis, V.; Arnalds, U.B.; Caballero, B.; García-Martín, A.; Cuevas, J.C.; *et al.* Influence of the magnetic field on the plasmonic properties of transparent Ni anti-dot arrays. *Appl. Phys. Lett.* **2012**, *101*, 063107.

62. Chetvertukhin, A.V.; Grunin, A.A.; Baryshev, A.V.; Dolgova, T.V.; Uchida, H.; Inoue, M.; Fedyanin, A.A. Magneto-optical Kerr effect enhancement at the Wood's anomaly in magnetoplasmonic crystals. *J. Magn. Magn. Mater.* **2012**, *324*, 3516–3518.

63. Hessel, A.; Oliner, A.A. A New Theory of Wood's Anomalies on Optical Gratings. *Appl. Opt.* **1965**, *4*, 1275–1297.

64. Galli, M.; Portalupi, S.L.; Belotti, M.; Andreani, L.C.; O'Faolain, L.; Krauss, T.F. Light scattering and Fano resonances in high-Q photonic crystal nanocavities. *Appl. Phys. Lett.* **2009**, *94*, 071101.

65. Gallinet, B.; Martin, O.J.F. *Ab initio* theory of Fano resonances in plasmonic nanostructures and metamaterials. *Phys. Rev. B* **2011**, *83*, 235427.

66. Maksymov, I.S.; Miroshnichenko, A.E. Active control over nanofocusing with nanorod plasmonic antennas. *Opt. Express* **2011**, *19*, 5888–5894.

67. Ichi Tanaka, S. Longitudinal Kerr Magneto-Optic Effect in Permalloy Film. *Jap. J. Appl. Phys.* **1963**, *2*, 548.

68. Robinson, C.C. Longitudinal Kerr Magneto-Optic Effect in Thin Films of Iron, Nickel, and Permalloy. *J. Opt. Soc. Am.* **1963**, *53*, 681–687.

69. Krinchik, G.; Chepurova, E.; Kraeva, T. Magneto-optical effects in nickel and Permalloy associated with frustrated total internal reflection. *Zh. Eksp. Teor. Fiz* **1985**, *88*, 277–285.

70. Belotelov, V.I.; Kreilkamp, L.E.; Akimov, I.A.; Kalish, A.N.; Bykov, D.A.; Kasture, S.; Yallapragada, V.J.; Gopal, A.V.; Grishin, A.M.; Khartsev, S.I.; *et al.* Plasmon-mediated magneto-optical transparency. *Nat. Commun.* **2013**, *4*, 2128.

71. Krinchik, G.; Gushchin, G.S. Magnetooptical effect of change of electronic structure of a ferromagnetic metal following rotation of the magnetization vector. *JETP Lett.* **1969**, *10*, 35–39.

72. Kurp, P. Green Computing. *Commun. ACM* **2008**, *51*, 11–13.

73. Samad, J.; Kumar, R. Computers and global warming. *Int. J. Eng. Innov. Technol.* **2013**, *2*, 241–244.

74. Hertel, R.; Wulfhekel, W.; Kirschner, J. Domain-Wall Induced Phase Shifts in Spin Waves. *Phys. Rev. Lett.* **2004**, *93*, 257202.

75. Kostylev, M.P.; Serga, A.A.; Schneider, T.; Leven, B.; Hillebrands, B. Spin-wave logical gates. *Appl. Phys. Lett.* **2005**, *87*, 153501.

76. Hertel, R. *Guided Spin Waves. Handbook of Magnetism and Advanced Magnetic Materials*; Wiley: New York, NY, USA, 2007.

77. Schneider, T.; Serga, A.A.; Leven, B.; Hillebrands, B.; Stamps, R.L.; Kostylev, M.P. Realization of spin-wave logic gates. *Appl. Phys. Lett.* **2008**, *92*, 022505.

78. Khitun, A.; Bao, M.; Wang, K.L. Magnonic logic circuits. *J. Phys. D* **2010**, *43*, 264005.

79. Ding, J.; Kostylev, M.; Adeyeye, A.O. Realization of a mesoscopic reprogrammable magnetic logic based on a nanoscale reconfigurable magnonic crystal. *Appl. Phys. Lett.* **2012**, *100*, 073114.

80. Demokritov, S.O. *Spin Wave Confinement*; Pan Stanford Publishing: Singapore, 2009.

81. Maksymov, I.S.; Kostylev, M. Broadband stripline ferromagnetic resonance spectroscopy of ferromagnetic films, multilayers and nanostructures. *Phys. E* **2015**, *69*, 253–293.

82. Wehrspohn, H.S.K.; Busch, K. *Nanophotonic Materials: Photonic Crystals, Plasmonics, and Metamaterials*; Wiley-VCH Verlag: Weinheim, Germany, 2008.

83. The International Technology Roadmap for Semiconductors (ITRS); Semiconductor Industry Association (SIA): Washington, DC, USA, 2011.

84. Kirilyuk, A.; Kimel, A.V.; Rasing, T. Ultrafast optical manipulation of magnetic order. *Rev. Mod. Phys.* **2010**, *82*, 2731–2784.

85. Wang, Z.; Wu, M. Chirped-microwave assisted magnetization reversal. *J. Appl. Phys.* **2009**, *105*, 093903.

86. Ding, J. *Novel Magnonic Crystals and Devices: Fabrication, Static and Dynamic Behaviors*; Department of Electrical and Computer Engineering, National University of Singapore: Singapore, 2013.

87. Ding, J.; Kostylev, M.; Adeyeye, A.O. Broadband ferromagnetic resonance spectroscopy of permalloy triangular nanorings. *Appl. Phys. Lett.* **2012**, *100*, 062401.

88. Ding, J.; Singh, N.; Kostylev, M.; Adeyeye, A.O. Static and dynamic magnetic properties of $Ni_{80}Fe_{20}$ anti-ring nanostructures. *Phys. Rev. B* **2013**, *88*, 014301.

89. Kostylev, M.; Zhong, S.; Ding, J.; Adeyeye, A.O. Resonance properties of bi-component arrays of magnetic dots magnetized perpendicular to their planes. *J. Appl. Phys.* **2013**, *114*, 113910.

90. Raether, H. *Surface Plasmons on Smooth and Rough Surfaces and on Gratings*; Springer Verlag: Berlin, Germany, 1988.

91. Adeyeye, A.O.; Singh, N. Large area patterned magnetic nanostructures. *J. Phys. D* **2008**, *41*, 153001.

92. Verellen, N.; Sonnefraud, Y.; Sobhani, H.; Hao, F.; Moshchalkov, V.V.; Dorpe, P.V.; Nordlander, P.; Maier, S.A. Fano Resonances in Individual Coherent Plasmonic Nanocavities. *Nano Lett.* **2009**, *9*, 1663–1667.

93. Santos, A.D.; Melo, L.G.C.; Martins, C.S.; Missell, F.P.; Souche, Y.; Machado, F.L.A.; Rezende, S.M. Domains and giant magneto-impedance in amorphous ribbons by magneto-optical Kerr effect. *J. Appl. Phys.* **1996**, *79*, 6546–6548.

94. Vavassori, P. Polarization modulation technique for magneto-optical quantitative vector magnetometry. *Appl. Phys. Lett.* **2000**, *77*, 1605–1607.

95. Gubbiotti, G.; Tacchi, S.; Carlotti, G.; Vavassori, P.; Singh, N.; Goolaup, S.; Adeyeye, A.O.; Stashkevich, A.; Kostylev, M. Magnetostatic interaction in arrays of nanometric permalloy wires: A magneto-optic Kerr effect and a Brillouin light scattering study. *Phys. Rev. B* **2005**, *72*, 224413.

96. Willig, L. Magnetisation Dynamics and Magneto-Optics of Nanostructures. Master Thesis, School of Physics, University of Western Australia, Perth, Australia, 2014.

97. Demokritov, S.; Tsymbal, E. Light scattering from spin waves in thin films and layered systems. *J. Phys.* **1994**, *6*, 7145.

98. Zhu, Y. *Modern Techniques for Characterizing Magnetic Materials*; Springer: Berlin, Germany, 2005.

99. Kronmüller, H.; Parkin, S. *Handbook of Magnetism and Advanced Magnetic Materials*; Wiley-Interscience: New York, NY, USA, 2007.

100. Borovik-Romanov, A.S.; Kreines, N.M. Brillouin-Mandelstam scattering from thermal and excited magnons. *Phys. Rep.* **1982**, *81*, 351–408.

101. Fukui, M.; Tada, O.; So, V.C.Y.; Stegeman, G.I. Enhanced Brillouin scattering involving surface plasmon polaritons. *J. Phys. C* **1981**, *14*, 5591.

102. Lee, S.; Hillebrands, B.; Dutcher, J.R.; Stegeman, G.I.; Knoll, W.; Nizzoli, F. Dispersion and localization of guided acoustic modes in a Langmuir-Blodgett film studied by surface-plasmon-polariton-enhanced Brillouin scattering. *Phys. Rev. B* **1990**, *41*, 5382–5387.

103. Utegulov, Z.N.; Shaw, J.M.; Draine, B.T.; Kim, S.A.; Johnson, W.L. Surface-plasmon enhancement of Brillouin light scattering from gold-nanodisk arrays on glass. *Proc. SPIE* **2007**, *6641*, 66411M:1–66411M:10.

104. Johnson, W.L.; Kim, S.A.; Utegulov, Z.N.; Shaw, J.M.; Draine, B.T. Optimization of Arrays of Gold Nanodisks for Plasmon-Mediated Brillouin Light Scattering. *J. Phys. Chem. C* **2009**, *113*, 14651–14657.

105. Van Exter, M.; Lagendijk, A. Ultrashort Surface-Plasmon and Phonon Dynamics. *Phys. Rev. Lett.* **1988**, *60*, 49–52.

106. Wang, J.; Wu, J.; Guo, C. Resolving dynamics of acoustic phonons by surface plasmons. *Opt. Lett.* **2007**, *32*, 719–721.

107. Khokhlov, N.E. Resonant Optical Effects Resulting from Optical, Magnetic and Acoustic Control of Plasmon-Polaritons in Multilayered Structures (in Russian). Ph.D. Thesis, Department of Photonics and Microwave Physics, Lomonosov Moscow State University, Moscow, Russia, 2015.

108. Wang, S.S.; Magnusson, R. Theory and applications of guided-mode resonance filters. *Appl. Opt.* **1993**, *32*, 2606–2613.

109. Naik, G.V.; Shalaev, V.M.; Boltasseva, A. Alternative Plasmonic Materials: Beyond Gold and Silver. *Adv. Mater.* **2013**, *25*, 3264–3294.

110. Kuznetsov, A.I.; Miroshnichenko, A.E.; Fu, Y.H.; Zhang, J.; Luk'yanchuk, B. Magnetic light. *Sci. Rep.* **2012**, *2*, 492.

111. Bai, B.; Tervo, J.; Turunen, J. Polarization conversion in resonant magneto-optic gratings. *New J. Phys.* **2006**, *8*, 205.

112. Marinchio, H.; Carminati, R.; García-Martín, A.; Sáenz, J.J. Magneto-optical Kerr effect in resonant subwavelength nanowire gratings. *New J. Phys.* **2014**, *16*, 015007.

113. Qin, J.; Deng, L.; Xie, J.; Tang, T.; Bi, L. Highly sensitive sensors based on magneto-optical surface plasmon resonance in Ag/CeYIG heterostructures. *AIP Adv.* **2015**, *5*, 017118.

114. Souche, Y.; Novosad, V.; Pannetier, B.; Geoffroy, O. Magneto-optical diffraction and transverse Kerr effect. *J. Magn. Magn. Mater.* **1998**, *177-181*, 1277–1278.

115. Amemiya, T.; Ishikawa, A.; Shoji, Y.; Hai, P.N.; Tanaka, M.; Mizumoto, T.; Tanaka, T.; Arai, S. Three-dimensional nanostructuring in YIG ferrite with femtosecond laser. *Opt. Lett.* **2014**, *39*, 212–215.

116. Maksymov, I.S.; Hutomo, J.; Kostylev, M. Transverse magneto-optical Kerr effect in subwavelength dielectric gratings. *Opt. Express* **2014**, *22*, 8720–8725.

117. Khokhlov, N.E.; Prokopov, A.R.; Shaposhnikov, A.N.; Berzhansky, V.N.; Kozhaev, M.A.; Andreev, S.N.; Ravishankar, A.P.; Achanta, V.G.; Bykov, D.A.; Zvezdin, A.K.; *et al.* Photonic crystals with plasmonic patterns: Novel type of the heterostructures for enhanced magneto-optical activity. *J. Phys. D* **2015**, *48*, 095001.

118. Kimel, A.V.; Kirilyuk, A.; Usachev, P.A.; Pisarev, R.V.; Balbashov, A.M.; Rasing, T. Ultrafast non-thermal control of magnetization by instantaneous photomagnetic pulses. *Nature* **2005**, *435*, 655–657.

119. Vabishchevich, P.P.; Frolov, A.Y.; Shcherbakov, M.R.; Grunin, A.A.; Dolgova, T.V.; Fedyanin, A.A. Magnetic field-controlled femtosecond pulse shaping by magnetoplasmonic crystals. *J. Appl. Phys.* **2013**, *113*, 17A947.

120. Shcherbakov, M.R.; Vabishchevich, P.P.; Frolov, A.Y.; Dolgova, T.V.; Fedyanin, A.A. Femtosecond intrapulse evolution of the magneto-optic Kerr effect in magnetoplasmonic crystals. *Phys. Rev. B* **2014**, *90*, 201405.

121. Satoh, T.; Terui, Y.; Moriya, R.; Ivanov, B.A.; Ando, K.; Saitoh, E.; Shimura, T.; Kuroda, K. Directional control of spin wave emission by spatially shaped light. *Nat. Photon.* **2012**, *6*, 662–666.

122. Pitaevskii, L.P. Electric Forces in a Transparent Dispersive Medium. *Sov. Phys. JETP* **1961**, *12*, 1008–1013.

123. Van der Ziel, J.P.; Pershan, P.S.; Malmstrom, L.D. Optically-Induced Magnetization Resulting from the Inverse Faraday Effect. *Phys. Rev. Lett.* **1965**, *15*, 190–193.

124. Rizzo, C.; Dupays, A.; Battesti, R.; Fouché, M.; Rikken, G.L.J.A. Inverse Cotton-Mouton effect of the vacuum and of atomic systems. *Europhys. Lett.* **2010**, *90*, 64003.

125. Popov, S.V.; Svirko, Y.P.; Zheludev, N.I. Coherent and incoherent specular inverse Faraday effect: $\chi(^3)$ measurements in opaque materials. *Opt. Lett.* **1994**, *19*, 13–15.

126. Perroni, C.A.; Liebsch, A. Coherent control of magnetization via inverse Faraday effect. *J. Phys.* **2006**, *18*, 7063.

127. Hertel, R. Microscopic theory of the inverse Faraday effect. *ArXiv E-Prints*, **2005.**

128. Satoh, T. Opto-magnonics: Light pulses manipulating spin waves. *SPIE Newsroom* **2012**, doi:10.1117/2.1201212.004631.

129. Stipe, B.C.; Strandl, T.C.; Poon, C.C.; Balamane, H.; Boone, T.D.; Katine, J.A.; Li, J.L.; Rawat, V.; Nemoto, H.; Hirotsune, A.; *et al.* Magnetic recording at 1.5 Pb m^2 using an integrated plasmonic antenna. *Nat. Photon.* **2010**, *4*, 484–488.

130. Guyader, L.L.; Savoini, M.; Moussaoui, S.E.; Buzzi, M.; Tsukamoto.; Itoh, A.; Kirilyuk, A.; Rasing, T.; Kimel, A.; Nolting, F. Nanoscale sub-100 picosecond all-optical magnetization switching in GdFeCo microstructures. *Nat. Commun.* **2015**, *6*, 5839.

131. Maksymov, I.S.; Staude, I.; Miroshnichenko, A.E.; Kivshar, Y.S. Optical Yagi-Uda nanoantennas. *Nanophotonics* **2012**, *1*, 65–81.

132. Stannigel, K.; Komar, P.; Habraken, S.J.M.; Bennett, S.D.; Lukin, M.D.; Zoller, P.; Rabl, P. Optomechanical Quantum Information Processing with Photons and Phonons. *Phys. Rev. Lett.* **2012**, *109*, 013603.

133. Belotelov, V.I.; Zvezdin, A.K. Inverse transverse magneto-optical Kerr effect. *Phys. Rev. B* **2012**, *86*, 155133.

134. Pohl, M.; Kreilkamp, L.E.; Belotelov, V.I.; Akimov, I.A.; Kalish, A.N.; Khokhlov, N.E.; Yallapragada, V.J.; Gopal, A.V.; Nur-E-Alam, M.; Vasiliev, M.; *et al.* Tuning of the transverse magneto-optical Kerr effect in magneto-plasmonic crystals. *New J. Phys.* **2013**, *15*, 075024.

135. Mittleman, D.M. Frontiers in terahertz sources and plasmonics. *Nat. Photon.* **2013**, *7*, 666–669.

136. Wang, K.; Mittleman, D.M. Metal wires for terahertz wave guiding. *Nature* **2004**, *432*, 376–379.

137. Han, J.; Lakhtakia, A.; Tian, Z.; Lu, X.; Zhang, W. Magnetic and magnetothermal tunabilities of subwavelength-hole arrays in a semiconductor sheet. *Opt. Lett.* **2009**, *34*, 1465–1467.

138. Lan, Y.C.; Chang, Y.C.; Lee, P.H. Manipulation of tunneling frequencies using magnetic fields for resonant tunneling effects of surface plasmons. *Appl. Phys. Lett.* **2007**, *90*, 171114.

139. Kumar, G.; Li, S.; Jadidi, M.M.; Murphy, T.E. Terahertz surface plasmon waveguide based on a one-dimensional array of silicon pillars. *New J. Phys.* **2013**, *15*, 085031.

140. Cong, L.; Cao, W.; Tian, Z.; Gu, J.; Han, J.; Zhang, W. Manipulating polarization states of terahertz radiation using metamaterials. *New J. Phys.* **2012**, *14*, 115013.

141. Tsiatmas, A.; Fedotov, V.A.; de Abajo, F.J.G.; Zheludev, N.I. Low-loss terahertz superconducting plasmonics. *New J. Phys.* **2012**, *14*, 115006.

142. Berry, C.W.; Jarrahi, M. Terahertz generation using plasmonic photoconductive gratings. *New J. Phys.* **2012**, *14*, 105029.

143. Hu, B.; Wang, Q.J.; Zhang, Y. Broadly tunable one-way terahertz plasmonic waveguide based on nonreciprocal surface magneto plasmons. *Opt. Lett.* **2012**, *37*, 1895–1897.

144. Fan, F.; Chang, S.J.; Gu, W.H.; Wang, X.H.; Chen, A.Q. Magnetically Tunable Terahertz Isolator Based on Structured Semiconductor Magneto Plasmonics. *IEEE Photon. Technol. Lett.* **2012**, *24*, 2080–2083.

145. Jablan, M.; Buljan, H.; Soljačić, M. Plasmonics in graphene at infrared frequencies. *Phys. Rev. B* **2009**, *80*, 245435.

146. Grigorenko, A. N.; Polini, M.; Novoselov, K.S. Graphene plasmonics. *Nat. Photon.* **2012**, *6*, 749–758.

147. Low, T.; Avouris, P. Graphene Plasmonics for Terahertz to Mid-Infrared Applications. *ACS Nano* **2014**, *8*, 1086–1101.

148. García de Abajo, F.J. Graphene Plasmonics: Challenges and Opportunities. *ACS Photon.* **2014**, *1*, 135–152.

149. Pimenov, A.; Shuvaev, A.; Loidl, A.; Schrettle, F.; Mukhin, A.A.; Travkin, V.D.; Ivanov, V.Y.; Balbashov, A.M. Magnetic and Magnetoelectric Excitations in $TbMnO_3$. *Phys. Rev. Lett.* **2009**, *102*, 107203.

150. Nakajima, M.; Namai, A.; Ohkoshi, S.; Suemoto, T. Ultrafast time domain demonstration of bulk magnetization precession at zero magnetic field ferromagnetic resonance induced by terahertz magnetic field. *Opt. Express* **2010**, *18*, 18260–18268.

151. Kampfrath, T.; Sell, A.; Klatt, G.; Pashkin, A.; Mährlein, S.; Dekorsy, T.; Wolf, M.; Fiebig, M.; Leitenstorfer, A.; Huber, R. Coherent terahertz control of antiferromagnetic spin waves. *Nat. Photon.* **2011**, *5*, 31–34.

152. Yamaguchi, K.; Kurihara, T.; Minami, Y.; Nakajima, M.; Suemoto, T. Terahertz Time-Domain Observation of Spin Reorientation in Orthoferrite $ErFeO_3$ through Magnetic Free Induction Decay. *Phys. Rev. Lett.* **2013**, *110*, 137204.

153. Wienholdt, S.; Hinzke, D.; Nowak, U. THz Switching of Antiferromagnets and Ferrimagnets. *Phys. Rev. Lett.* **2012**, *108*, 247207.

154. Constable, E.; Cortie, D.L.; Horvat, J.; Lewis, R.A.; Cheng, Z.; Deng, G.; Cao, S.; Yuan, S.; Ma, G. Complementary terahertz absorption and inelastic neutron study of the dynamic anisotropy contribution to zone-center spin waves in a canted antiferromagnet $NdFeO_3$. *Phys. Rev. B* **2014**, *90*, 054413.

155. Kurihara, T.; Nakamura, K.; Yamaguchi, K.; Sekine, Y.; Saito, Y.; Nakajima, M.; Oto, K.; Watanabe, H.; Suemoto, T. Enhanced spin-precession dynamics in a spin-metamaterial coupled resonator observed in terahertz time-domain measurements. *Phys. Rev. B* **2014**, *90*, 144408.

156. Davoyan, A.R.; Engheta, N. Nonreciprocal Rotating Power Flow within Plasmonic Nanostructures. *Phys. Rev. Lett.* **2013**, *111*, 047401.

157. Davoyan, A.R.; Engheta, N. Theory of Wave Propagation in Magnetized Near-Zero-Epsilon Metamaterials: Evidence for One-Way Photonic States and Magnetically Switched Transparency and Opacity. *Phys. Rev. Lett.* **2013**, *111*, 257401.

158. Chettiar, U.K.; Davoyan, A.R.; Engheta, N. Hotspots from nonreciprocal surface waves. *Opt. Lett.* **2014**, *39*, 1760–1763.

159. Hadad, Y.; Steinberg, B.Z. Magnetized Spiral Chains of Plasmonic Ellipsoids for One-Way Optical Waveguides. *Phys. Rev. Lett.* **2010**, *105*, 233904.

160. Hadad, Y.; Steinberg, B.Z. One way optical waveguides for matched non-reciprocal nanoantennas with dynamic beam scanning functionality. *Opt. Express* **2013**, *21*, A77–A83.

161. Mazor, Y.; Steinberg, B.Z. Longitudinal chirality, enhanced nonreciprocity, and nanoscale planar one-way plasmonic guiding. *Phys. Rev. B* **2012**, *86*, 045120.

162. Mazor, Y.; Steinberg, B.Z. Metaweaves: Sector-Way Nonreciprocal Metasurfaces. *Phys. Rev. Lett.* **2014**, *112*, 153901.

163. Hadad, Y.; Mazor, Y.; Steinberg, B.Z. Green's function theory for one-way particle chains. *Phys. Rev. B* **2013**, *87*, 035130.

164. Lozovik, Y.E. Plasmonics and magnetoplasmonics based on graphene and a topological insulator. *Phys. Usp.* **2012**, *55*, 1035.

165. Crassee, I.; Orlita, M.; Potemski, M.; Walter, A.L.; Ostler, M.; Seyller, T.; Gaponenko, I.; Chen, J.; Kuzmenko, A.B. Intrinsic Terahertz Plasmons and Magnetoplasmons in Large Scale Monolayer Graphene. *Nano Lett.* **2012**, *12*, 2470–2474.

166. Ferreira, A.; Peres, N.M.R.; Castro Neto, A.H. Confined magneto-optical waves in graphene. *Phys. Rev. B* **2012**, *85*, 205426.

167. Yan, H.; Li, Z.; Li, X.; Zhu, W.; Avouris, P.; Xia, F. Infrared Spectroscopy of Tunable Dirac Terahertz Magneto-Plasmons in Graphene. *Nano Lett.* **2012**, *12*, 3766–3771.

168. Zhou, Y.; Xu, X.; Fan, H.; Ren, Z.; Bai, J.; Wang, L. Tunable magnetoplasmons for efficient terahertz modulator and isolator by gated monolayer graphene. *Phys. Chem. Chem. Phys.* **2013**, *15*, 5084–5090.

169. Chamanara, N.; Caloz, C. Tunable Terahertz Graphene Magnetoplasmons: Non-Reciprocal Components and Applications. In Proceedings of the 2014 8th European Conference on Antennas and Propagation (EuCAP), Hague, The Netherlands, 6–11 April 2014; pp. 670–671.

170. Hu, B.; Tao, J.; Zhang, Y.; Wang, Q.J. Magneto-plasmonics in graphene-dielectric sandwich. *Opt. Express* **2014**, *22*, 21727–21738.

171. Hadad, Y.; Davoyan, A.R.; Engheta, N.; Steinberg, B.Z. Extreme and Quantized Magneto-Optics with Graphene Meta-Atoms and Metasurfaces. *ACS Photon.* **2014**, *1*, 1068–1073.

172. Ebbesen, T.W.; Lezec, H.J.; Ghaemi, H.F.; Thio, T.; Wolff, P.A. Extraordinary optical transmission through sub-wavelength hole arrays. *Nature* **1998**, *391*, 667–669.

173. Bergman, D.J.; Strelniker, Y.M. Anisotropic ac Electrical Permittivity of a Periodic Metal-Dielectric Composite Film in a Strong Magnetic Field. *Phys. Rev. Lett.* **1998**, *80*, 857–860.

174. Strelniker, Y.M.; Bergman, D.J. Optical transmission through metal films with a subwavelength hole array in the presence of a magnetic field. *Phys. Rev. B* **1999**, *59*, R12763–R12766.

175. Strelniker, Y.M.; Bergman, D.J. Transmittance and transparency of subwavelength-perforated conducting films in the presence of a magnetic field. *Phys. Rev. B* **2008**, *77*, 205113.

176. Strelniker, Y.M.; Bergman, D.J. Strong angular magneto-induced anisotropy of Voigt effect in metal-dielectric metamaterials with periodic nanostructures. *Phys. Rev. B* **2014**, *89*, 125312.

177. Battula, A.; Chen, S.; Lu, Y.; Knize, R.J.; Reinhardt, K. Tuning the extraordinary optical transmission through subwavelength hole array by applying a magnetic field. *Opt. Lett.* **2007**, *32*, 2692–2694.

178. García-Martín, A.; Armelles, G.; Pereira, S. Light transport in photonic crystals composed of magneto-optically active materials. *Phys. Rev. B* **2005**, *71*, 205116.

179. Helseth, L.E. Tunable plasma response of a metal/ferromagnetic composite material. *Phys. Rev. B* **2005**, *72*, 033409.

180. Khanikaev, A.B.; Baryshev, A.V.; Fedyanin, A.A.; Granovsky, A.B.; Inoue, M. Anomalous Faraday effect of a system with extraordinary optical transmittance. *Opt. Express* **2007**, *15*, 6612–6622.

181. Zhou, R.; Li, H.; Zhou, B.; Wu, L.; Liu, X.; Gao, Y. Transmission through a perforated metal film by applying an external magnetic field. *Solid State Commun.* **2009**, *149*, 657–661.

182. Ou, N.; Shyu, J.; Wu, J.; Wu, T. Extraordinary Optical Transmission Through Dielectric Hole-Array Coated With TbFeCo Thin Film. *IEEE Trans. Magnet.* **2009**, *45*, 4027–4029.

183. Bergman, D.J.; Stockman, M.I. Surface Plasmon Amplification by Stimulated Emission of Radiation: Quantum Generation of Coherent Surface Plasmons in Nanosystems. *Phys. Rev. Lett.* **2003**, *90*, 027402.

184. Baranov, D.G.; Vinogradov, A.P.; Lisyansky, A.A.; Strelniker, Y.M.; Bergman, D.J. Magneto-optical spaser. *Opt. Lett.* **2013**, *38*, 2002–2004.

185. Xu, X.W.; Solanki, S.; Liang, X.A.; Pan, Y.C.; Chong, T.C. Full high-definition digital 3D holographic display and its enabling technologies. *Proc. SPIE* **2010**, *7730*, 77301C.

186. Butykai, A.; Orbán, A.; Kocsis, V.; Szaller, D.; Bordács, S.; Tátrai-Szekeres, E.; ans A. Bóta, L.F.K.; Vértessy, B.G.; Zelles, T.; Kézsmárki, I. Malaria pigment crystals as magnetic micro-rotors: key for high-sensitivity diagnosis. *Sci. Rep.* **2013**, *3*, 1431.

6

Dealloying of Cu-Based Metallic Glasses in Acidic Solutions: Products and Energy Storage Applications

Zhifeng Wang [1,2,3,†], **Jiangyun Liu** [1,†], **Chunling Qin** [1,†,*], **Hui Yu** [1], **Xingchuan Xia** [1], **Chaoyang Wang** [1], **Yanshan Zhang** [1], **Qingfeng Hu** [1] **and Weimin Zhao** [1,3,*]

[1] School of Materials Science and Engineering, Hebei University of Technology, Tianjin 300130, China; E-Mails: wangzf@hebut.edu.cn (Z.W.); liujiangyun1991@163.com (J.L.); yuhuidavid@hebut.edu.cn (H.Y.); xc_xia@hebut.edu.cn (X.X.); wangchaoyang92@126.com (C.W.); 15222812590@163.com (Y.Z.); huqingfeng_hebut@163.com (Q.H.)

[2] Key Laboratory for New Type of Functional Materials in Hebei Province, Hebei University of Technology, Tianjin 300130, China

[3] CITIC Dicastal Co. Ltd., Qinhuangdao 066011, China

† These authors contributed equally to this work.

* Authors to whom correspondence should be addressed; E-Mails: clqin@hebut.edu.cn (C.Q.); wmzhao@yahoo.com (W.Z.)

Academic Editor: Jiye Fang

Abstract: Dealloying, a famous ancient etching technique, was used to produce nanoporous metals decades ago. With the development of dealloying techniques and theories, various interesting dealloying products including nanoporous metals/alloys, metal oxides and composites, which exhibit excellent catalytic, optical and sensing performance, have been developed in recent years. As a result, the research on dealloying products is of great importance for developing new materials with superior physical and chemical properties. In this paper, typical dealloying products from Cu-based metallic glasses after dealloying in hydrofluoric acid and hydrochloric acid solutions are summarized. Several potential application fields of these dealloying products are discussed. A promising application of nanoporous Cu (NPC) and NPC-contained composites related to the energy storage field is introduced. It is expected that more promising dealloying products could be developed for practical energy storage applications.

Keywords: dealloying; metallic glass; nanoporous copper; energy storage; application

1. Introduction

Dealloying, which is a well known etching technique, refers to selective dissolution of one or more components out of an alloy [1], leaving residual noble metal nanoporous structure. Such a technique, initially known as depletion gilding [2], has been used by metalsmiths to gold-coat artifacts for millennia, dating back to the Nagada period (prehistory 4000–3100 BC) [2]. However, some speculate that a nanoporous structure is formed during alloy corrosion owing to the limitation of observation at the nanoscale in the past. This idea has recently been receiving renewed attention because various dealloying products, including nanoporous metals, nanoporous alloys, metallic oxide and composites with excellent physical and chemical properties, can be fabricated by this method [3–6] and can be clearly observed using a high performance scanning electron microscope (SEM) and transmission electron microscopy (TEM), in recent decades [7]. As a result, research on products and applications of an alloy after dealloying has become a hot topic and is drawing increasing attention.

Nanoporous metals and alloys are known as easily obtained dealloying products after corrosion. With the development of dealloying techniques and theories, nanoporous metals such as Au [8], Pd [9], Pt [10], Ag [11] and Cu [12], as well as nanoporous alloys including Au–Ag [13], Pt–Ru [14], Pd–Ag [15] and Pt–Au [16] have been developed in the latest 15 years. They have been revealed for potential applications in catalysis [17], heat exchangers [18], actuators [19], energy storage [20], biosensors [21] and surface-enhanced Raman scattering [22]. Besides this applied research, some basic research related to dealloying has been carried out. Erlebacher and coworkers [23] proposed an atomistic model to explain the formation mechanism of the nanoporous structure during dealloying. Delogu's group [24] developed theoretical models to relate structure and mechanical properties in nanoporous metals. Hosson *et al.* [25,26] studied strain in nanoporous gold. Biener [27], Lian [28] and Weissmüller *et al.* [29] studied mechanical properties of nanoporous metal and composites. It was found that after mixing nanoporous gold with epoxy, the composite became stronger and harder than each constituent phase individually. Thus, as a kind of engineering material, the application field of nanoporous metal can be extended. However, extensive studies have focused on the fabrication and properties of nanoporous noble metals and alloys. Research pertaining to inexpensive nanoporous metals and other fascinating dealloying products is quite rare. As we know, the cost must be taken into account for practical commercial application of the dealloyed products. Recently, a new series of dealloyed products including nanoporous Cu (NPC) and nano/micro-Cu oxides with low cost and unique properties have been developed in our group [30–34]. This paper presents an overview of our recent advancement of dealloyed products via free-dealloying Cu-based metallic glasses (MGs) in hydrofluoric acid (HF) and hydrochloric acid (HCl) solutions, as well as a brief review of others' results related to dealloying. The prospective potential applications of the dealloyed NPC and composites are also introduced.

2. Fabrication of NPC

The fabrication of inexpensive NPC has attracted the attention of numerous researchers. Many direct factors, such as alloy composition, dealloying solutions, etching time, dealloying temperatures and electrochemical potential, have been widely studied [35–39]. These factors are confirmed to influence the specific structural characteristics of the NPC.

Based on a large number of experiments, the microstructure and chemical composition of a precursor alloy are the key parameters to determine the dealloying process and the morphology of nanopores. To prepare uniform nanoporous metal, a homogeneous single phase of dealloying precursor is necessary. Due to their single phase nature, representative single phase solid solution and metallic glasses are good candidates as dealloying precursors. For example, Au–Ag alloy [40] is an ideal and classical precursor to produce uniform nanoporous Au by selectively dissolving the constituent element Ag, because Au and Ag are completely miscible through the entire composition range and no phase separation occurs during the dealloying process. For two-phase or multiphase alloy systems, however, the dealloying results are complicated. In the case of a two-phase system, if one phase can be removed from the alloy and another remains, porous/nonporous metal composites can be fabricated. If the dealloying takes place within each phase, then nanoporous metal composite with two kinds of nanoporous metals or two different pore size distributions can be obtained [41,42]. Up to now, Mn–Cu [43], Al–Cu [44], Mg–Cu [45] and Zn–Cu [46] crystalline precursors have been used to produce NPC with different porous structures. Aside from crystal precursors, recently, amorphous materials such as MGs have also been demonstrated to be a new type of precursor for dealloying to nanoporous metals. MGs are known for their chemically homogeneous single-phase nature which lack crystalline defects and large-scaled phase segregations. They have thousands of compositions and can be used to fabricate various nanoporous metals which cannot be achieved from conventional crystalline precursors. Nowadays, MG precursors including Ti–Cu(–Au) [47,48], Mg–Cu–Y [49], Mg–Cu–Gd [50], Cu–Zr–Ti [51], Cu–Hf–Al [30], Cu–Zr(–Al) [31] and Al–Cu–Mg(–Ni) [52] have been used to obtain uniform NPC through chemical/electrochemical dealloying processes. As shown in Figure 1, a NPC ribbon, which obtained by dealloying of Cu–Hf–Al MGs [30], presented a typical three-dimensional (3D) continuous nanoporous structure. For a binary MG, different atomic ratios influence the pore size of the resulting 3D nanoporous structure. For example, as atomic ratios of the Cu in Ti–Cu ribbons increased, the pore size of NPC decreased [47]. Moreover, the ligament/pore size of dealloyed NPC can be tailored by the addition of different third components into binary MGs. For instance, the pore size of NPC dealloyed from the $Ti_{60}Cu_{39}Ag_1$ alloy was smaller than that of the $Ti_{60}Cu_{40}$ alloy [53]. A noble Au additive in Ti–Cu MGs resulted in the formation of an NPC with ultrafine nanoporous structure [48]. An active Al additive in Cu–Zr MGs, however, led to the formation of a wider ligament of dealloyed NPC [31].

For dealloying study, a proper etching solution must be correctly selected. For obtaining NPC using a dealloying method, the constituent elements of the precursor alloy exhibiting a large difference in the galvanic series [54] in an etching solution is a necessary requirement. Figure 2 shows open-circuit potentials *vs.* time curves of the metal Cu, Hf and Al in 0.5 M HF solution at 298 K open to air. Among the constituent elements, Cu metal shows much higher stability in HF solution due to its much nobler potential, whereas Hf and Al metals exhibit high electrochemical activity. It can be also seen from Figure 2 that the potential differences between Cu and other two elements are noticeable and greater than 0.8 V, which provides an advantageous driving force for the dissolution of less noble elements Hf and Al under free corrosion conditions. As a result, a uniform NPC can be obtained by dealloying the $Cu_{52.5}Hf_{40}Al_{7.5}$ MG in 0.5 M HF solution [30]. Various corrosive solutions, such as HCl [55], H_2SO_4 [56], HF [57], H_3PO_4 [58] and HBF_4 [59] solutions, NaOH aqueous alkali [60] and NaCl saline solution [61], were successfully used to produce NPC. Representative NPC obtained by dealloying of MGs in different electrolytes is listed in Table 1.

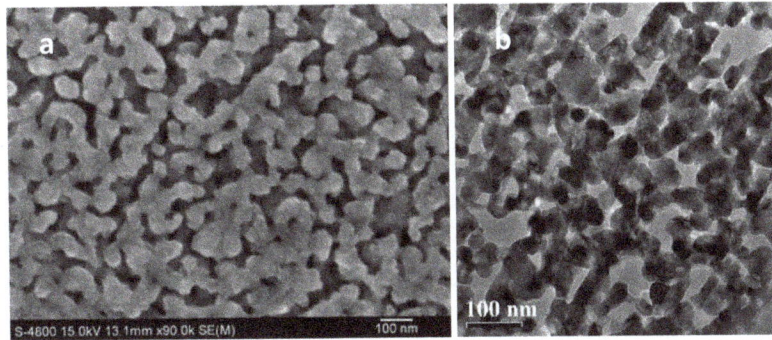

Figure 1. Typical nanoporous Cu (NPC) obtained by dealloying of $Cu_{52.5}Hf_{40}Al_{7.5}$ MGs in 0.5 M hydrofluoric acid (HF) solution for 300 s [30]. (**a**) Scanning electron microscope (SEM) image; (**b**) Transmission electron microscopy (TEM) image.

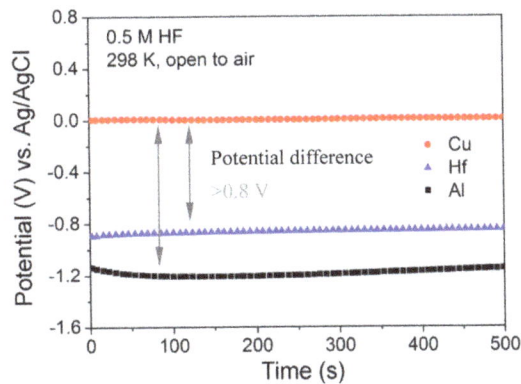

Figure 2. Open-circuit potentials of metals Cu, Hf and Al in 0.5 M HF solution at 298 K open to air [30].

Table 1. Representative NPC obtained via dealloying of MGs.

Dealloying Solution	Precursors	References
HCl	$Al_{70}Cu_{18}Mg_{12}$, $(Al_{75}Cu_{17}Mg_8)_{97}Ni_3$	[52,62]
H_2SO_4	$Mg_{90-x}Cu_xY_{10}$ (x = 20, 25, 30, 40 at.%)	[49,56]
	$Mg_{65}Cu_{25}Gd_{10}$	[50]
H_2SO_4 + PVP	$Mg_{65}Cu_{25}Y_{10}$	[12]
	$Cu_{52.5}Hf_{40}Al_{7.5}$	[30]
	$Cu_{50}Zr_{50-x}Al_x$ (x = 0, 5 at.%)	[31]
HF	$Cu_{60}Zr_{30}Ti_{10}$	[51]
	$Ti_{100-x}Cu_x$ (x = 33, 40, 50, 60, 70 at.%)	[47,63,64]
	Ti–Cu–(Ag, Au, Ni, Pd, Pt)	[48,53,57,65–67]
HF + PVP	$Ti_{60}Cu_{40}$	[68]

Diverse acidic solutions have been widely used for dealloying to produce NPC. It was reported that the ligament/pore size of dealloyed NPC was strongly influenced by the corrosive solution. Generally, the length scale of the NPC ligament and pore size turned to be greater in more concentrated solutions [47]. In addition, an introduction of organic macromolecules into dealloying solution was conducive to refined nanoporous structures [12,68]. It was reported that the pore and ligament sizes decreased in a mixed solution of poly-vinylpyrrolidone (PVP) and sulfuric acid compared to those in the PVP-free H_2SO_4 solution. With increasing concentration of PVP in the dealloying solutions, the sizes of nanopores and ligaments decreased.

Besides the precursor composition and dealloying solution, etching conditions such as dealloying time and temperature also have great effects on the ligament/pore size of the nanoporous metals. Controlling etching conditions is a simple and effective way to tailor porous structures, which determines the final physical and chemical properties of the NPC. Since the dealloying takes place layer by layer, the thickness of the nanoporous metal increases with increasing the etching time. By choosing a proper etching time, we can produce a complete 3D nanoporous structure material without further coarsening of the ligament/pore. Meanwhile, inefficient dealloying of a MG precursor resulted in the fabrication of a nanoporous metal/MG/nanoporous metal composite with a sandwich-like structure [30]. The sandwich-like structure composites containing a MG ductile interlayer showed good bendability [30]. It was presented that the interplay between ligament width d of NPC and dealloying time t can be deduced [56] as:

$$\ln d = \frac{1}{n}\ln t - \frac{E}{nRT} + \frac{1}{n}\ln(KD_0) \tag{1}$$

where n is the coarsening exponent, T is the temperature, E is the activation energy for the ligament growth, R is the gas constant, D_0 and K are constants. According to the formula, a linear law between $\ln d$ and $\ln t$ can be established. In addition, the linear laws were revealed between logarithm of the pore sizes and logarithm of the dealloying time [57], as well as between logarithm of the ligament width and reciprocal of the dealloying temperature [48]. Generally, the ligament/pore size increases with the dealloying time/temperature. This corollary has been verified by some studies [22,56].

There are also some indirect factors to influence the microstructure of dealloyed nanoporous metals. These indirect factors include the parting limit (for most dealloying precursors, the critical range of the noble component in an alloy is between 20 and 60 at.%), the diffusivity of a noble metal at alloy/electrolyte interfaces, the critical potential (for the electrochemical dealloying), volume shrinks and surface cracks, and so on [69]. In general, the microstructure of nanoporous metals fabricated in dealloying process is influenced by many factors, which should be controlled carefully.

3. Synthesis of NPC/Metal Oxides Composites

The NPC/metal oxides composites integrate NPC and other functional metal oxides together. They can be created through various methods and have potential for application in different fields. Relatively active NPC obtained by dealloying can be easily oxidized into NPC/Cu_2O or NPC/CuO composites by annealing in air [58,70]. It was found that heat treatments toward NPC between 200 and 600 °C led to the formation of Cu oxide (CuO and Cu_2O) layers [58]. These oxide layers with a thickness of ten to dozens of nanometers can be produced on the surface of the 3D NPC via *in situ* thermal oxidation process. For NPC with relative density of 20%~30%, the oxide layers were mainly made up of Cu_2O after annealing at 200 °C for 30 min, while those that were composed of CuO after annealing between 400 and 600 °C for 30 min. It was reported that the oxide layer growth thicknesses in a temperature range of 100~600 °C can be estimated using the following formula [58]:

$$d_{oxide}(t) = A \exp\left(\frac{-Q}{R \times T}\right) \times t^{1/2} + d_0 \qquad (2)$$

where $d_{oxide}(t)$ is the thickness of the formed copper oxide as a function of time, R is the gas constant, T is the temperature, t is time in minutes, A is the initial coefficient with values ranging from 5.518×10^5 to 6.658×10^7 Å·min$^{-1/2}$, Q is activation energy and d_0 is the initial copper oxide thickness. Thus, the thickness of Cu oxide film on the surface of NPC can be controlled by adjusting oxidation temperature and annealing time.

Ding and co-workers [71] revealed a general corrosion strategy for the straightforward fabrication of a variety of nano-structured metal oxide through a dealloying and spontaneous oxidation method. The approach mainly involves the alloying consisting of the targeted transition metals and more active metal species, and a subsequent selective leaching of active metals in proper etching liquid. During this corrosion process, the transition metal atoms left behind will undergo spontaneous oxidation at the metal/electrolyte interface to form metal oxides. By following this route, nano-structured Co_3O_4, Fe_3O_4, WO_3, TiO_2 and Mn_3O_4 with intricate structural properties have been successfully synthesized [5,71–75]. To develop this corrosion strategy, Zhang et al. [76] reported an attempt to dealloy Cu–Fe–Al ternary alloy in alkaline solutions. As a result, NPC/$(Cu,Fe)_3O_4$ composites were obtained by a direct one-step dealloying process. These composites were composed of a NPC matrix with ligament/channel sizes of 20–40 nm and octahedral $(Fe,Cu)_3O_4$ embedded particles 600–800 nm in size. The formation of these composites can be explained by the surface diffusion of Cu adatoms (to form a NPC matrix) and oxidation of the active Fe/Cu adatoms (to form metal oxides) during dealloying.

By drawing on Ding's corrosion strategy [71], here we report a successful fabrication of a NPC/Cu_2O composite using the oxygen-assisted dealloying method. Unlike in our previous study [30], the HF etching solution is replenished with oxygen during the dealloying process. Then, a part of active Cu adatoms can be smoothly oxidated to Cu_2O nanoparticles. It can be seen in Figure 3a that the color of the ribbons before and after dealloying in oxygen-enriched 0.65 M HF solutions obviously changes. With the increase of the dealloying time, the ribbon color turns from argenteous to a typical Cu metallic luster and finally changes to dark red. During the dealloying with etching time from 0 to 420 s, the ribbons keep their mechanical integrity, which is important for the subsequent applications. Figure 3b shows X-ray diffraction (XRD) patterns of $Cu_{52.5}Hf_{40}Al_{7.5}$ MGs dealloying in oxygen-enriched 0.65 M HF with different etching times. The diffraction pattern of the as-spun ribbon shows a characteristic broad halo peak without appreciable crystal phase, indicating a single homogeneous amorphous structure. The SEM image (Figure 4a) of the ribbon treated in 0.65 M HF for 180 s presents nanoporous structure. The XRD pattern of the ribbon exhibits sharp crystal peaks which match with (111), (200), (220) crystal planes of Cu (JCPDS No.04-0836). However, no remaining amorphous phase is recognized in the XRD pattern. It is indicated that most of the Hf and Al elements are selectively removed from the Cu–Hf–Al precursor alloy after dealloying in 0.65 M HF for 180 s, leaving much nobler Cu element behind. As a result, the as-obtained porous metal can be identified as

NPC. When the dealloying time is prolonged to 300 s, it is observed that many Cu_2O nanocubes are formed on surfaces of the NPC (Figure 4b). When the dealloying time reaches 420 s, more Cu_2O nanocubes coat or embed into the NPC structure to form the NPC/Cu_2O nanocube composite (Figure 4c). These are also verified by the increasing intensity of Cu_2O peaks in Figure 3b. Moreover, it can be seen that the dealloying product exhibits a new nanoporous composite structure.

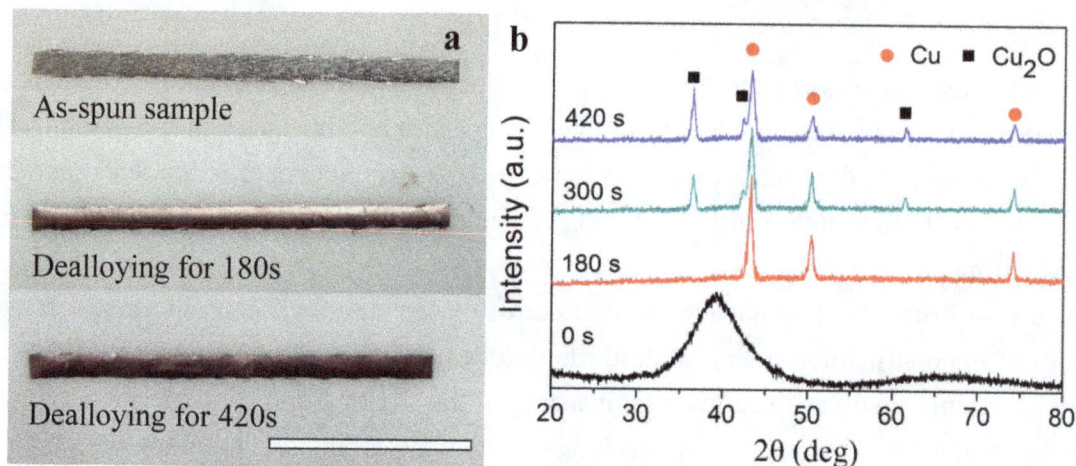

Figure 3. Macrophotograph (**a**) and X-ray diffraction (XRD) patterns (**b**) of the melt-spun $Cu_{52.5}Hf_{40}Al_{7.5}$ ribbon before and after dealloying in oxygen-enriched 0.65 M HF for different time at 298 K open to air. Scale bar, 1 cm.

From the above typical example, a general method to produce NPC/Cu_2O composite structure is schematically illustrated in Figure 4d–i. The method is based on dealloying in an oxygen-enriched corrosive solution. Firstly, a 3D NPC is obtained in dealloying process. Secondly, a portion of copper on the surfaces of NPC ligaments reacts with dissolved oxygen to form Cu_2O nanoparticles. Through the long dealloying process, a mass of Cu_2O nanoparticles are formed on the surfaces of NPC ligaments by reaction between copper and oxygen. At last, a Cu_2O particle layer with a certain thickness is formed on NPC surfaces. Thus, NPC/Cu_2O composites exhibiting new nanoporous structure can be produced and can be further controlled by adjusting the dealloying time and the oxygen content in the etching solution. A similar method is also used to produce porous CuO nanoplate-films with an oxidation-assisted dealloying method [77]. It was observed that the Cu component in an alloy was oxidized preferentially into Cu_2O nanocubes due to free oxidation by dissolved oxygen in electrolytes. Then, the Cu_2O nanocubes are further oxidized into CuO nanoplates mainly owing to primary-cell-induced oxygen consuming corrosion. Such a method can be used universally to fabricate various porous metal oxide nanostructural films on flexible substrates for future nanostructure-based integrated circuit, sensor and solar cell applications.

Besides the above methods, NPC/metal oxides composites can also be synthesized through chemical deposition of metal oxides on prefabricated NPC. Recently, our group produced a new NPC-supported MnO_2 composite (MnO_2/NPC/MnO_2 sandwich structure) [30]. We firstly synthesized a monolithic NPC ribbon with good mechanical integrity and bendability by designing a ductile MG-containing interlayer in the ribbon (Figure 5a). Then, the NPC was used as the substrate for the MnO_2 deposition. It can be seen from Figure 5b that the as-obtained MnO_2 prepared through the classical chemical reaction between KMnO$_4$ and ethanol was composed of nanosized globular particles, which showed a serious particle aggregation phenomenon. By using a ductile NPC support, however, MnO$_2$ nanoflakes (Figure 5c) were homogeneously deposited on the surface of the NPC substrate. This result indicated that the NPC with large specific surface areas and excellent electrical conductivity can effectively promote the morphological change of MnO$_2$ from globular particles to nanoflakes for larger specific surface area and improve the utilization of MnO$_2$ surface active sites. This method can be extended to develop more NPC/metal oxide composites with distinctive functional properties.

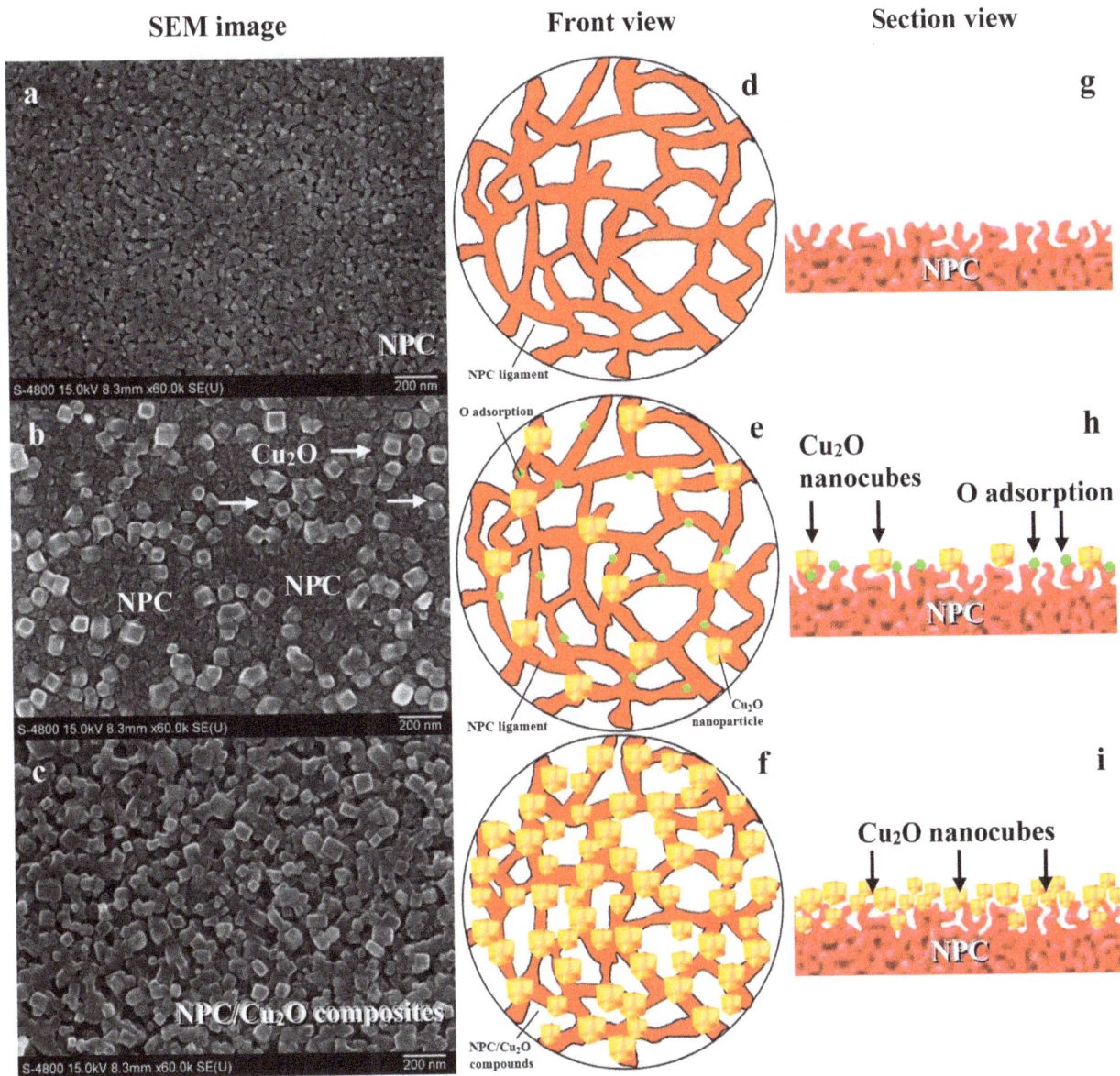

Figure 4. SEM (**a–c**) and schematic images (**d–i**) of Cu$_{52.5}$Hf$_{40}$Al$_{7.5}$ MGs dealloying in O-enriched 0.65 M HF with different time. (**a,d,g**) 180 s; (**b,e,h**) 300 s; (**c,f,i**) 420 s.

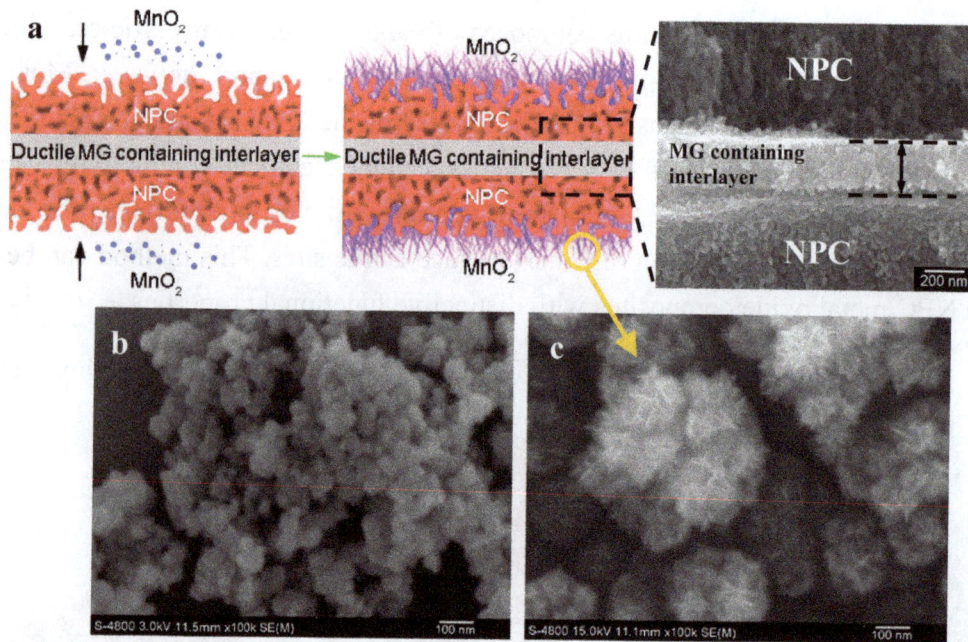

Figure 5. Fabrication of NPC/MnO$_2$ composite [30]. (**a**) Schematic of fabrication process; (**b**) as-prepared MnO$_2$ powders; (**c**) NPC/MnO$_2$ composite with 50 wt.% MnO$_2$.

4. Synthesis of Cu$_2$O Particles on Surface of MGs

So far as we know, Cu–Hf–Al MGs presented relatively good corrosion resistance in many electrolytes [78]. When dealloying Cu–Hf–Al MGs in different etching solutions, the dealloying products are diverse. As discussed in the last section, by dealloying Cu–Hf–Al MGs in 0.5 M HF solution, NPC as the general dealloying product can be fabricated. when dealloying Cu–Hf–Al MGs in HCl solutions, however, dealloying products are mainly Cu$_2$O crystals with different and interesting morphologies, which are reviewed in this section. The reason why NPC cannot be produced in Cu–Hf–Al MGs by using HCl electrolytes can be explained as follows. The corrosion rates of the Cu$_{52.5}$Hf$_{40}$Al$_{7.5}$ MG both in 0.5 M HCl and 0.5 M HF solutions can be estimated using the following formula [79]:

$$R = \frac{87600\Delta w}{s \cdot \rho \cdot t}$$

(3)

where R is corrosion rate (mm·y^{-1}), Δw is weight loss (g), s is surface area of specimen (cm^2), ρ is density of specimen (g·cm^{-3}) and t is immersion time (h). This shows that the corrosion rate of the Cu$_{52.5}$Hf$_{40}$Al$_{7.5}$ MG in 0.5 M HCl solution is about 0.2 mm·y^{-1}, while it is more than 500 mm·y^{-1} in 0.5 M HF electrolyte [80]. Consequently, it is very difficult to etch Cu$_{52.5}$Hf$_{40}$Al$_{7.5}$ MGs in HCl solution. On the other hand, Cu, Hf and Al elements showed noticeable potential difference in HF electrolytes, while the potential difference is small in HCl solution. For example, the potential difference between Cu and Hf in 0.5 M HF electrolytes was distinct and more than 0.8 V, which provides an advantageous driving force for the dissolution of less noble elements Hf under free corrosion condition. The potential difference between Cu and Hf in 0.5 M HCl solution, however, was less than 0.2 V [80]. Thus, it is difficult to remove Hf in HCl solution under free etching conditions. Considering the above two aspects, NPC cannot be easily synthesized by dealloying of Cu–Hf–Al MGs in HCl electrolytes.

4.1. Synthesis of Regular Cu₂O Particles on Surface of MGs

Xue *et al.* [81] found that Cu_2O, the initially formed product of Cu oxidation in air or oxygen atmosphere, can spontaneously grow at the surface of Cu foil in Cl^- solutions at room temperature. By adjusting concentrations of Cl^- ions, Cu_2O crystals with various shapes including octahedra, rhombic dodecahedra and spheres are produced and can be tailored on the surface of Cu foil. Using the salutary experience of this route, we found that various and regular Cu_2O particles can also be fabricated on surfaces of MGs by dealloying Cu-based MG ribbons in the HCl solution with low concentrations for different times. For example, by dealloying $Cu_{52.5}Hf_{40}Al_{7.5}$ MG ribbons in 0.05 M diluted HCl solution for 4 h, 5 h, 6 h, 8 h, 14 h, 20 h and 24 h, Cu_2O crystals with truncated tetrahedron, cube, cuboctahedron, truncated octahedron, octahedron, hexapod and octahedron-detached hexapod shapes [32,33] were synthesized, respectively. So, regular Cu_2O particles with designable morphology can be tailored in the diluted HCl solution by simply controlling dealloying time.

The characteristics of these regular Cu_2O crystals are listed in Table 2. It can be seen that the 3D sizes of Cu_2O crystals do not change much in the first 8 h of dealloying though they show different edge lengths. Then, the sizes of Cu_2O crystals increase obviously after dealloying for 14 h, and get bigger with the increase in dealloying time. It should be noted that the volume fraction of Cu_2O particles on MG surface is low (less than 20%) and does not change much with the extension of etching time. Since the Cu_2O crystals with different morphologies possess important electrical and optical properties [82,83], it is necessary to enhance the volume fraction of Cu_2O particles on the glassy surface.

Table 2. Characteristics of regular Cu_2O crystals produced by free dealloying of $Cu_{52.5}Hf_{40}Al_{7.5}$ MG in 0.05 M HCl solution for different times at 298 K open to air [32,33].

Dealloying Time/h	4	5	6	8	14	20	24
Morphology	Truncated tetrahedron	Cube	Cuboctahedron	Truncated octahedron	Octahedron	Hexapods	Octahedron-detached hexapods
Edge length/nm	~300	~300	~300	~150	~450	~500	~1100
Volume fraction/%	10.6	13.3	12.2	13.9	15.8	14.4	19.8 (mixed with other shapes)

4.2. Preparation of Cu₂O Micro-flowers on Surface of MGs

In order to improve the volume fraction of Cu_2O particles on glassy surfaces, Cu-based MG ribbons dealloying in HCl solutions with enhanced concentration were studied [34]. After $Cu_{52.5}Hf_{40}Al_{7.5}$ glassy ribbons were dealloyed in 0.1 M, 0.2 M and 0.4 M HCl solution for 8 h, Cu_2O particles formed on MG surfaces exhibited a flower-like shape. The mean surface coverage rate of Cu_2O micro-flowers increased from 17.2% to 33.1% with the increase in HCl concentration. The Cu_2O coverage rate was improved compare with the above work (13.9%, 0.05 M HCl for 8 h) [32]. The improvement in the Cu_2O coverage rate arose from the increased HCl concentration that promoted the dealloying reaction including the reaction speed and reaction product. As a result, the sizes of Cu_2O crystals gradually increased and Cu_2O crystals in regular polyhedral shapes could not be maintained but rather grow into micro-flowers. On the other hand, $Cu_{52.5}Hf_{40}Al_{7.5}$ glassy ribbons dealloying in 0.5 M HCl solution for 8 h, 14 h and 20 h were also studied [34]. When the dealloying time was extended from 8 h to 14 h, some cracks were formed on the glassy surface. Furthermore, deeper cracks were observed in the ribbon surface dealloyed for 20 h. It was found that plentiful Cu_2O/CuO particles grew up from these crack walls. With the increase of the dealloying time, the mean surface coverage rate of Cu_xO ($x = 1,2$) crystals increased gradually.

From the above results, we can conclude that higher concentration of HCl solutions and longer dealloying time are beneficial for improving the volume fraction of Cu_2O particles on glassy surfaces. However, the cracks formed on the Cu–Hf–Al glassy surfaces in concentrated HCl solutions are unavoidable. They are harmful to the mechanical integrity of an alloy ribbon. To solve this problem, dealloying attempts to Cu–Hf–Al–Nb glassy ribbons are made. For $Cu_{52.5}Hf_{40}Al_{7.5}$ glassy ribbon dealloying in 1.2 M HCl for only 1.5 h (Figure 6a,b), many cracks form on the ribbon surface. Cu_2O micro-flowers with diameter about 600 nm grow up from the crack walls. For $Cu_{52.5}Hf_{40}Al_5Nb_{2.5}$ glassy ribbon dealloying in 1.2 M HCl for 14 h (Figure 6c,d), dimples instead of cracks are present on the ribbon surface. Cu_2O micro-flowers with the biggest size to 1.1 μm grow along the edges of the dimples. For $Cu_{50}Hf_{40}Al_5Nb_5$ glassy ribbon dealloying in 1.2 M HCl for 14 h (Figure 6e,f), however, a smooth surface is retained. Cu_2O micro-flowers with diameter less than 700 nm are created on the glassy surface. The volume fraction of Cu_2O crystals on glassy surfaces reaches 39% ± 5 %. Thus, a composite with an ideal glassy surface plus a relatively high volume fraction of Cu_2O micro-flowers is successfully synthesized.

Figure 7 shows potentiodynamic polarization curves of Cu–Hf–Al(–Nb) glassy ribbons in 1.2 M HCl solution. With the increase of the Nb content, MG samples present lower corrosion current density and more positive corrosion potential, indicating a big improvement in the corrosion resistance of the MG samples. So, $Cu_{50}Hf_{40}Al_5Nb_5$ glassy ribbon with good corrosion resistance in 1.2 M HCl solution can maintain its smooth surface during the long corrosion process. To our knowledge, MG is a good carrier for Cu_2O particles, because MG exhibits high strength, high toughness and high corrosion resistance. In addition, the Cu_2O crystals formed on MG surfaces are easier to store or extract as compared to traditional chemical methods. Consequently, the MG/Cu_2O micro-flower compounds contain multiple potential properties which are well worth developing.

Figure 6. SEM images of Cu-based MG dealloying in 1.2 M HCl solution at 298 K. (**a,b**) $Cu_{52.5}Hf_{40}Al_{7.5}$ MG, dealloying for 1.5 h; (**c,d**) $Cu_{52.5}Hf_{40}Al_5Nb_{2.5}$ MG, dealloying for 14 h; (**e,f**) $Cu_{50}Hf_{40}Al_5Nb_5$ MG, dealloying for 14 h.

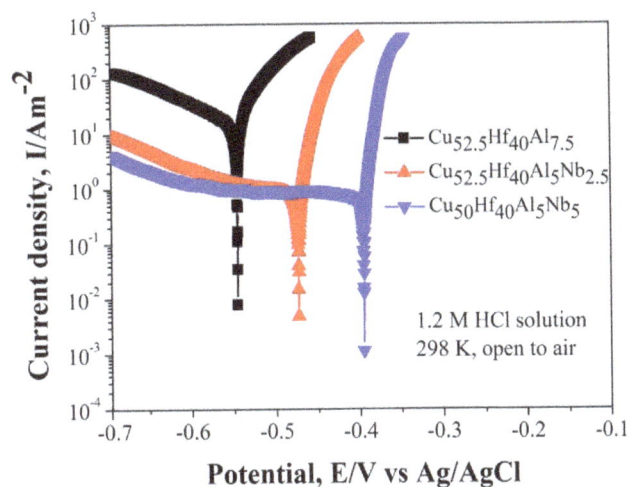

Figure 7. Potentiodynamic polarization curves of Cu–Hf–Al(–Nb) MG in 1.2 M HCl solution at 298 K open to air.

Dealloying products of Cu-based MG ribbons in HF and HCl solutions are summarized in Table 3. By choosing different Cu-based MGs and etching conditions, various dealloying products with different surface and inner compositions can be tailored. These dealloying products with multiple functionalities could be applied in wide fields in near future.

Table 3. Summary of dealloying products of Cu-based MG in acidic solutions at 298 K.

Etching Solution	Dealloying Products		MG	Dealloying Conditions	References
	Inner	Surface			
HF	NPC	NPC	$Cu_{52.5}Hf_{40}Al_{7.5}$	0.5 M HF, 300 s	[30]
	NPC	Regular Cu_2O particles	$Cu_{52.5}Hf_{40}Al_{7.5}$	oxygen-enriched 0.65 M HF, 420 s	This study
HCl	MG	Regular Cu_2O particles	$Cu_{52.5}Hf_{40}Al_{7.5}$	0.05 M HCl, 4~24 h	[32,33]
	MG	Cu_2O micro-flowers	$Cu_{50}Hf_{40}Al_5Nb_5$	1.2 M HCl, 14 h	This study

5. Applications of Dealloying Products

The paper reviews representative dealloying products of Cu-based MGs in acidic solutions, such as NPC, NPC/Cu_2O composites and metallic glass-supported Cu_2O composites with tunable Cu_2O shapes. These new dealloying products with unique structures and multiple properties would provide a wide array of possible applications in many areas.

Nanoporous metals, as the most famous dealloying products, have been the focus of much attention due to their potential in various applications. So far, the study on nanoporous gold (NPG) makes up a high proportion of dealloying works. Just in recent years, fabrication and application development of inexpensive NPC have attracted considerable attention from many researchers. Representative applications of NPC and composites are listed in Table 4. The first significant development toward NPC was reported by Chen's group [22]. It was found that the tunable nanoporosity leads to a remarkable improvement in surface-enhanced Raman scattering (SERS) of NPC, which was helpful in developing inexpensive SERS substrates for sensitive instrumentations in molecular diagnostics. After that, diversified reports about NPC applications were unveiled. For example, NPC was considered to be a good support for horseradish peroxidase immobilization [84]. Moreover, NPC-contained composites presented superior catalytic and sensitive performance in oxidation of hydrazine and alcohols, degrading organic compounds and detecting glucose [50,85–88].

Table 4. Representative applications of NPC and composites.

Products	Application Field	References
NPC	Surface-enhanced Raman scattering (SERS)	[22,89]
	Immobilization of horseradish peroxidase	[84]
NPC/Ag core–shell composite	SERS	[90]
NPC/Pt(Pd) core–shell composite	Methanol electro-oxidation	[91]
NPC/Au core-shell composite	Electrocatalysis and nonenzymatic biosensing	[87]
Ni-B amorphous nanoparticles modified NPC	Ethanol oxidation	[88,92]
NPC/Cu composite	Electro-oxidation of hydrazine	[85]
Nanoporous Cu–O system	Catalysts towards CO oxidation	[93]
NPC/Cu_2O nanocomposite	Adsorption of methyl orange	[94]

Table 4. *Cont.*

Products	Application Field	References
NPC/MG composite	Degradation of azo dye	[86]
	Degradation of phenol	[50]
NPC/Si composite		[95]
NPC/Cu$_2$O composite	Lithium-ion battery anodes	[70]
Cu/NPC/MnO$_2$ composite		[96]
NPC/MnO$_2$ composite	Supercapacitor electrodes	[30]
NPC/(Fe,Cu)$_3$O$_4$ composites	Excellent magnetic/electrical properties for potential applications in sensors, information storage, and so on	[76]

Besides these developments, an important application field for NPC is energy storage, especially for electrochemical supercapacitors (ECs). As charge-storage devices, ECs possess a unique combination of high power, high energy and long lifetimes [97,98]. They were widely used in portable electronics and hybrid electric vehicles [99,100]. There are two basic kinds of ECs: one is double-layer capacitance relying on surface ion adsorption, such as carbon based materials (e.g., carbon nanotubes, graphene, porous carbon, carbon aerogel and active carbon fiber) [101–105]; another is pseudo-capacitance relying on surface redox reactions, including metal oxides (e.g., RuO_2, MnO_2, Co_3O_4 and NiO) [106–109] and conducting polymers (e.g., polyaniline, polypyrrole and polythiophene) [110–112]. It should be noted that the capacitive performances of ECs depend on the effective surface area and conductivity of electrode materials [98]. Thus, NPG with a high specific surface area and high conductivity has been firstly tried as a current collector or a substrate support for the electrode materials in supercapacitors. It was reported that bare NPG-based double-layer supercapacitors had excellent charge-discharge cycling stability, although their capacitance was still too low (~5 $F \cdot g^{-1}$) for real applications [113,114]. For enhancing capacitive performance, pseudo-capacitive materials with high theoretical values such as MnO_2, polyaniline and polypyrrole (PPy) were deposited on the high-surface-area and highly conductive NPG surface [115–117]. These nanocomposite-based supercapacitors exhibit high capacitance properties (e.g., a specific capacitance of ~1145 $F \cdot g^{-1}$ for NPG/MnO_2, a power density of 296 $kW \cdot kg^{-1}$ for NPG/PPy). However, Au is a well-known noble metal, which would limit its large-scale application. With the aim of reducing the cost and promoting the practical application of nanoporous metals in the EC field, nanoporous Ni has been fabricated by dealloying Mn–Ni, Al–Ni, Cu–Ni, Fe–Ni, Mg–Ni, Zn–Ni and Mn–Ni–Cu alloy in different solutions [118–121]. The nanoporous Ni as an electrode substrate presented a stable areal capacitance (1.7 $F \cdot cm^{-2}$) [122]; however, a very small pore size (usually less than 10 nm) inhibited the growth of the pseudo-capacitive materials inside the nanopores. Therefore, it is expected that the inexpensive nanoporous Cu (NPC) with tunable pore size in a wide range from 10 to 60 nm could be considered as an excellent substrate support for pseudo-capacitive materials.

In our recent work, a new NPC-supported MnO_2 (NPC/MnO_2) composite for ECs was produced [30]. It can be clearly seen from the cyclic voltammogram (CV) curves (Figure 8a) that the capacitive current of the NPC in the voltage window was negligibly low, indicating NPC was only used as a stable substrate. On the other hand, it was surprisingly found that the closed area of CV curves (Figure 8a) and the specific capacitance values (Figure 8b) of the NPC/MnO$_2$ composite were remarkably enhanced as compared to that of the pure MnO$_2$ powders. When combining the electrochemical performance of pure MnO$_2$ and NPC/MnO$_2$ composites with their corresponding surface morphologies and distribution of MnO$_2$ (Figure 5b,c), it was indicated that NPC substrate can efficiently improve the utilization of MnO$_2$ surface active sites and promote MnO$_2$ chemical reactions. The cycling stability of NPC/MnO$_2$ composites with different weight ratios is presented in Figure 8c. The specific capacitance for the two composites remained greater than 97% and 82% of the initial value after 500 and 1000 cycles, respectively. All the results showed that the as-obtained NPC/MnO$_2$ composite had a commendable potential for EC application. In future works, the optimization of the coated capacitive materials, including their amount and structure, could further enhance their capacitive performance and accelerate these inexpensive ECs for a practical application.

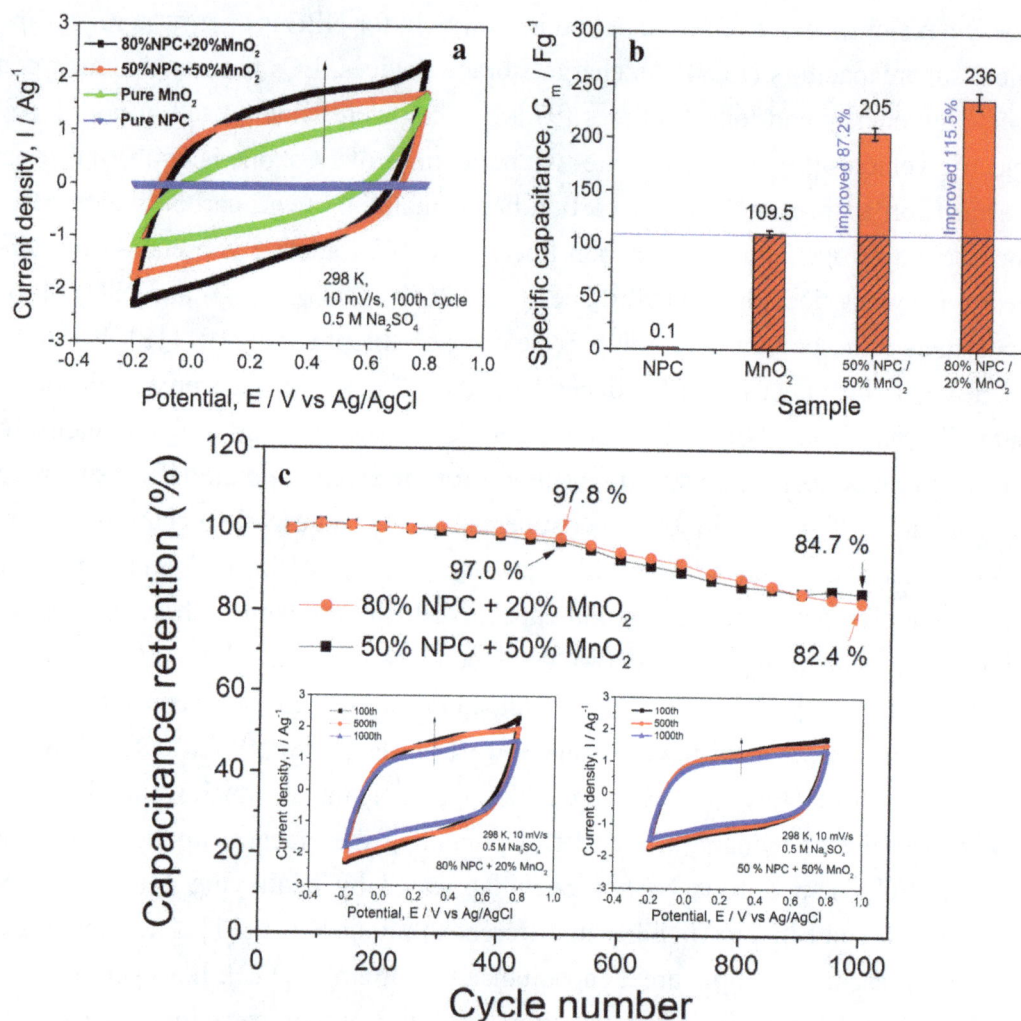

Figure 8. NPC/MnO$_2$ composite for ECs. (**a**) CV curves of monolithic NPC, pure MnO$_2$ and NPC/MnO$_2$ composites with different weight ratios in 0.5 M Na$_2$SO$_4$ solution; (**b**) specific capacitance of monolithic NPC, pure MnO$_2$ and NPC/MnO$_2$ composites with different weight ratios; (**c**) cycling performance of the NPC/MnO$_2$ composite at the scan rate of 10 mV·s^{-1}; the inserts show CV curves of NPC/MnO$_2$ composites with different weight ratios in 0.5 M Na$_2$SO$_4$ solution at different scan cycles. Adapted with permission from Elsevier, Copyright 2015 [30].

In NPC/MnO$_2$ composite supercapacitor electrodes, the NPC, as a current collector, was the support for MnO$_2$ nanoparticles. The presence of NPC changed the morphology of MnO$_2$ and improved the utilization of surface active sites of MnO$_2$. This is of great significance to develop electrode materials with high electrochemical performance. On the other hand, it is known that NPC is unstable and has a tendency to form Cu$_2$O and CuO under high voltage. Using this, Cu/Cu$_2$O composites [123,124] and Cu/CuO composites [125] were successfully developed for pseudocapacitance electrodes. The pseudocapacitance effect of these composites came from the redox transformation between the Cu$^+$ and Cu^{2+} species in the process of charging and discharging. As a result, the synergistic effect of Cu oxide and MnO$_2$ contributed to the pseudocapacitance properties of the composites, showing a low cost method for improvement of pseudocapacitance properties. Moreover, in comparison with carbon-based supercapacitor, the NPC-based supercapacitor exhibits distinct merits. Although carbon is a kind of low cost material and exhibits ultra-high specific capacitance, the production of carbon nanotubes, graphene, *etc.* includes a complex preparation process, expensive equipment, and low yield. Thus, it is difficult to achieve mass production of carbon-based supercapacitors at low cost. In contrast, the preparation of NPC is based on free dealloying, which is a very simple and rapid fabrication method. That makes NPC-based supercapacitors economical and practical. As a result, the development of NPC-based supercapacitors with low cost, stable properties and long lifetimes is of great importance.

An NPC used in lithium-ion batteries is another energy storage material. The fabrication of nanoporous metal-based composites for lithium-ion batteries is still in its early stages. Lang's team [96] uncovered a flexible Cu/NPC/MnO$_2$ hybrid bulk electrode for high-performance lithium-ion batteries, which showed a capacity as high as ~1100 mA·h·g^{-1} for 1000 cycles. In addition, it was reported that an NPC supported cuprous oxide prepared by the direct oxidation of NPC at high temperatures was applied in high-performance lithium ion battery anodes [70].

Up to now, the majority of research about applications of Cu-containing alloys for dealloying products has focused on NPC and NPC-based composites. Application development in broader fields for other dealloying products, such as NPC/Cu$_2$O composites and MG/Cu$_2$O composites with tailored Cu$_2$O shapes, needs to be carried out in the near future. The corresponding experimental results will be presented soon by our group.

6. Conclusions and Outlook

The fabrication and applications of nanoporous metals and composites with dealloying techniques have earned more and more attention from academic research. These dealloying products show good performance in many fields. Although the history of research into dealloying is very short, the achievements gained in such a short period of time are amazing. According to the data from the "Web of Science" website (Figure 9), there have been 852 papers published in the dealloying field in the last 10 years (1 January 2004–31 March 2015). Every paper is cited nearly 20 times. In addition, the "H-index" is 63, which is far above average. The number of published papers and citations has risen over the last several years. In 2014, there were more than 150 published papers and over 4000 total citations. All the statistics show dealloying studies to be a hot topic in academia. As for papers published in countries/territories, the number of papers related to dealloying published by China, the USA, Japan and Germany is significantly higher than those of other countries/territories. In particular, papers in China account for more than forty percent. Per organization, Shandong University and Tohoku University, as leaders in the study of dealloying, have published the most papers in the world. The popularity of study in the dealloying field is thanks to rapid developments in the fields of energy, catalysis, biosensing, *etc*. Dealloying has become a multidisciplinary and interdisciplinary research. Dealloying research is also attracting more and more researchers from different disciplines. With sufficient cooperation, there will be more surprises in the near future.

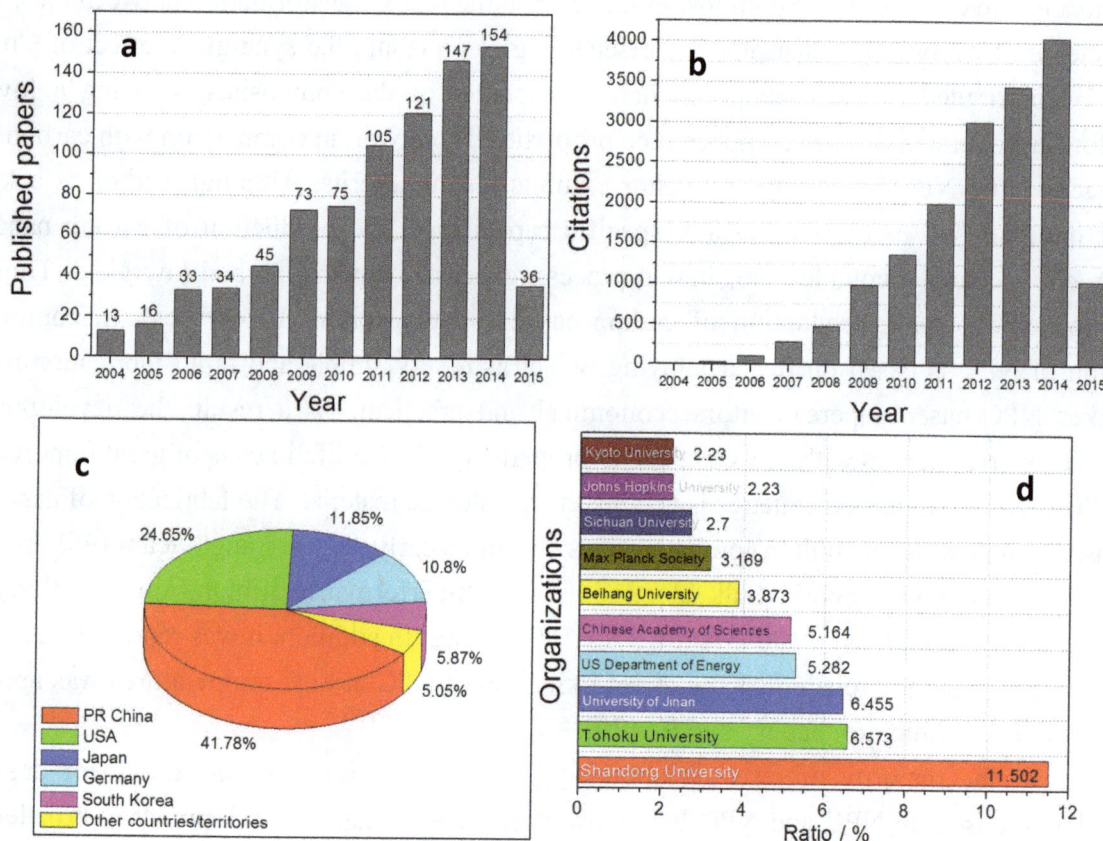

Figure 9. Analyses of publications in the dealloying field using the topic of "dealloying" (1 January 2004–31 March 2015). (**a**) Papers published in each year; (**b**) citations in each year; (**c**) pie chart showing the paper publication ratio of different countries/territories; (**d**) bar graph showing the paper publication ratio of different organizations.

In this paper, the dealloying products via free-dealloying of Cu-based metallic glasses in acid solutions are summarized. With different dealloying conditions, various dealloying products including NPC, NPC/Cu$_2$O composites and MG/Cu$_2$O composites are fabricated. NPC with uniform pore structure can be obtained by dealloying because of homogeneous structure of MG precursors. The ligament/pore size of dealloyed NPC can be tailored by the addition of different trace elements into MGs. A noble element such as Ag additive in MGs results in the formation of a nanoprous Cu–Ag alloy with ultrafine porous structure. An active element such as Al additive, however, leads to the formation of a wider ligament of NPC. Besides alloy compositions, many factors, such as dealloying solutions, etching time, dealloying temperatures and so on, have significant influence on the ligament/pore size of the resulting NPC. The change in ligament/pore size will influence the performance of the NPC in many application fields. Consequently, the application properties of NPC can be tailored by simply changing these factors. In addition, when dealloying in an oxygen-enrichment solution, metal oxide can be formed. In this study, NPC/Cu_2O composites are synthesized by this method, which is called the dealloying and spontaneous oxidation method. The electrochemical properties of the composites need to be further developed. Meanwhile, MG presents excellent corrosion resistance performance in an appropriate corrosion environment. In this situation, MG ribbon with good mechanical integrity can be retained after corrosion, which provides an opportunity for the formation of MG-based compounds. The application fields of these compounds deserve to be further studied.

Energy storage is one of the most important and meaningful application fields of dealloying products. NPC possesses excellent electrical conductivity, and can be used as a current collector and support material for metal oxide. Thus, the as-obtained NPC/metal oxide composite material can be used as a low-cost supercapacitor and lithium ion battery electrodes. Through the combination of dealloying with other pore-forming technology, nano/mirco-porous Cu or so called bimodal porous Cu can be prepared. This type of composite with porous structure is expected to have a more outstanding performance in the field of energy storage. In lithium-air batteries, for example, the big pores can accommodate more discharge product (Li_2O_2) to enhance the energy density, while the small ligament/pore with high catalytic activity can effectively catalyze the reduction of Li_2O_2 into Li during the charge process to enhance the cycling stability [98]. At the same time, the porous structured composite can be used as a substrate to form nano/micro-scaled metal oxides with different morphologies using diverse techniques, such as anodization, heat treatment, chemical deposition and so on. The series of compounds are expected to show improved performance in energy storage.

Acknowledgments

The authors would like to acknowledge the financial support from both the "100 Talents Project" of Hebei Province, China (E2012100009), the Youth Fund of Key Laboratory for New Type of Functional Materials in Hebei Province, China (201503) and the Natural Science Foundation of Hebei Province, China (E2015202081).

Author Contributions

Zhifeng Wang and Chunling Qin designed this study and wrote the paper; Zhifeng Wang and Jiangyun Liu carried out the experiments and analyzed the data; Hui Yu, Xingchuan Xia, Chaoyang Wang, Yanshan Zhang, Qingfeng Hu and Weimin Zhao performed auxiliary operations in experiments. All authors discussed the results and commented on the manuscript.

Conflicts of Interest

The authors declare no conflict of interest.

References

1.   Zhang, C.; Sun, J.; Xu, J.; Wang, X.; Ji, H.; Zhao, C.; Zhang, Z. Formation and microstructure of nanoporous silver by dealloying rapidly solidified Zn–Ag alloys. *Electrochim. Acta* **2012**, *63*, 302–311.

2. Darque-Ceretti, E.; Felder, E.; Aucouturier, M. Foil and leaf gilding on cultural artifacts: Forming and adhesion. *Matéria* **2011**, *16*, 540–559.

3. Ding, Y.; Kim, Y.; Erlebacher, J. Nanoporous gold leaf: "Ancient technology"/advanced material. *Adv. Mater.* **2004**, *16*, 1897–1900.

4. Xu, C.; Liu, Y.; Wang, J.; Geng, H.; Qiu, H. Nanoporous PdCu alloy for formic acid electro-oxidation. *J. Power Sources* **2012**, *199*, 124–131.

5. Xu, C.; Liu, Y.; Zhou, C.; Wang, L.; Geng, H.; Ding, Y. An *in situ* dealloying and oxidation route to Co_3O_4 nanosheets and their ambient-temperature CO oxidation activity. *ChemCatChem* **2011**, *3*, 399–407.

6. Lang, X.; Fu, H.; Hou, C.; Han, G.; Yang, P.; Liu, Y.; Jiang, Q. Nanoporous gold supported cobalt oxide microelectrodes as high-performance electrochemical biosensors. *Nat. Commun.* **2013**, *4*, doi:10.1038/ncomms3169.

7. Pickering, H.; Wagner, C. Electrolytic dissolution of binary alloys containing a noble metal. *J. Electrochem. Soc.* **1967**, *114*, 698–706.

8. Seker, E.; Reed, M.; Begley, M. Nanoporous gold: Fabrication, characterization, and applications. *Materials* **2009**, *2*, 2188–2215.

9. Hakamada, M.; Nakano, H.; Furukawa, T.; Takahashi, M.; Mabuchi, M. Hydrogen storage properties of nanoporous palladium fabricated by dealloying. *J. Phys. Chem. C* **2010**, *114*, 868–873.

10. Jin, H.; Kramer, D.; Ivanisenko, Y.; Weissmüller, J. Macroscopically strong nanoporous Pt prepared by dealloying. *Adv. Eng. Mater.* **2007**, *9*, 849–854.

11. Li, Z.; Wang, X.; Lu, X. Refinement of nanoporous silver by adding surfactant to the electrolyte. *ECS Electrochem. Lett.* **2014**, *3*, C13–C16.

12. Dan, Z.; Qin, F.; Yamaura, S.; Xie, G.; Makino, A.; Hara, N. Refinement of nanoporous copper by dealloying MgCuY amorphous alloys in sulfuric acids containing polyvinylpyrrolidone. *J. Electrochem. Soc.* **2014**, *161*, C120–C125.

13. Gu, X.; Xu, L.; Tian, F.; Ding ,Y. Au–Ag alloy nanoporous nanotubes. *Nano Res.* **2009**, *2*, 386–393.

14. Koczkur, K.; Yi, Q.; Chen, A. Nanoporous Pt–Ru networks and their electrocatalytical properties. *Adv. Mater.* **2007**, *19*, 2648–2652.

15. Xu, C.; Liu, Y.; Su, F.; Liu, A.; Qiu, H. Nanoporous PtAg and PtCu alloys with hollow ligaments for enhanced electrocatalysis and glucose biosensing. *Biosens. Bioelectron.* **2011**, *27*, 160–166.

16. Zhang, Z.; Wang, Y.; Wang, X. Nanoporous bimetallic Pt–Au alloy nanocomposites with superior catalytic activity towards electro-oxidation of methanol and formic acid. *Nanoscale* **2011**, *3*, 1663–1674.

17. Chen, X.; Jiang, Y.; Sun, J.; Jin, C.; Zhang, Z. Highly active nanoporous Pt-based alloy as anode and cathode catalyst for direct methanol fuel cells. *J. Power Sources* **2014**, *267*, 212–218.

18. Tang, Y.; Tang, B.; Qing, J.; Li, Q.; Lu, L. Nanoporous metallic surface: Facile fabrication and enhancement of boiling heat transfer. *Appl. Surf. Sci.* **2012**, *258*, 8747–8751.

19. Biener, J.; Wittstock, A.; Zepeda-Ruiz, L.A.; Biener, M.M.; Zielasek, V.; Kramer, D.; Viswanath, R.N.; Weissmüller, J.; Bäumer, M.; Hamza, A.V. Surface-chemistry-driven actuation in nanoporous gold. *Nat. Mater.* **2009**, *8*, 47–51.

20. Kang, J.; Hirata, A.; Qiu, H.; Chen, L.; Ge, X.; Fujita, T.; Chen, M. Self-grown oxy-hydroxide@nanoporous metal electrode for high-performance supercapacitors. *Adv. Mater.* **2014**, *26*, 269–272.

21. Chen, L.; Fujita, T.; Chen, M. Biofunctionalized nanoporous gold for electrochemical biosensors. *Electrochim. Acta* **2012**, *67*, 1–5.

22. Chen, L.; Yu, J.; Fujita, T.; Chen, M. Nanoporous copper with tunable nanoporosity for SERS applications. *Adv. Funct. Mater.* **2009**, *19*, 1221–1226.

23. Erlebacher, J.; Aziz, M.; Karma, A.; Dimitrov, N.; Sieradzki, K. Evolution of nanoporosity in dealloying. *Nature* **2001**, *410*, 450–453.

24. Pia, G.; Delogu, F. Nanoporous Au: Statistical analysis of morphological features and evaluation of their influence on the elastic deformation behavior by phenomenological modeling. *Acta Mater.* **2015**, *85*, 250–260.

25. Detsi, E.; Punzhin, S.; Rao, J.; Onck, P.; Hosson, J. Enhanced strain in functional nanoporous gold with a dual microscopic length scale structure. *ACS Nano* **2012**, *6*, 3734–3744.

26. Detsi, E.; Chen, Z.; Vellinga, W.; Onck, P.; Hosson, J. Reversible strain by physisorption in nanoporous gold. *Appl. Phys. Lett.* **2011**, *99*, 083104.

27. Biener, J.; Hodge, A.; Hayes, J.; Volkert, C.; Zepeda-Ruiz, L.; Hamza, A.; Abraham, F. Size effects on the mechanical behavior of nanoporous Au. *Nano Lett.* **2006**, *6*, 2379–2382.

28. Kong, Q.; Lian, L.; Liu, Y.; Zhang, J. Fabrication and compression properties of bulk hierarchical nanoporous copper with fine ligament. *Mater. Lett.* **2014**, *127*, 59–62.

29. Weissmüller, J.; Wang, K. Composites of nanoporous gold and polymer. *Adv. Mater.* **2013**, *25*, 1280–1284.

30. Wang, Z.; Liu, J.; Qin, C.; Liu, L.; Zhao, W.; Inoue, A. Fabrication and new electrochemical properties of nanoporous Cu by dealloying amorphous Cu–Hf–Al alloys. *Intermetallics* **2015**, *56*, 48–55.

31. Qin, C.; Wang, Z.; Liu, H.; Liu, L.; Wang, H.; Ding, J.; Zhao, W. Monolithic nanoporous copper with novel electrochemical properties fabricated by dealloying Cu–Zr(–Al) metallic glasses. *Mater. Sci. Forum* **2014**, *783–786*, 1925–1930.

32. Wang, Z.; Qin, C.; Zhao, W.; Jia, J. Tunable Cu_2O nanocrystals fabricated by free dealloying of amorphous ribbons. *J. Nanomater.* **2012**, *2012*, 126715.

33. Wang, Z.; Wang, L.; Qin, C.; Liu, J.; Li, Y.; Zhao, W. Tailored dealloying products of Cu-based metallic glasses in hydrochloric acid solutions. *Mater. Res.* **2014**, *17*, 1003–1009.

34. Wang, Z.; Qin, C.; Liu, L.; Wang, L.; Ding, J.; Zhao, W. Synthesis of Cu_xO ($x = 1,2$)/amorphous compounds by dealloying and spontaneous oxidation method. *Mater. Res.* **2014**, *17*, 33–37.

35. Li, M.; Zhou, Y.Z.; Geng, H. Fabrication of nanoporous copper ribbons by dealloying of Al–Cu alloys. *J. Porous Mater.* **2012**, *19*, 791–796.

36. Liu, W.; Zhang, S.; Li, N.; Zheng, J.; An, S.; Li, G. Influence of dealloying solution on the microstructure of monolithic nanoporous copper through chemical dealloying of Al 30 at.% Cu alloy. *Int. J. Electrochem. Sci.* **2012**, *7*, 7993–8006.

37. Tan, X.; Li, K.; Niu, G.; Yi, Z.; Luo, J.; Liu, Y.; Han, S.; Wu, W.; Tang, Y. Effect of heat treatment of Mn–Cu precursors on morphology of dealloyed nanoporous copper. *J. Cent. South Univ.* **2012**, *19*, 17–21.

38. Rizzi, P.; Scaglione, F.; Battezzati, L. Nanoporous gold by dealloying of an amorphous precursor. *J. Alloy. Compd.* **2014**, *586*, S117–S120.

39. Scaglione, F.; Rizzi, P.; Celegato, F.; Battezzati, L. Synthesis of nanoporous gold by free corrosion of an amorphous precursor. *J. Alloy. Compd.* **2014**, *615*, S142–S147.

40. Qiu, H.; Peng, L.; Li, X.; Xu, H.; Wang, Y. Using corrosion to fabricate various nanoporous metal structures. *Corros. Sci.* **2015**, *92*, 16–31.

41. Feng, Y.; Zhang, S.; Xing, Y.; Liu, W. Preparation and characterization of nanoporous Cu_6Sn_5/Cu composite by chemical dealloying of Al–Cu–Sn ternary alloy. *J. Mater. Sci.* **2012**, *47*, 5911–5917.

42. Liu, W.; Zhang, S.; Li, N.; Zheng, J.; An, S.; Xing, Y. Formation of nanoporous copper with hybrid-modal pore size distributions related to surface diffusion of copper atoms during dealloying of Mg 13.5 at.% Cu alloy in an acidic solution. *Int. J. Electrochem. Sci.* **2012**, *7*, 6365–6377.

43. Hayes, J.; Hodge, A.; Biener, J.; Hamza, A.; Sieradzki, K. Monolithic nanoporous copper by dealloying Mn–Cu. *J. Mater. Res.* **2006**, *21*, 2611–2616.

44. Qi, Z.; Zhao, C.; Wang, X.; Lin, J.; Shao, W.; Zhang, Z.; Bian, X. Formation and characterization of monolithic nanoporous copper by chemical dealloying of Al–Cu alloys. *J. Phys. Chem. C* **2009**, *113*, 6694–6698.

45. Zhao, C.; Qi, Z.; Wang, X.; Zhang, Z. Fabrication and characterization of monolithic nanoporous copper through chemical dealloying of Mg–Cu alloys. *Corros. Sci.* **2009**, *51*, 2120–2125.

46. Lin, B.; Kong, L.; Hodgson, P.; Dumée, L. Impact of the de-alloying kinetics and alloy microstructure on the final morphology of de-alloyed meso-porous metal films. *Nanomaterials* **2014**, *4*, 856–878.

47. Dan, Z.; Qin, F.; Sugawara, Y.; Muto, I.; Hara, N. Fabrication of nanoporous copper by dealloying amorphous binary Ti–Cu alloys in hydrofluoric acid solutions. *Intermetallics* **2012**, *29*, 14–20.

48. Dan, Z.; Qin, F.; Sugawara, Y.; Muto, I.; Hara, N. Dependency of the formation of Au-stabilized nanoporous copper on the dealloying temperature. *Microporous Mesoporous Mater.* **2014**, *186*, 181–186.

49. Luo, X.; Li, R.; Liu, Z.; Huang, L.; Shi, M.; Xu, T.; Zhang, T. Three-dimensional nanoporous copper with high surface area by dealloying Mg–Cu–Y metallic glasses. *Mater. Lett.* **2012**, *76*, 96–99.

50. Deng, Z.; Zhang, C.; Liu, L. Chemically dealloyed MgCuGd metallic glass with enhanced catalytic activity in degradation of phenol. *Intermetallics* **2014**, *52*, 9–14.

51. Abe, H.; Sato, K.; Nishikawa, H.; Takemoto, T.; Fukuhara, M.; Inoue, A. Dealloying of Cu–Zr–Ti bulk metallic glass in hydrofluoric acid solution. *Mater. Trans.* **2009**, *50*, 1255–1258.

52. Aburada, T.; Fitz-Gerald, J.; Scully, J. Synthesis of nanoporous copper by dealloying of Al–Cu–Mg amorphous alloys in acidic solution: The effect of nickel. *Corros. Sci.* **2011**, *53*, 1627–1632.

53. Dan, Z.; Qin, F.; Makino, A.; Sugawara, Y.; Muto, I.; Hara, N. Fabrication of nanoporous copper by dealloying of amorphous Ti–Cu–Ag alloys. *J. Alloy. Compd.* **2014**, *586*, S134–S138.

54. Zhang, Q.; Zhang, Z. On the electrochemical dealloying of Al-based alloys in a NaCl aqueous solution. *Phys. Chem. Chem. Phys.* **2010**, *12*, 1453–1472.

55. Wang, Y.; Zhang, W.; Inoue, A. Nanoporous Cu wide ribbons with good mechanical integrity. *Mater. Sci. Eng. B* **2012**, *177*, 532–535.

56. Luo, X.; Li, R.; Huang, L.; Zhang, T. Nucleation and growth of nanoporous copper ligaments during electrochemical dealloying of Mg-based metallic glasses. *Corros. Sci.* **2013**, *67*, 100–108.

57. Dan, Z.; Qin, F.; Sugawara, Y.; Muto, I.; Makino, A.; Hara, N. Nickel-stabilized nanoporous copper fabricated from ternary TiCuNi amorphous alloys. *Mater. Lett.* **2013**, *94*, 128–131.

58. Cheng, I.; Hodge, A. Morphology, oxidation, and mechanical behavior of nanoporous Cu foams. *Adv. Eng. Mater.* **2012**, *14*, 219–226.

59. Zhang, X.; Li, Y.; Liu, Y.; Zhang, H. Fabrication of a bimodal micro/nanoporous metal by the Gasar and dealloying processes. *Mater. Lett.* **2013**, *92*, 448–451.

60. Liu, W.; Zhang, S.; Li, N.; Zheng, J.; Xing, Y. Influence of phase constituent and proportion in initial Al–Cu alloys on formation of monolithic nanoporous copper through chemical dealloying in an alkaline solution. *Corros. Sci.* **2011**, *53*, 809–814.

61. Zhao, C.; Wang, X.; Qi, Z.; Ji, H.; Zhang, Z. On the electrochemical dealloying of Mg–Cu alloys in a NaCl aqueous solution. *Corros. Sci.* **2010**, *52*, 3962–3972.

62. Aburada, T.; Unlu, N.; Fitz-Gerald, J.; Scully, J. Effect of Ni as a minority alloying element on the corrosion behavior in Al–Cu–Mg–(Ni) metallic glasses. *Scr. Mater.* **2008**, *58*, 623–626.

63. Dan, Z.; Qin, F.; Sugawara, Y.; Muto, I.; Hara, N. Uniform evolution of nanoporosity on amorphous Ti–Cu alloys. *J. Nanosci. Nanotechnol.* **2014**, *14*, 7879–7883.

64. Lan, G.; Xie, Z.; Huang, Z.; Yang, S.; Zhang, X.; Zeng, Y.; Jiang, J. Amorphous alloy: Promising precursor to form nanoflowerpot. *Adv. Mater. Sci. Eng.* **2014**, *2014*, 263681.

65. Dan, Z.; Qin, F.; Hara, N. Refinement of nanoporous copper: A summary of micro-alloying of Au-group and Pt-group elements. *Mater. Trans.* **2014**, *55*, 796–800.

66. Dan, Z.; Qin, F.; Sugawara, Y.; Muto, I.; Hara, N. Fabrication of ultrafine nanoporous copper by the minor addition of gold. *Mater. Trans.* **2012**, *53*, 1765–1769.

67. Dan, Z.; Qin, F.; Sugawara, Y.; Muto, I.; Hara, N. Elaboration of nanoporous copper by modifying surface diffusivity by the minor addition of gold. *Microporous Mesoporous Mater.* **2013**, *165*, 257–264.

68. Dan, Z.H.; Qin, F.X.; Hara, N. Polyvinylpyrrolidone macromolecules function as a diffusion barrier during dealloying. *Mater. Chem. Phys.* **2014**, *146*, 277–282.

69. Xu, Q. *Nanoporous Materials: Synthesis and Applications*; CRC Press: Boca Raton, FL, USA, 2013; Chapter 5.

70. Liu, D.; Yang, Z.; Wang, P.; Li, F.; Wang, D.; He, D. Preparation of 3D nanoporous copper-supported cuprous oxide for high-performance lithium ion battery anodes. *Nanoscale* **2013**, *5*, 1917–1921.

71. Xu, C.; Wang, R.; Zhang, Y.; Ding, Y. A general corrosion route to nanostructured metal oxides. *Nanoscale* **2010**, *2*, 906–909.

72. Hao, Q.; Li, M.; Jia, S.; Zhao, X.; Xu, C. Controllable preparation of Co_3O_4 nanosheets and their electrochemical performance for Li-ion batteries. *RSC Adv.* **2013**, *3*, 7850–7854.

73. Jia, S.; Song, T.; Zhao, B.; Zhai, Q.; Gao, Y. Regular Fe_3O_4 octahedrons with excellent soft magnetic properties prepared by dealloying technique. *J. Alloy. Compd.* **2014**, *585*, 580–586.

74. Liu, Z.; Yamazaki, T.; Shen, Y.; Meng, D.; Kikuta, T.; Nakatani, N.; Kawabata, T. Dealloying derived synthesis of W nanopetal films and their transformation into WO_3. *J. Phys. Chem. C* **2008**, *112*, 1391–1395.

75. Zhao, Z.; Xu, J.; Liaw, P.; Wu, B.; Wang, Y. One-step formation and photocatalytic performance of spindle-like TiO_2 nanorods synthesized by dealloying amorphous $Cu_{50}Ti_{50}$ alloy. *Corros. Sci.* **2014**, *84*, 66–73.

76. Qi, Z.; Gong, Y.; Zhang, C.; Xu, J.; Wang, X.; Zhao, C.; Ji, H.; Zhang, Z. Fabrication and characterization of magnetic nanoporous Cu/(Fe,Cu)$_3$O$_4$ composites with excellent electrical conductivity by one-step dealloying. *J. Mater. Chem.* **2011**, *21*, 9716–9724.

77. Wang, H.; Jiang, M.; Su, J.; Liu, Y. Fabrication of porous CuO nanoplate-films by oxidation-assisted dealloying method. *Surf. Coat. Technol.* **2014**, *249*, 19–23.

78. Zhang, B.; Chen, Y.; Guo, H. Electrochemical behavior of Cu–Hf–Al amorphous films. *ECS Electrochem. Lett.* **2014**, *3*, C1–C3.

79. Zahrani, E.; Alfantazi, A. Molten salt induced corrosion of Inconel 625 superalloy in PbSO$_4$–Pb$_3$O$_4$–PbCl$_2$–Fe$_2$O$_3$–ZnO environment. *Corros. Sci.* **2012**, *65*, 340–359.

80. Wang, Z. *Fabrication of Nanoporous Copper and Nano/Micro Cuprous Oxide Particles by Dealloying Method*; Hebei University of Technology: Tianjin, China, 2013.

81. Chen, K.; Xue, D. Cu-based materials as high-performance electrodes toward electrochemical energy storage. *Funct. Mater. Lett.* **2014**, *7*, 1430001.

82. Poizot, P.; Laruelle, S.; Grugeon, S.; Dupont, L.; Tarascon, J. Nano-sized transition-metal oxides as negative-electrode materials for lithium-ion batteries. *Nature* **2000**, *407*, 496–499.

83. Xu, H.; Wang, W.; Zhu, W. Shape evolution and size-controllable synthesis of Cu$_2$O octahedra and their morphology-dependent photocatalytic properties. *J. Phys. Chem. B* **2006**, *110*, 13829–13834.

84. Qiu, H.; Lu, L.; Huang, X.; Zhang, Z.; Qu, Y. Immobilization of horseradish peroxidase on nanoporous copper and its potential applications. *Bioresour. Technol.* **2010**, *101*, 9415–9420.

85. Jia, F.; Zhao, J.; Yu, X. Nanoporous Cu film/Cu plate with superior catalytic performance toward electro-oxidation of hydrazine. *J. Power Sources* **2013**, *222*, 135–139.

86. Luo, X.; Li, R.; Zong, J.; Zhang, Y.; Li, H.; Zhang, T. Enhanced degradation of azo dye by nanoporous-copper-decorated Mg–Cu–Y metallic glass powder through dealloying pretreatment. *Appl. Surf. Sci.* **2014**, *305*, 314–320.

87. Chen, L.; Fujita, T.; Ding, Y.; Chen, M. A three-dimensional gold-decorated nanoporous copper core–shell composite for electrocatalysis and nonenzymatic biosensing. *Adv. Funct. Mater.* **2010**, *20*, 2279–2285.

88. Zhang, S.; Zheng, Y.; Yuan, L.; Zhao, L. Ni–B amorphous alloy nanoparticles modified nanoporous Cu toward ethanol oxidation in alkaline medium. *J. Power Sources* **2014**, *247*, 428–436.

89. Li, M.; Su, Y.; Zhao, J.; Geng, H.; Zhang, J.; Zhang, L.; Yang, C.; Zhang, Y. One-pot preparation of thin nanoporous copper foils with enhanced light absorption and SERS properties. *CrystEngComm* **2015**, *17*, 1296–1304.

90. Chen, L.; Zhang, L.; Fujita, T.; Chen, M. Surface-enhanced Raman scattering of silver@nanoporous copper core–shell composites synthesized by an *in situ* sacrificial template approach. *J. Phys. Chem. C* **2009**, *113*, 14195–14199.

91. Xu, C.; Liu, Y.; Wang, J.; Geng, H.; Qiu, H. Fabrication of nanoporous Cu–Pt(Pd) core/shell structure by galvanic replacement and its application in electrocatalysis. *ACS Appl. Mater. Interfaces* **2011**, *3*, 4626–4632.

92. Zhang, S.; Zheng, Y.; Yuan, L.; Wang, X.; Zhao, L. In situ synthesis of nickel–boron amorphous alloy nanoparticles electrode on nanoporous copper film/brass plate for ethanol electro-oxidation. *Int. J. Hydrog. Energy* **2014**, *39*, 3100–3108.

93. Kou, T.; Si, C.; Gao, Y.; Frenzel, J.; Wang, H.; Yan, X.; Bai, Q.; Eggeler, G.; Zhang, Z. Large-scale synthesis and catalytic activity of nanoporous Cu–O system towards CO oxidation. *RSC Adv.* **2014**, *4*, 65004.

94. Kou, T.; Wang, Y.; Zhang, C.; Sun, J.; Zhang, Z. Adsorption behavior of methyl orange onto nanoporous core–shell Cu@Cu$_2$O nanocomposite. *Chem. Eng. J.* **2013**, *223*, 76–83.

95. Li, G.; Song, Y.; Zhang, L.; Wei, X.; Song, X.; Sun, Z. Nanoporous copper silicon composite prepared by chemical dealloying as anode material for Lithium-ion batteries. *Funct. Mater. Lett.* **2013**, *6*, 1350033.

96. Hou, C.; Lang, X.; Han, G.; Li, Y.; Zhao, L.; Wen, Z.; Zhu, Y.; Zhao, M.; Li, J.; Lian, J.; *et al.* Integrated solid/nanoporous copper/oxide hybrid bulk electrodes for high-performance lithium-ion batteries. *Sci. Rep.* **2013**, *3*, doi:10.1038/srep02878.

97. Winter, M.; Brodd, R. What are batteries, fuel cells, and supercapacitors? *Chem. Rev.* **2004**, *104*, 4245–4270.

98. Qiu, H.; Xu, H.; Liu, L.; Wang, Y. Correlation of the structure and applications of dealloyed nanoporous metals in catalysis and energy conversion/storage. *Nanoscale* **2015**, *7*, 386–400.

99. Simon, P.; Gogotsi, Y. Materials for electrochemical capacitors. *Nat. Mater.* **2008**, *7*, 845–854.

100. Miller, J.; Simon, P. Electrochemical capacitors for energy management. *Science* **2008**, *321*, 651–652.

101. Huang, H.; Zhang, W.; Fu, Y.; Wang, X. Controlled growth of nanostructured MnO$_2$ on carbon nanotubes for high-performance electrochemical capacitors. *Electrochim. Acta* **2015**, *152*, 480–488.

102. Miller, J.; Outlaw, R.; Holloway, B. Graphene electric double layer capacitor with ultra-high-power performance. *Electrochim. Acta* **2011**, *56*, 10443–10449.

103. Guo, Y.; Shi, Z.; Chen, M.; Wang, C. Hierarchical porous carbon derived from sulfonated pitch for electrical double layer capacitors. *J. Power Sources* **2014**, *252*, 235–243.

104. Liu, D.; Shen, J.; Liu, N.; Yang, H.; Du, A. Preparation of activated carbon aerogels with hierarchically porous structures for electrical double layer capacitors. *Electrochim. Acta* **2013**, *89*, 571–576.

105. Du, X.; Zhao, W.; Wang, Y.; Wang, C.; Chen, M.; Qi, T.; Hua, C.; Ma, M. Preparation of activated carbon hollow fibers from ramie at low temperature for electric double-layer capacitor applications. *Bioresour. Technol.* **2013**, *149*, 31–37.

106. Kuratani, K.; Kiyobayashi, T.; Kuriyama, N. Influence of the mesoporous structure on capacitance of the RuO$_2$ electrode. *J. Power Sources* **2009**, *189*, 1284–1291.

107. Cheng, S.; Yang, L.; Chen, D.; Ji, X.; Jiang, Z.; Ding, D.; Liu, M. Phase evolution of an alpha MnO$_2$-based electrode for pseudo-capacitors probed by in operando Raman spectroscopy. *Nano Energy* **2014**, *9*, 161–167.

108. Zhang, Y.; Li, L.; Shi, S.; Xiong, Q.; Zhao, X.; Wang, X.; Gu, C.; Tu, J. Synthesis of porous Co$_3$O$_4$ nanoflake array and its temperature behavior as pseudo-capacitor electrode. *J. Power Sources* **2014**, *256*, 200–205.

109. Yuan, C.; Hou, L.; Feng, Y.; Xiong, S.; Zhang, X. Sacrificial template synthesis of short mesoporous NiO nanotubes and their application in electrochemical capacitors. *Electrochim. Acta* **2013**, *88*, 507–512.

110. Bavio, M.; Acosta, G.; Kessler, T. Polyaniline and polyaniline–carbon black nanostructures as electrochemical capacitor electrode materials. *Int. J. Hydrog. Energy* **2014**, *39*, 8582–8589.

111. Lee, H.; Cho, M.; Kim, I.; Nam, J.; Lee, Y. RuO_x/polypyrrole nanocomposite electrode for electrochemical capacitors. *Synth. Met.* **2010**, *160*, 1055–1059.

112. Aradilla, D.; Estrany, F.; Casellas, F.; Iribarren, J.; Alemán, C. All-polythiophene rechargeable batteries. *Org. Electron.* **2014**, *15*, 40–46.

113. Cortie, M.; Maaroof, A.; Smith, G. Electrochemical capacitance of mesoporous gold. *Gold Bull.* **2005**, *38*, 14–22.

114. Snyder, J.; Asanithi, P.; Dalton, A.; Erlebacher, J. Stabilized nanoporous metals by dealloying ternary alloy precursors. *Adv. Mater.* **2008**, *20*, 4883–4886.

115. Lang, X.; Hirata, A.; Fujita, T.; Chen, M. Nanoporous metal/oxide hybrid electrodes for electrochemical supercapacitors. *Nat. Nanotechnol.* **2011**, *6*, 232–236.

116. Lang, X.; Zhang, L.; Fujita, T.; Ding, Y.; Chen, M. Three-dimensional bicontinuous nanoporous Au/polyaniline hybrid films for high-performance electrochemical supercapacitors. *J. Power Sources* **2012**, *197*, 325–329.

117. Meng, F.; Ding, Y. Sub-micrometer-thick all-solid-state supercapacitors with high power and energy densities. *Adv. Mater.* **2011**, *23*, 4098–4102.

118. Wang, L.; Balka, T. Synthesis of nanoporous nickel thin films from various precursors. *Philos. Mag. Lett.* **2014**, *94*, 573–581.

119. Dan, Z.; Qin, F.; Sugawara, Y.; Muto, I.; Hara, N. Bimodal nanoporous nickel prepared by dealloying $Ni_{38}Mn_{62}$ alloys. *Intermetallics* **2012**, *31*, 157–164.

120. Fukumizu, T.; Kotani, F.; Yoshida, A.; Katagiri, A. Electrochemical formation of porous nickel in zinc chloride-alkali chloride melts. *J. Electrochem. Soc.* **2006**, *153*, C629–C633.

121. Hakamada, M.; Mabuchi, M. Preparation of nanoporous Ni and Ni–Cu by dealloying of rolled Ni–Mn and Ni–Cu–Mn alloys. *J. Alloy. Compd.* **2009**, *485*, 583–587.

122. Qiu, H.; Kang, J.; Liu, P.; Hirata, A.; Fujita, T.; Chen, M. Fabrication of large-scale nanoporous nickel with a tunable pore size for energy storage. *J. Power Sources* **2014**, *247*, 896–905.

123. Dong, C.; Bai, Q.; Cheng, G.; Zhao, B.; Wang, H.; Gao, Y.; Zhang, Z. Flexible and ultralong-life cuprous oxide microsphere-nanosheets with superior pseudocapacitive properties. *RSC Adv.* **2015**, *5*, 6207–6214.

124. Dong, C.; Wang, Y.; Xu, J.; Cheng, G.; Yang, W.; Kou, T.; Zhang, Z.; Ding, Y. 3D binder-free Cu_2O@Cu nanoneedle arrays for high-performance asymmetric supercapacitors. *J. Mater. Chem. A* **2014**, *2*, 18229–18235.

125. Momeni, M.; Nazari, Z.; Kazempour, A.; Hakimiyan, M.; Mirhoseini, S. Preparation of CuO nanostructures coating on copper as supercapacitor materials. *Surf. Eng.* **2014**, *30*, 775–778.

7

DNA-Protected Silver Clusters for Nanophotonics

Elisabeth Gwinn [1,*], Danielle Schultz [2], Stacy M. Copp [1] and Steven Swasey [2]

[1] Department of Physics, The University of California, Santa Barbara, Santa Barbara, CA 93106, USA; E-Mail: shiffler@physics.ucsb.edu

[2] Department of Chemistry and Biochemistry, The University of California, Santa Barbara, Santa Barbara, CA 93106, USA; E-Mails: dschultz@chem.ucsb.edu (D.S.); sswasey@chem.ucsb.edu (S.S.)

* Author to whom correspondence should be addressed; E-Mail: bgwinn@physics.ucsb.edu

Academic Editor: Stephen Ralph

Abstract: DNA-protected silver clusters (Ag_N-DNA) possess unique fluorescence properties that depend on the specific DNA template that stabilizes the cluster. They exhibit peak emission wavelengths that range across the visible and near-IR spectrum. This wide color palette, combined with low toxicity, high fluorescence quantum yields of some clusters, low synthesis costs, small cluster sizes and compatibility with DNA are enabling many applications that employ Ag_N-DNA. Here we review what is known about the underlying composition and structure of Ag_N-DNA, and how these relate to the optical properties of these fascinating, hybrid biomolecule-metal cluster nanomaterials. We place Ag_N-DNA in the general context of ligand-stabilized metal clusters and compare their properties to those of other noble metal clusters stabilized by small molecule ligands. The methods used to isolate pure Ag_N-DNA for analysis of composition and for studies of solution and single-emitter optical properties are discussed. We give a brief overview of structurally sensitive chiroptical studies, both theoretical and experimental, and review experiments on bringing silver clusters of distinct size and color into nanoscale DNA assemblies. Progress towards using DNA scaffolds to assemble multi-cluster arrays is also reviewed.

Keywords: ligand-protected metal clusters; DNA templates; silver cluster; machine learning; DNA nanotechnology; fluorescence

1. Introduction

In this article, we review the properties of a compelling new class of nanoscale optical element: fluorescent silver clusters that are stabilized by single-stranded (*ss*) "binding pockets" in oligonucleotide strands and strand assemblies. Such DNA-stabilized Ag clusters (Ag_N-DNA), first reported by Petty [1], are now emerging in applications that range from biological imaging [2], to molecular logic schemes [3] and strand-exchange on-off switches [4], to sensors for single base mutations [5–7], microRNAs [8], and DNA target strands [9]. These applications, recently reviewed in [10], all rely on the special sensitivity of fluorescence color and intensity to the specific nucleic acid environment of the silver cluster. Here we will focus on what our group has learned about the underlying composition and structure of pure Ag_N-DNA that lead to these fascinating properties (Figure 1a,b) and the prospects for realizing metal cluster nanophotonics on DNA scaffolds that self-assemble by Watson-Crick pairing of synthetic DNA strands, such as DNA origami [11], tile-based assemblies [12], and nanoscale three-dimensional shapes [13].

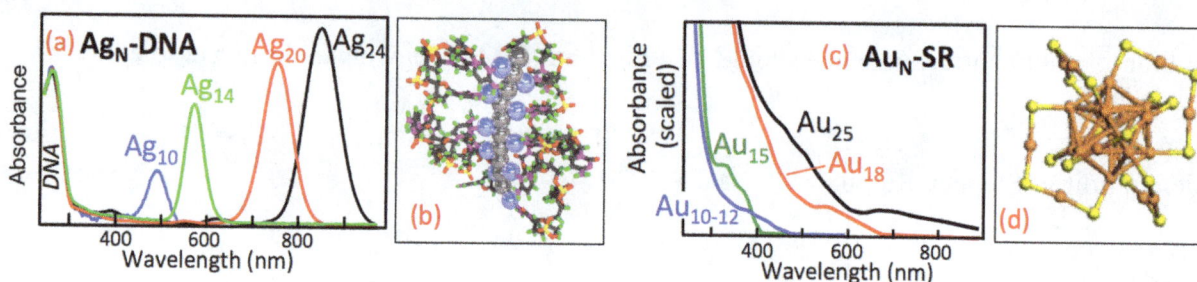

Figure 1. Experimental absorbance spectra are strikingly different for ligand-protected clusters composed of the same number of metal atoms but with different shapes and type of metal. (**a**) Ag_N clusters protected by different DNA template strands each display a single, narrow absorbance transition beyond the 260 nm DNA peak [14]; (**b**) A proposed Ag_N-DNA structure [14], whose neutral cluster core (gray) is held by base-bound Ag^+ (blue). The depicted *rod* cluster shape reproduces the striking spectral purity and size-dependent colors that are characteristic of Ag_N-DNA; (**c**) The broad spectra of Au_N-SR clusters protected by thiolate (SR) ligands reflect their *globular* cluster shape and Au composition. (Adapted from reference [15] and reprinted with permission); (**d**) The globular core of Au_N-SR is protected by SR ligands that incorporate additional Au atoms. Orange: Au. Yellow: S. (R groups not shown). (Adapted with permission from reference [16]).

Several groups have already demonstrated intriguing optical phenomena from the interaction of nanoscale optical elements arranged at close proximity on DNA scaffolds, including fluorescent dyes, metal nanoparticles and semiconductor quantum dots [17–20]. Due to their unique combination of metallic and molecular attributes, ligand-protected metal clusters of up to some tens of atoms in size are beginning to be examined as a distinct class of nanophotonic element. These clusters can exhibit metal-like optical response, due to the collective excitations of delocalized valence electrons, and molecule-like high fluorescence quantum yields, related to the sparse density of states in the cluster regime.

Ag$_N$-DNA exemplify the special properties of metal clusters [14] and are especially exciting for nanophotonics because they integrate naturally with DNA assemblies [21] and because the DNA template can be varied to produce fluorescent silver clusters at blue-green through infrared wavelengths (Figure 1a). While atomic-level structures for Ag$_N$-DNA are still not established, recent work reviewed here suggests that the selection of particular fluorescence colors by the DNA template arises largely from cluster size selection, with additional color tuning by the specific base and silver cation environment.

2. Results and Discussion

Our initial work on Ag$_N$-DNA established the sensitivity of silver cluster fluorescence to the sequence and secondary structure of the host DNA [21]. Using select DNA oligomers, we found that visibly fluorescent silver clusters form preferentially in single-stranded (*ss*) regions of DNA hosts, rather than on double-stranded (*ds*) regions. This selectivity opened the possibility of precisely positioning silver clusters on *ss*DNA extrusions from *ds*DNA scaffolds. Additionally we used hairpin DNA templates, with *ds* stems and *ss* loops in their native state, to show that both G-and C-rich regions within the *ss* segment favor formation of fluorescent clusters, but runs of several consecutive T or A bases suppress formation of fluorescent clusters [21]. The sensitivity of silver cluster color to changes in sequence has been confirmed and expanded upon by others [22–24], and shown by us to extend to Ag$_N$-RNAs [25].

Silver clusters formed on different oligonucleotide templates display tremendous variation in emission wavelengths (~400–800 nm), as summarized in [26], as well as widely varying quantum yields [27]. The strong variation in cluster properties with the specific details of host strands is an exciting aspect of Ag$_N$-DNA that is beginning to be addressed systematically, as discussed below. However the challenge of isolating pure Ag$_N$-DNA has impeded understanding of how optical properties relate to the silver content, as opposed to the base composition of the DNA template.

Ag$_N$-DNA can be synthesized quite simply by borohydride reduction of solutions of Ag$^+$ mixed with DNA template strands that have sequences selected to produce fluorescent solutions. This facile synthesis procedure typically produces heterogeneous solutions of many different Ag$_N$-DNA species, with different numbers N of silver atoms attached to the DNA, as well as different numbers n_s of the DNA oligomers held together by silver atoms [28]. Typically most of the silver-bearing DNA products are non-fluorescent ("dark"), with the fluorescent Ag$_N$-DNA comprising only a few to some tens of percent of all silver-DNA products. This heterogeneity produced confusion in the early literature regarding the silver content, N, in fluorescent Ag$_N$-DNA, with claims of quite small $N = 2$–3 from studies of impure solutions [21,23,29]. In contrast, in studies of pure material $N = 10$ is the smallest silver content identified to date for a fluorescent Ag$_N$-DNA [14,28].

2.1. Isolation of Pure Ag$_N$-DNA and Identification of Composition

2.1.1. HPLC Separation of Ag$_N$-DNA and Sizing by in-Line Mass Spectrometry

Figure 2a schematically illustrates the system we use to isolate pure Ag$_N$-DNA for identification of composition and optical properties [28]. The as-synthesized, impure samples are injected into a core-shell C18 column for reverse-phase, ion pair (IP) HPLC. Because column affinities depend on the DNA conformation and composition, under optimized conditions different species of Ag_N-DNA desorb from the column at different solvent compositions, resulting in fully or partly purified "plugs" concentrated with a particular Ag_N-DNA. The plugs travel past an absorbance detector set to 260 nm to monitor passage of all DNA-containing material, and then past a fluorescence array detector. This method of HPLC separation is effective for more robust Ag_N-DNA; however in several cases of strands known to produce fluorescent solutions, we have found that the Ag_N-DNA responsible for fluorescence are too fragile to survive HPLC under the solution conditions we have employed [28].

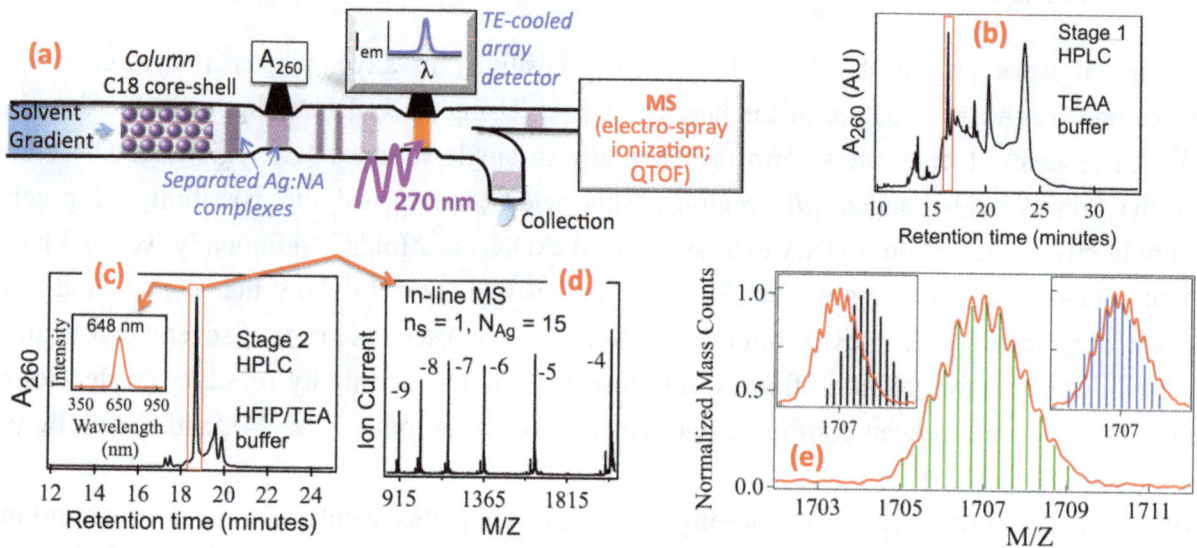

Figure 2. (a) Schematic of the HPLC-mass spectrometry (MS) system used to determine composition and optical properties of pure Ag_N-DNA [14,28]; (b) Absorbance chromatogram, A_{260}, from the 1st stage of purification by HPLC, with TEAA/TEA buffer, shows multiple peaks corresponding to many different Ag-bearing DNA species in the as-synthesized solution. The boxed peak is for the red fluorescent Ag_N-DNA; (c) A_{260} chromatogram for the 2nd HPLC stage, with HFIP/TEA buffer, which separates out remaining impurities. This enables inline MS (d) to identify the red emitter as an Ag_{15}-DNA containing just one DNA strand ($n_s = 1$); (e) We separately identify M and Z by resolving the isotope peak envelope in high resolution MS. Alignment of measured (red) and calculated isotope patterns (bars) determines the total (N), neutral (N_0) and cationic (N_+) Ag content in Ag_N-DNA, where the total Ag content $N = N_0 + N_+$. (Adapted with permission from references [14,28]).

Composition is analyzed by coupling into the electrospray ionization (ESI) source of a high resolution mass spectrometer (MS). In the ideal case of complete separation, only one species is detected in MS, identifying the composition of the Ag_N-DNA whose properties were measured by the upstream optical detectors.

Figure 2b shows the 260 nm absorbance, A_{260}, plotted *versus* time for an HPLC separation of the unpurified Ag-DNA solution formed on a typical 22-base host strand, $T_2CGC_6GC_4AG_2CGT_2$, selected because it forms a nearly *spectrally* pure, red fluorescent solution [28]. However, having a single dominant *emission* peak does not indicate that the corresponding Ag_N-DNA is the majority product in solution. In fact, as shown by the absorbance chromatogram in Figure 2b, the fluorescent Ag_N-DNA comprises just a small fraction of the many dark Ag-DNA products formed by the synthesis. Here, the time axis corresponds to increasing methanol concentration in the water-methanol HPLC running buffer, which contains triethyl ammonium acetate (TEAA)/triethyl amine (TEA) as the ion pairing (IP) buffer. (An IP buffer is necessary for solvent-dependent retention to the C18 column.)

The many chromatogram peaks in Figure 2b each correspond to a different Ag_N-DNA species. The red boxed peak marks the *fluorescent* Ag_N-DNA. To identify composition, we collected the eluent plug corresponding to this peak and reinjected this partially purified material into the HPLC system for a second round of purification using a different IP system, 1,1,1,3,3,3 hexafluoro-2-propanol (HFIP)/triethylamine (TEA). Use of HFIP/TEA is crucial because it greatly enhances electrospray ionization rates of DNA relative to TEAA [30,31], providing sufficient sensitivity for small analyte quantities; and because the column retention has a different dependence on the DNA composition, which causes residual impurities from the 1st round of HPLC purification to separate (Figure 2c). For this example case, in the 2nd round HFIP/TEA separation the in-line ESI-MS shows various charge states of a *single* species with 15 Ag atoms attached to the DNA strand, identifying this red emitter as an Ag_{15}-DNA (Figure 2d).

In addition to such strand monomer Ag_N-DNA with N silver atoms on $n_s = 1$ DNA strand, we have also identified strand dimer products where $n_s = 2$ copies of the strand are glued together by the silver content [14,26,28]. Formation of fluorescent, strand dimer Ag_N-DNA appears to dominate for short (<16 base) DNA strands [26].

Another method that has been used in attempts to determine composition of fluorescent Ag_N-DNA is inductively coupled plasma atomic emission spectroscopy (ICP-AES) [32]. In this technique, aliquots were collected during one stage of HPLC purification, during the chromatogram peak for elution of a fluorescent product. After acid digestion, ICP-AES determines the ratio of total silver to total DNA content. If high purity is achieved in HPLC, this can be an accurate way to determine stoichiometry. However, ICP-AES provides no information on the degree of purity achieved in HPLC, and in the case of strand dimer Ag_N-DNA can lead to a factor of two underestimation of size. Petty *et al.* discovered an IR emitting Ag_N-DNA that forms with the template strand $C_3AC_3AC_3TC_3A$ [32]. HPLC-MS identified this Ag_N-DNA as a strand *dimer* ($n_s = 2$) with $N = 20$ [28], consistent with the N/n_s~10 stoichiometry found by ICP-AES, but twice as large as the composition inferred from assuming $n_s = 1$. HPLC-MS also revealed a *non*-fluorescent product composed of just one DNA strand and 10 silver atoms: the same 10Ag/DNA strand stoichiometry as the fluorescent product, but with half the mass [28]. This dark Ag_{10}-DNA had a substantially earlier column elution time than the fluorescent, strand dimer Ag_{20}-DNA, consistent with the smaller DNA content [28].

Subsequent studies of other IR-emitting Ag_N-DNA have also found N~20, in some cases with the silver content gluing two DNA strands together [33,34]. These papers [33,34] expressed concern that the electrospray ionization process might cause aggregation of Ag_N-DNA. We view this as unrealistic given the pronounced physical differences between increasing the concentration of analytes in an equilibrium solution and the non-equilibrium Coulomb explosion mechanism for desolvation and ionization in ESI-MS, which is routinely used to generate and size individual ions of large biomolecules (such as proteins with molecular weights 10–20 times larger than those of Ag_N-DNA), despite the propensity for aggregation of such large molecules in equilibrium solutions.

A third technique that has been used to estimate stoichiometry of fluorescent Ag_N-DNA that are present in unpurified solutions is related to the "Job plot" analysis [35]. The Job plot obtains binding stoichiometry graphically from the dependence of product yields on the relative concentrations of two species that bind to form a complex, at fixed total concentration, and is known to be valid for certain limiting cases of binding reactions [35]. Studies of unpurified, red-emitting silver-DNA solutions formed on partial dsDNA duplexes with abasic sites used this indirect method to infer a total silver content, N, of approximately 2 silver atoms per DNA duplex [36]. This is a much smaller size regime than found in the above studies of purified materials, and might indicate a qualitatively different type of emissive structure; however due to use of just one total concentration, the applicability of the assumptions that underlie Job plot analyses was not tested [35].

In summary, most prior work on Ag_N-DNA have employed unpurified solutions, which typically contain a mixture of many silver-bearing DNA products in which the fluorescent Ag_N-DNA is a minor component. Often this may not be an issue; for example, in development of sensing schemes that are insensitive to silver-DNA byproducts. However for information about composition and structure, conclusions derived from studies of unpurified solutions [37] may be of questionable relevance.

2.1.2. High Resolution ESI-MS of Ag_N-DNA Reveals both Neutral and Cationic Silver Content

We use negative ion electrospray ionization mass spectrometry (ESI-MS) to determine how silver is incorporated into pure Ag_N-DNA. Due to the ease of deprotonating DNA, negative ion ESI-MS is a well-established technique for determining the composition of weakly bound, noncovalent complexes with dsDNA or ssDNA in solution [38–40]. Use of high resolution MS (HRMS) is important because HRMS separately determines the ion mass, M, and charge, Z, rather than just the ratio M/Z. With both M and Z, we can determine how the total silver content, N, in Ag_N-DNA, divides into cationic (Ag^+) content, N_+, and neutral (Ag°) content, N_0. This is accomplished in HRMS by resolving the isotope peak envelope that arises from the natural variation in isotopic abundances of the elements.

Figure 2e shows an example for a particular, pure Ag_N-DNA [14]. The finger pattern of peaks is the isotope envelope. The separation in M/Z between peaks must correspond to ΔM of precisely one atomic mass unit (amu), and thus identifies the charge state, Z. Here $Z = -en_{pr} + eN_+$, n_{pr} is the number of protons removed from the DNA by ESI, N_+ is the number of attached Ag^+ in the Ag_N-DNA, and e is the fundamental charge. With the measured Z, the M/Z of the peaks determines M, where $M = M_{DNA} + m_{Ag}(N_+ + N_0) - n_{pr}$, M_{DNA} is the total mass of the unionized DNA and m_{Ag} is the silver atom mass in amu. To determine N_+ and N_0, we vary N_+ to obtain agreement between the calculated and measured isotope peak envelope (Figure 2e).

Studies of many different Ag_N-DNA, discussed below, have shown that they contain roughly equal numbers of Ag^+ and Ag°. The Ag^+ appear to act as part of the DNA base ligands that support a neutral, metallically bonded cluster.

Identification of composition by ESI-HRMS can be complicated by fragmentation of delicate assemblies during ionization. For example, ESI-MS of short Watson-Crick paired DNA oligomers display peaks for the individual ssDNA strands as well as for the dsDNA duplex, due to partial disruption of hydrogen bonding in ESI. Fragmentation is also evident in MS for more delicate Ag_N-DNA. Figure 3 shows an example that identifies a parent Ag_{21}-DNA that contains $n_s = 2$ DNA strands of the 15-base template, CGC$_6$T$_2$G$_2$CGT [28]. ESI-MS also shows peaks for two fragments of the parent strand, an Ag$_{10}$-DNA and an Ag$_{11}$-DNA, each a strand monomer containing just one ($n_s = 1$) strand. The non-overlapping HPLC retention times of strand monomer and strand dimer products establish that the monomer peaks detected in ESI-MS are generated by fragmentation, rather than being residual impurities. We infer that this Ag$_{21}$-DNA contains two weakly bound strands, each primarily associated with $N = 10$ or 11 silver atoms. In analysis of optical properties of pure Ag$_N$-DNA, discussed below, we have excluded strand dimers with this fragmentation behavior.

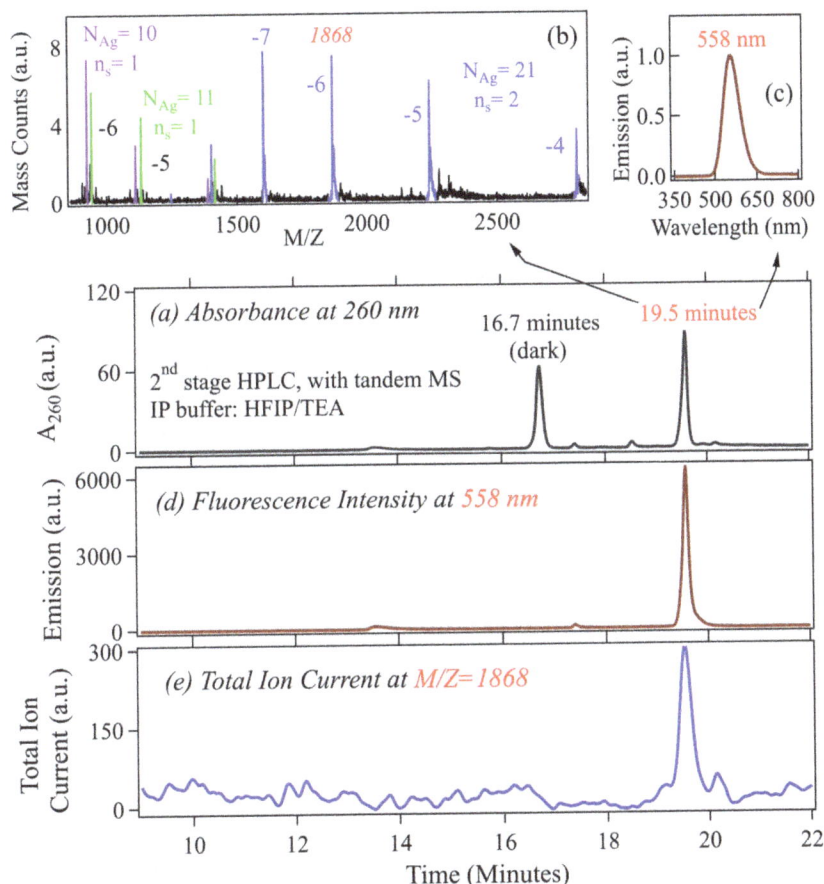

Figure 3. Identification of the composition of a weakly-bound strand dimer ($n_s = 2$) Ag$_N$-DNA. Chromatograms are from the 2nd stage of HPLC purification (HFIP/TEA) with inline MS. Reprinted with permission from reference [28]. (**a**) A$_{260}$ peaks for elution of all DNA-containing species. The fluorescent Ag$_N$-DNA elutes at 19.5 minutes; (**b,c**) MS and emission spectra corresponding to the 19.5 min A$_{260}$ peak in (**a**); (**b**) The $Z = -4, -5,$ -6 and -7 charge ladder (blue) identifies $N = 21$ as the total silver content and $n_s = 2$ as the number of DNA strands. The additional strand monomer peaks correspond to $n_s = 1$ and $N = 10$ (purple) and 11 (green) fragments (as expected, the integrated fragment counts are equal within experimental error); (**d**) Fluorescence chromatogram at 558 nm; and (**e**) Mass chromatogram of the $N_{Ag} = 21$, $n_s = 2$ IR emitter, for $M/Z = 1868$.

2.1.3. Universal Excitation of Ag$_N$-DNA via the Bases

Fluorescent Ag$_N$-DNA are universally excited *via* the DNA bases, regardless of emission color [41]. This property is very useful for purification of Ag$_N$-DNA and for rapid characterization of products formed on many different DNA hosts. In both cases we use a UV LED to excite all fluorescent Ag$_N$-DNA present in solution. While the specific mechanism for the energy transfer from the initial UV-excited state to the visible (or IR) emissive state has not yet been identified, it is facilitated by the close proximity of the DNA bases and the silver clusters they bind, and may be promoted by overlap of transition energies for the bases with higher energy transitions of metal clusters, followed by decay into the lower energy emissive state.

2.1.4. Native Secondary Structure of DNA and Ag$_N$-DNA

Current DNA nanotechnology is built on the canonical Watson-Crick (WC) hydrogen bonding of T to A and C to G. For integration of Ag$_N$-DNA into DNA nanostructures, it is important to know whether WC-paired DNA structure is maintained upon introduction of silver cations and the subsequent reduction to form silver clusters. For *ss*DNA hosts, circular dichroism studies of pure Ag$_N$-DNA and of the same DNA mixed with Ag$^+$, without reduction, show that Ag$^+$ and Ag$_N$ clusters reshape the DNA from the random coil form of the bare DNA to a more structured form [42]. One known mechanism for such structural change is the formation of Ag$^+$-bridged cytosine base pairs [43], though other base-dependent pairings mediated by Ag$^+$ may also be important.

A number of other Ag$_N$-DNA studies have employed DNA oligomers with partial or complete WC pairing in the native DNA [10,21,36,44,45]. However, depending on the relative stabilities of WC pairing and silver-mediated interactions between the bases, the native base pairing may be disrupted by the introduction of silver cations and subsequent reduction. The assumption that native WC pairing will persist after Ag$_N$ synthesis appears to hold in some cases. For 19-base, complementary C-rich and G-rich strands, the fluorescence observed for synthesis on the *individual* strands was quenched for synthesis on the *ds*DNA formed by the mixed, annealed strands, indicating that sufficient WC pairing persisted to suppress the formation of fluorescent Ag$_N$-DNA [21]. Later studies of *ss*DNA homo-base strands of all T, all A, all C and all G bases showed that homobase runs of T and A do not produce fluorescent clusters at neutral pH, while homobase runs of C or G bases do template fluorescent silver clusters [25]. This is consistent with the persistence of base pairing in the G,C-rich *ds* stems of native DNA hairpins with 5 base loops, which showed only very weak emission for T and A loops, relative to the stronger fluorescence for G or C loops [21].

Other studies of native hairpins with C loops of varying length found two dominant fluorescent products, one green-emitting and one red-emitting [44]. This evidence for different structural forms of clusters formed on the same DNA host was later confirmed in electrophoretic mobility studies that found significant differences in hydrodynamic radius for the green and red emitters. The differences in diffusion constants were attributed to disruption of base pairing in the *ds* stem by the incorporation of silver to form the green emitter [45]. The necks of hairpin loops [21,44,45] and mismatches and abasic sites [36,46] in *ds*DNA provide lesions through which Ag$^+$ can potentially invade to form fluorescent clusters, despite primarily *ds*DNA surroundings. Even in fully base-paired, duplex DNA, fluorescent products have been observed [46], indicating the disruption of WC pairing by silver incorporation. Successful prediction of whether a region of planned WC pairing will in fact retain native structure after silver cluster synthesis will require a better understanding of the base-specific interactions of silver with DNA.

2.2. Optical Properties of Pure Ag$_N$-DNA and Comparison to Other Ligand-Stabilized Metal Clusters

2.2.1. Sensitivity of Metal Cluster Optical Absorbance Spectra to Cluster Shape, Size and Composition

Metal *clusters* [47] are distinguished from metal *complexes* by the metal-metal bonding in clusters, which delocalizes the valence electron density over the entire cluster. Metal clusters are distinguished from metal *nanoparticles* by the small physical sizes of clusters, on the order of up some tens of atoms, corresponding to dimensions up to a few Fermi wavelengths. Because bare metal clusters agglomerate on contact, for use in solution or solid state they must be stabilized by a surface covering of protecting ligands. Most studied are small molecule ligands that cap nearly spherical, "globular"-shaped metal clusters. A well-studied example is the case of Au clusters protected by thiolate (SR) ligands (Figure 1c,d) [15,16,48].

DNA provides the protective ligand environment for Ag$_N$-DNA, which are remarkable for their unique combination of metallic and molecular attributes, with metal-like behavior of their optical spectra yet molecule-like fluorescence quantum yields that can approach 100% [14]. For much larger metal nanoparticles with $N > \sim 10^2 - 10^3$ atoms, the plasmon resonances that dominate optical spectra arise from the collective response of all the valence electrons to electromagnetic fields, which are well known to lie at energies sensitive to the shape of the metal nanoparticle [49].

At the small sizes of metal clusters, optical transitions can still retain plasmonic (collective) character, depending on cluster composition, shape and size. In the cluster size regime, a *rod* shape uniquely concentrates the absorbance spectrum into strong collective transitions. This is illustrated in Figure 4a,b in calculated results for atomic chains of silver atoms [50]. The longitudinal "L" collective excitation of valence electrons that oscillate parallel to the rod (Figure 4a) has exceptional color tunability by rod length (Figure 4b), that strikingly resembles the Ag$_N$-DNA data in Figure 1 a,b. In contrast, the weaker "T" plasmon at higher energy, with valence electrons oscillating perpendicular to the rod (Figure 4a,b), shifts relatively little with cluster size. Additional, much smaller peaks have primarily single-electron (rather than collective) character.

To display the profound effect of cluster shape, Figure 4c compares the calculated absorbance spectra of rod (red) and globular (green) Ag$_{10}$ clusters. In the globule, scattered transitions at higher energy replace the rod's concentrated "L" plasmon [51]. Data on globular silver clusters stabilized by *p*-MBA ligands exhibit a qualitatively similar forest of many absorbance peaks, rising above a broad background [52].

In addition to cluster shape, the choice of metal also has a profound effect on the optical spectra of clusters. In gold, *d* orbital transitions extend to lower energies than in silver. The greater mixing with valence electron transitions in gold produces congested spectra with weak absorbance peaks, as shown in calculations for Ag$_{12}$ and Au$_{12}$ *rods* in Figure 4d [53]. In addition to producing more complex optical spectra, the stronger *d* orbital mixing in gold may also contribute to lower fluorescence quantum yields of gold clusters, below 10% for Au_N-SR [54], in comparison to values above 90% reported for specific Ag_N-DNA [14].

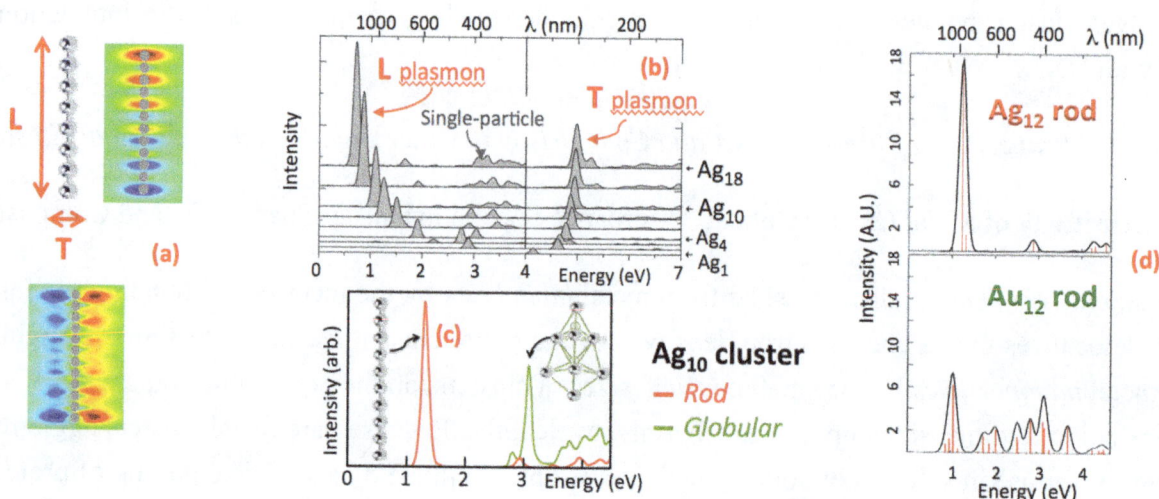

Figure 4. Sensitivity of cluster absorbance spectra to the cluster shape, size and composition. (**a**) An atomic silver rod, with the 2 types of collective excitations illustrated by their electron density profiles (color). Valence electrons slosh longitudinal to the rod in the "L" plasmon, and transverse to the rod in the "T" plasmon; (**b**) These L and T collective excitations dominate the optical absorbance spectra of Ag rods. Much weaker transitions are single particle-like. (Adapted with permission from reference [50]); (**c**) When the same number of Ag atoms is rearranged from a *rod* (red) into a *globule* (green), the absorbance shifts to higher energy and becomes more scattered. (Adapted with permission from reference [51]); (**d**) Calculated absorbance spectra of Ag_{12} (top) and Au_{12} (bottom) *rod* clusters. The congested spectrum for Au_{12} arises from mixing of valence electron and *d* orbital transitions. (Adapted with permission from reference [53]).

2.2.2. Evidence for Rod-Shaped Clusters in Fluorescent Ag_N-DNA

If Ag_N-DNA contained globular silver clusters, their absorbance spectra should show multiple peaks in the near-UV to blue, as exemplified by the green curve in Figure 4c. But instead we find that spectra of solutions of pure, fluorescent Ag_N-DNA have a single, dominant absorbance peak (Figure 5a) at visible to near-IR wavelengths. This peak shifts to lower energy/longer wavelength with increasing neutral cluster size, N_0 (Figure 1a, Figure 4b). This is the behavior predicted for rod-shaped clusters [50,53]. (Figure 5b, green line).

In cryogenic spectroscopy of individual Ag_N-DNA emitters (Figure 4f), we found broad emission lines even at 2K (−271 °C) [55], in contrast to the dramatic line narrowing typical of deep-cooled, single organic molecules. This qualitative difference from conventional molecular behavior is another metal-like aspect of clusters, expected to arise from dephasing processes of collective excitations [56,57]. Our cryogenic microscopy also found a strong polarization dependence of individual Ag_N-DNA, as expected for cluster rods (Figure 4g) [58].

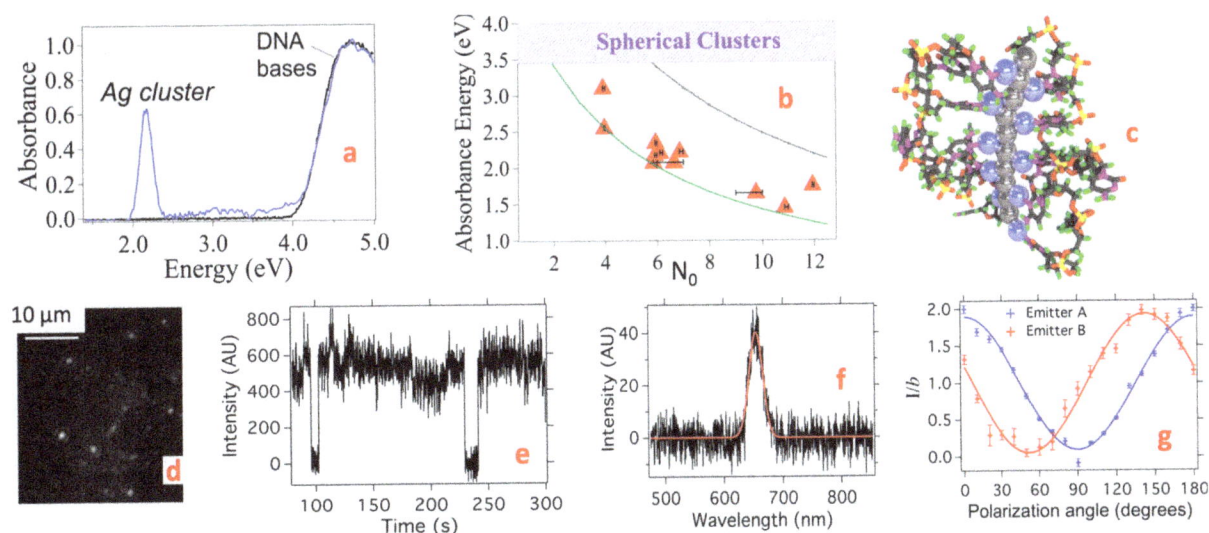

Figure 5. Optical properties of pure Ag_N-DNA. (Adapted with permission from references [14,55]). (**a**) Fluorescent Ag_N-DNA show a single absorbance peak (blue) at energies below the DNA peak. Black: DNA alone; (**b**) Peak absorbance energies of fluorescent Ag_N-DNA (red) follow the same trend as predicted for Ag cluster rods with 1-atom cross section (green curve) [14]. Blue shading: Range of main absorbance peaks for same-size globular clusters. Gray curve: predictions for a thicker Ag cluster rod [53]; (**c**) Another example rod structure that is consistent with optical data and the measured N_0 and N_+. The Ag° rod (gray) is shown attached to DNA bases via Ag^+ (blue) [14]; (**d–g**) Single Ag_{15}-DNA at 2K (excited at 590 nm) [55]; (**d**) Wide field image shows bright spots from individual Ag_{15}-DNA; (**e**) Observation of 1-step blinking between bright and dark states confirms emission from a single Ag_{15}-DNA; (**f**) Spectra for individual Ag_{15}-DNA remain spectrally broad at 2K, consistent with collective transitions [55]; (**g**) Dependence of emission on polarization for 2 individual Ag_{15}-DNA. The high modulation index is expected for rod clusters [58].

2.2.3. Magic Colors from Magic Number Cluster Sizes in Ag_N-DNA

DNA-stabilized silver clusters are remarkable for the selection of fluorescence color by the sequence of the stabilizing DNA oligomer. Yet despite a growing number of applications that exploit this property, large-scale studies to probe the origins of Ag_N-DNA color, and whether certain colors occur more frequently than others, have been lacking. We examined a set of 684 randomly chosen 10-base DNA oligomers to address these questions [26]. Rather than a flat distribution, we found that specific color bands dominate. Cluster size data indicate that these "magic colors" originate from the existence of magic number sizes for Ag_N-DNA that are different from the magic number sizes characteristic of globular gold clusters stabilized by small-molecule ligands.

Cluster physics predicts enhanced abundances at certain "magic numbers" of metal atoms that are bonded together in a connected cluster, due to enhanced stabilities from electronic shell closings at these special sizes [47]. To investigate the possibility of magic numbers, and associated "magic colors", in Ag_N-DNA, we used robotic synthesis in a well plate format that enabled rapid read-out of emission wavelengths using a well-plate fluorimeter, with UV light exciting all fluorescent Ag_N-DNA present in each well [26]. Of the 684 random 10-base strands, ~25% produced fluorescence with just one emission peak and a narrow enough peak width to indicate a single dominant type of fluorescent Ag_N-DNA [26].

These peak wavelengths exhibited a bimodal color distribution with enhanced ("magic") abundances of green-emitting Ag_N-DNA near 540 nm and red emitting Ag_N-DNA near 630 nm (Figure 6a; the plate reader's detector sensitivity at near-IR wavelengths was too poor to examine the possibility of magic color bands beyond ~750 nm) [26].

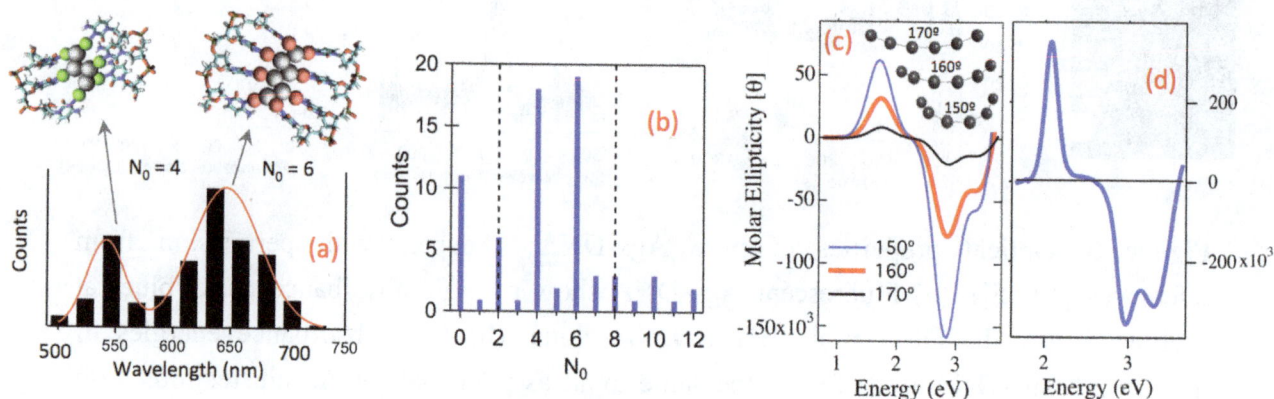

Figure 6. Magic colors and magic number neutral cluster sizes in (**a,b**) Ag_N-DNA and chiroptical properties of pure (**c,d**) Ag_N-DNA. (Adapted with permission from references [26,42]). (**a**) Fluorescence colors measured for Ag_N-DNA formed on many different DNA templates exhibit "magic colors": high green abundances near 540 nm and high red abundances near 630 nm [26]; (**b**) $N_0 = 4$ and 6 are magic (highly abundant) cluster sizes across all Ag_N-DNA with composition determined by ESI-MS [26]. Dashed lines: spherical "superatom" magic numbers 2 and 8 are ***not*** magic for Ag_N-DNA, indicating non-spherical (rod) shapes for the silver clusters; (**c**) Calculated circular dichroism spectra for bare, chiral Ag cluster rods show a consistent pattern of positive and negative peaks for different curvatures [42]; (**d**) CD data on pure Ag_N-DNA show the same peak pattern as theory [42].

To investigate the origin for these magic color bands, we used the system in Figure 2a to identify Ag_N-DNA products formed on many different strands. We found high abundances at "magic" $N_0 = 4$ (Figure 6b), which correspond predominantly to green magic colors, and also high abundances at magic $N_0 = 6$, which correspond predominantly to red magic colors. In contrast, numbers of silver cations, Ag^+, spread widely; e.g., $N_0 = 6$ clusters had $N_+ = 6$–10. Thus, neutral silver content, N_0, is magic, and cationic silver content, N_+, is not.

The known structures of ligand-stabilized Au clusters provide insight into why N_0 is magic, but total $N = N_+ + N_0$ is not [48,59]. In globular Au_N-SR, total gold content, N, is not magic because ligands incorporate some Au atoms, leaving cluster cores with smaller, magic number sizes that are predicted by the spherical "superatom" model. We infer that base-Ag^+ complexes act as ligands, analogous to the Au in SR units.

A crucial difference is that DNA presents multiple base ligands along a line-like backbone, which apparently enforces rod cluster shapes. This suggests that a perimeter of base-attached Ag^+ surrounds the cluster rod. Figure 5c shows a proposed type of structure compatible with this idea. While the structural details, such as specific sites and geometries of Ag^+-base attachment, are not known, the optical data discussed above indicate that a rod-like neutral cluster core is key. Variations in N_+ at fixed cluster core size, N_0, presumably contribute to the widths of the magic color bands in Figure 6a, which exhibit a population spread of ~60 nm in red peak emission wavelengths, and ~30 nm for green peak emission wavelengths (Figure 6a).

For globular clusters, the spherical "superatom" model predicts 2 and 8 as the smallest magic numbers of cluster valence electrons. But spherical symmetry breaking in rod-shaped clusters is expected to result in non-spherical magic numbers, with an even-odd oscillation of stability as a function of size. Consistent with a non-spherical shape, the most prominent magic numbers of neutral Ag atoms in Ag_N-DNA are 4 and 6 (Figure 6b), not 2 and 8 as for the spherical "superatom" model (dashed lines, Figure 6b).

2.2.4. Chiroptical Properties of Pure Ag_N-DNA

Measurement of cluster structure in Ag_N-DNA by AFM, TEM or X-ray studies is problematic due to N of just 10–30 Ag atoms in large DNA surroundings. To obtain better structural understanding we turned to circular dichroism (CD) spectroscopy, due to its high conformational sensitivity [42].

Prior CD studies by others that used as-synthesized silver-DNA solutions [1,60,61], which typically contain mostly non-fluorescent silver-DNA products, could not draw meaningful conclusions on the minor, fluorescent product. We used pure fluorescent Ag_N-DNA spanning the visible to near IR. Changes in the UV CD of pure cluster solutions showed that silver incorporation in fluorescent Ag_N-DNA restructures DNA conformation [42]. The NIR through visible wavelength CD probes the chirality of the clusters themselves. We compared data to quantum chemical calculations carried out in the group of Christine Aikens [42]. They found that a rod of silver atoms with a chiral twist reproduces the special pattern of positive and negative dichroic peaks in the data, and agree in overall magnitude with experiment (Figure 6c,d) [42]. This transparent relation between cluster shape and optical chirality is in qualitative contrast to the case of gold clusters, where mixing with d-orbital transitions produces complex CD spectra [62] with no simple correlation to cluster structure.

2.2.5. Equilibrium between Dark and Fluorescent forms of Ag_N-DNA

A fascinating aspect of Ag_N-DNA is the equilibrium between fluorescent and dark cluster forms with the same total silver content, N, that has been established for certain choices of DNA template [41,62]. This equilibrium can be manipulated by varying solution conditions, with potential for exploitation in diverse sensing applications, and may be the mechanistic underpinning for changes in fluorescence reported in some of the previously developed sensing schemes [10].

Petty and co-workers found reversible conversion between dark and fluorescent forms upon hybridization of an Ag_N-DNA with a partially complementary strand [63]. Reversible fluorescence quenching due to thermal strand unbinding, and an associated isosbestic point in the temperature-dependent absorption spectra (Figure 7a), indicated a temperature-controlled 2-species equilibrium between a dark form with peak absorbance near 400 nm and a fluorescent form with peak absorbance near 490 nm and peak emission near 550 nm (Figure 7b). Purification by size exclusion chromatography, followed by ICP-AES, indicated a stoichiometry of approximately 11 silver atoms per strand in the dark form, and the same silver content in the 2-strand bright form.

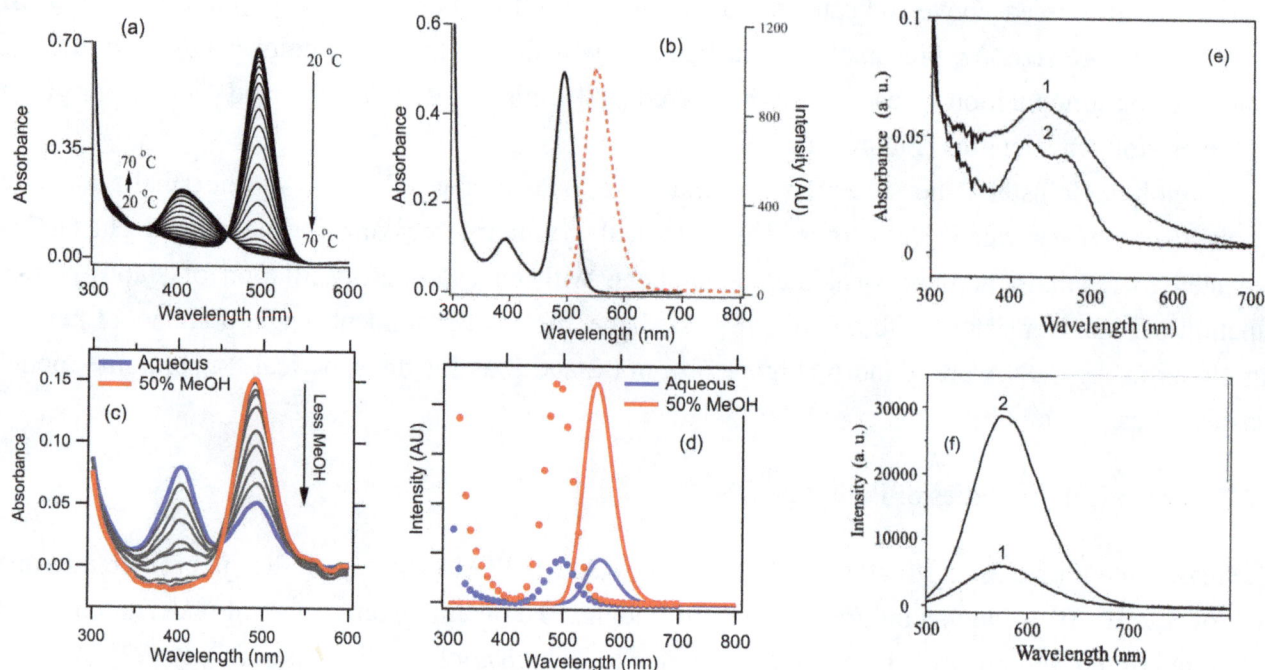

Figure 7. (Adapted with permission from [42,63,64]). (**a,b**) *Hybridization*-tuned equilibrium between dark and fluorescent forms of pure Ag_N-DNA with 10–12 Ag atoms per DNA assembly. (**a**) Varying temperature, T, varies hybridization extent, producing an absorbance isosbestic point (~460 nm), that indicates an equilibrium between isomers. Peaks are at 400 nm (dark form) and 490 nm (fluorescent form) [63]; (**b**) Emission (red) and absorbance at intermediate T, with both cluster forms present. Excitation at 490 nm produces emission near 570 nm, but excitation near 400 nm produces no emission [63]; (**c,d**) *Solvent*-tuned equilibrium between dark and fluorescent forms of pure Ag_{10}-DNA [41]; (**c**) Absorbance for MeOH fractions of 0% to 50% of the solvent, in 5% increments. The isosbestic point (near 450 nm) indicates equilibrium between two forms of the Ag_{10}-DNA. Absorbance peaks at 400 nm (dark form) and 490 nm (fluorescent form) [42], just as for the hybridization-tuned equilibrium in (**a,b**); (**d**) Fluorescence excitation spectra (dotted) and emission spectra at 0% and 50% MeOH. The increase in intensity is due to the higher fraction of the fluorescent form [42]; (**e,f**) Studies of Ag-DNA and Ag-Cu-DNA made by reducing Cu^{2+} on the fluorescent Ag-DNA solution [64]; (**e**) Absorbance of unpurified solutions shows two peaks at similar wavelengths for Ag-DNA (1) and Ag-Cu-DNA (2), close to those for the dark and fluorescent cluster forms of pure Ag_N-DNA in (**a**) and (**c**). The poorly resolved peak structure in the unpurified solutions in (**e**) indicate the presence of additional cluster species. (**f**) Emission from the unpurified Ag-DNA solution (1) increased upon reduction of Cu^{2+} (2). The 560 nm emission lies near the 560–570 nm emission peaks in the pure Ag_N-DNAs in (**a–d**), suggesting that Cu^{2+} may also be controlling an equilibrium between dark and fluorescent cluster forms.

A different means of controlling the equilibrium between between bright and dark forms of pure Ag_N-DNA with $N = 10$ was established for an Ag_{10}-DNA that formed on a particular 19-base DNA template [42]. In 50% methanol-50% aqueous solution, the absorbance spectrum exhibited a single cluster absorbance peak at 490 nm (Figure 7c) that produces an emission peak at 560 nm. Reducing the methanol fraction introduced an additional, dark form with a shorter wavelength absorbance peak near 400 nm, causing a decrease in emission intensity for excitation at 490 nm due to the decreased fraction of the fluorescent form (Figure 7d). Solvent exchange confirmed the reversibility of this bright-dark conversion, and the isosbestic point in the absorbance spectra collected for different methanol fractions indicated a 2-species equilibrium. Varying the concentration of sucrose produced similar behavior for this Ag_{10}-DNA [42].

The addition of alcohols and sucrose to aqueous solution alters DNA conformation through changes in hydration [65,66]. Thus it appears that solvent-mediated changes in DNA induce structural changes in the embedded silver clusters [42]. Hybridization also alters DNA conformation. The similarities in the changes in optical properties of these Ag_N-DNA with similar $N = 10$–11, under both solvent and hybridization-mediated control, suggest that other means of DNA structural manipulation, such as changes in ionic strength, may also induce such structural transitions of embedded clusters.

The similar stoichiometries, absorbance wavelengths of the dark and bright forms, and emission wavelengths of the bright form suggest that the different mixed-base templates used in these separate studies produced similarly structured dark and bright species. In both cases, the observed spectral switching between bright and dark forms indicates a change in bonding, perhaps due to a change in the preferred magic number cluster size that favors conversion of a larger cluster into two smaller clusters, and/or changes between neutral and cationic form of some of the silver atoms. The shorter peak absorbance wavelength of the dark form is consistent with separation of the fluorescent cluster form into smaller, dark subclusters.

2.2.6. Effects of Copper and Gold Reduction on Fluorescent Silver-DNA Solutions

Changes in emission intensity have also been produced by reduction of Cu^{2+} on pre-synthesized, fluorescent Ag-DNA solutions and by simultaneous reduction of Ag^+ and Au^{3+} on DNA [64,67]. Reduction of Cu^{2+} on fluorescent Ag-DNA solutions was reported to enhance emission near 570 nm, with almost no shift in peak wavelength. This was interpreted as evidence for incorporation of a $Cu°$ atom into a two atom $Ag°$ cluster, based on the most abundant peak detected in ESI-MS [64]. This is a much smaller silver content than the $N =10$–12 Ag atoms measured for pure Ag_N-DNA with nearly the same emission wavelength (560–570) nm, as discussed above. Attempts to purify by capillary and gel electrophoresis were made in the Cu-Ag-DNA studies [64] but it is unclear whether these separations had sufficient resolution to isolate fluorescent products from dark products with similar mobilities. In our own work, we have found that electrophoresis typically fails as a purification method due to similar electrophoretic mobilities of multiple Ag_N-DNA products.

We would expect significant shifts in wavelength, rather than the observed shift of just ~2 nm in peak emission, to result from the addition of one copper atom to a cluster of two silver atoms. Unlike Ag^+, which bind specifically to the DNA bases, Cu^{2+} binds also to the DNA backbone [68], and may attach to the DNA at separate sites from the silver cluster. An alternative explanation for the increase in fluorescence brightness observed in [64] may be a shift in equilibrium between bright and dark silver cluster forms with the same silver content, as discussed above. Such a shift could be induced by a conformational change in the DNA from incorporation of copper at sites not directly connected to the silver cluster.

The substantial changes in fluorescence color reported for co-reduction of gold and silver cations onto DNA, relative to silver alone [67], do suggest a change in the population of fluorescent cluster sizes, perhaps reflecting the formation of Au-Ag alloy clusters. Recent work on thiolate-protected metal clusters reported the replacement of 13 gold atoms in an Au_{25} clusters with silver atoms [69]. The thiolate-protected Au_{25} cluster exhibited extremely weak fluorescence (quantum yield ~0.1%) at ~820 nm, but the thiolate-protected alloy cluster $Au_{12}Ag_{13}$ was brightly fluorescent (quantum yield ~40%) with substantially blue-shifted peak emission to 680 nm. The studies of fluorescent Au/Ag-DNA solutions claimed a composition of $3Ag°$ per DNA oligomer, corresponding to a ~530 nm emission peak, and a composition of $2Ag°$ and $1 Au°$, corresponding to a ~630 nm emission peak. These assignments of composition appear to have been based on the most abundant peak detected in ESI-MS of filtered, but otherwise unpurified, solutions. Based on emission color alone, a comparison to studies of thiolate-protected Au and AuAg alloy clusters [69] and to Ag_N-DNA purified by multiple stages of HPLC [14,28] would suggest a higher total metal content of ~10–20 Au and Ag atoms.

2.2.7. The Sequence-Color Code for Ag_N-DNA

Understanding why certain DNA sequences produce brightly fluorescent Ag_N-DNA solutions, while other apparently similar sequences do not, is crucial for further development of these intriguing nanomaterials. This is a challenging problem due to the enormous sequence space at the strand lengths, L, typically used in applications: even for L of just 10 bases, there are over 10^6 different sequences $(4^{L=10})$. Therefore a "hit or miss" approach is unlikely uncover specific sequences that lead to high chemical yields of Ag_N-DNA with desired properties.

Our recent approach towards "cracking the code" for the sequence characteristics that govern formation of fluorescent Ag_N-DNA is to apply machine learning algorithms to large, strategically selected data sets (Figure 8) [70]. The data is acquired by robotic synthesis and rapid array optical characterization, which make feasible the query of hundreds of distinct template strands in parallel. The underlying hypothesis is that "bright" multi-base motifs within these templates select for formation of fluorescent silver clusters, while "dark" motifs favor non-fluorescent products.

The pattern recognition algorithm we employed, the support vector machine (SVM), is a classifier that learns to separate two classes of training data, which are represented in a high-dimensional feature space, by fitting an optimal hyperplane between the two classes [71]. SVMs are widely used in bioinformatics, for example in protein-protein binding site prediction [72] and gene classification [73]. For Ag_N-DNA, we chose the two training data classes to correspond to "bright" DNA templates that stabilize fluorescent Ag_N-DNA and "dark" templates that do not.

We used data from randomly generated $L = 10$ base templates to train the SVM to make predictions of the probability of brightness for new, untested DNA templates. The data sets were collected one week after synthesis to eliminate unstable fluorescent products. Part of the data set was used to train the SVM, and the remainder tested the accuracy, A, of predictions made by the SVM. A is simply the fraction of test templates that the SVM correctly selects as "bright" or "dark".

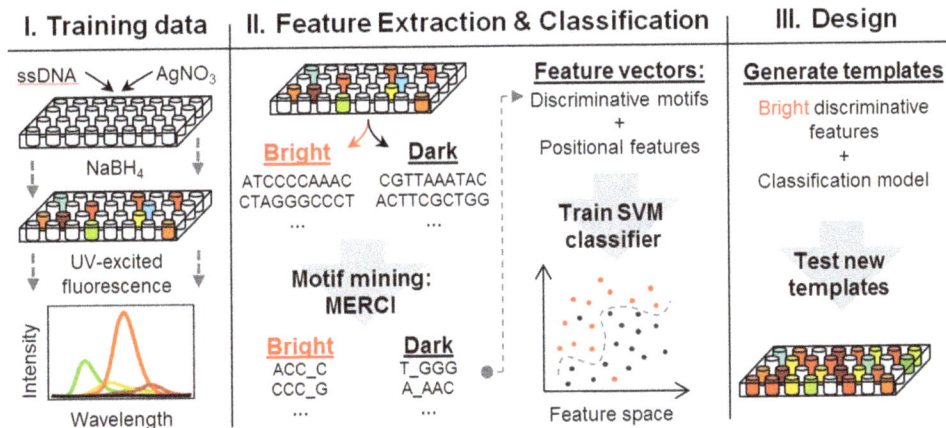

Figure 8. Methods used to establish that multi-base motifs are key to formation of fluorescent Ag$_N$-DNA, to recognize discriminative base motifs within DNA templates and to construct new templates for solutions with increased brightness. (Adapted with permission from reference [70]). (**a**) Robotic synthesis on nearly 700 randomly generated 10-base DNA templates, and fluorescence measurement one week later, created a large, unbiased data set without unstable fluorescent products; (**b**) By running support vector machine (SVM) classifiers with different choices of feature vector, we established that multibase motifs are key to formation of fluorescent Ag$_N$-DNA. A motif-miner algorithm, MERCI, was optimized to identify these discriminative motifs; (**c**) Combining the discovery of certain multi-base motifs important for determining fluorescence brightness with a simple generative algorithm, the probability of selecting DNA templates that stabilize fluorescent silver clusters was increased by a factor of >3 relative to random templates.

To convey information to the SVM, each bright and dark template must be represented by a "feature vector" that contains information on sequence. The most obvious choice is to use the entire template sequence, with each of the ten bases coded as an integer. However, trained SVMs using feature vectors composed of the entire sequence gave poor accuracy, $A \approx 60\%$, for bright-dark predictions. This indicates that the entire 10-base sequence is not what distinguishes bright from dark templates. By instead forming feature vectors that contain of the number of times certain multi-base motifs appear in the template, we achieved much improved accuracy, $A > 85\%$, for predicting whether a new 10-base template will be bright or dark after silver cluster synthesis [70].

To identify these bright-dark discriminative motifs, we optimized a motif-miner employed in bioinformatics, MERCI [74], to recognize bright motifs that favor fluorescent Ag$_N$-DNA and dark motifs that favor non-fluorescent products. The identified discriminative motifs contained 3 to 5 bases, with 4 and 5 base motifs making up 98% of those identified by MERCI [70]. Thus it appears that motifs of 4–5 base length are sufficient to define Ag$_N$ clusters with the requisite structure to emit at wavelengths within the detection bandwidth of our well plate reader. Several of the identified bright motifs contain consecutive C bases and/or consecutive G bases, consistent with previous findings of fluorescent Ag$_N$-DNA formed on C- and G-rich templates. We find that A bases are also common in bright motifs (the complete list of identified bright and dark motifs is given in reference [70]). Because consecutive runs of A bases were previously found not to produce fluorescent clusters [19,27], the presence of *A* bases in bright motifs suggests that having a C or G base flanking an A base may enable incorporation of silver cations in favorable modes for formation of fluorescent clusters upon reduction.

Aside from motif composition, the *number* of bright motifs required to stabilize an emissive Ag cluster is important for template design strategies. In studies of composition, we have found that for template strands with 16 or fewer bases, *two* copies of the same strand simultaneously stabilized the clusters. This implies that at least two bright motifs are required. In longer templates, this cluster "sandwiching" between bright motifs can be achieved by folding the strand around the cluster. With shorter templates, stiffness at length scales below the ~2 nm persistence length of *ss*DNA may preclude such folding, so that clusters instead engage multiple bright motifs by simultaneously attaching to two strands [70].

To create new DNA templates for bright Ag_N-DNA solutions, we used a simple generative algorithm that biases the template contents by including brightness-weighted discriminative motifs [70]. We classified each of the newly generated templates as bright or dark using our previously trained SVM and experimentally tested the effectiveness of this motif-based design method for the 374 template sequences to which the SVM assigned the highest brightness probabilities. The average fluorescence brightness of Ag_N-DNA solutions synthesized with this designed template set was much brighter, by a factor of >3 at one week after synthesis, relative to the random template set used to train the SVM. This inclusion of greater numbers of "bright" base motifs also red-shifted the Ag_N-DNA color distribution of the designed templates relative to the random templates. It appears that the higher average number of bright motifs in the designed templates increased the numbers of Ag^+ incorporated before reduction, causing an increase in the average silver cluster size formed by reduction and thus longer fluorescence wavelength. We expect that a more sophisticated generative algorithm, incorporating sparse dark motifs to limit byproducts, may further increase chemical yields of fluorescent products, and further anticipate that the motif miner-SVM approach can be generalized to prediction of templates that select for Ag_N-DNA color.

2.2.8. Stabilities of Ag_N-DNA

Just as Ag_N-DNA display widely varying colors and fluorescence quantum yields, their solution stabilities also vary widely for reasons that are not well understood. Temporal stability ranges from very poor, with fluorescence decaying over a matter of hours for storage at room temperature [75], to excellent: emission of a 530 nm emitting Ag_{11}-DNA [44] has persisted over several years in our labs, and a 615 nm emitting Ag_N-DNA with stability over at least one year of storage has also been reported [60]. For Ag_N-DNA that are stable enough to isolate by HPLC, we find that pure solutions are typically stable over weeks to months for storage at 10 °C at micromolar concentrations.

Another important issue for many applications is photostability. Early studies of fluorescent, ~700 nm emitting Ag_N-DNA that were immobilized in polyvinyl alcohol films (PVA) found high photostability, superior to cyanine dyes [29]. For many biomolecular applications, solution photostability is relevant rather than behavior in PVA. Solution photostability of Ag_N-DNA is poor in some cases, with wide variations in photobleaching rates amongst fluorescent solutions formed on different DNA templates [75,76]. These photostability studies proposed a redox-controlled mechanism for the photobleaching of a red-emitting Ag_N-DNA that simultaneously enhanced fluorescence from a green-emitting Ag_N-DNA. A redox reaction was also indicated by the conversion of green-emitting solutions back to the red form by addition of $NaBH_4$ [76]. Earlier studies had identified the green emitter as an Ag_{11}-DNA and the red emitter as an Ag_{13}-DNA [44] with a smaller hydrodynamic radius [45], so the interconversion changes DNA conformation in addition to adding or removing silver. The two silver atoms "lost" by the Ag_{13}-DNA to form the Ag_{11}-DNA may transfer to other strands in solution. In general, changes in solution conditions that alter fluorescence color presumably signal changes in cluster size by inter- or inter-strand transfer of silver content.

2.2.9. Electronic Properties of the Bases and Ag_N-DNA

The base-specific interactions of silver with the DNA template and competition of cluster formation with base pairing, and with agglomeration to form larger metal particles, presumably all underlie the wide variation in Ag_N-DNA properties found for seemingly minor variations in DNA template sequence. Early studies of Ag_N-DNA and Ag_N-RNA found that fluorescent clusters formed on strands of all C or all G bases, but not on strands of all T or all A bases at neutral pH [28]. The persistence of Ag^+ content as roughly half the total silver in pure Ag_N-DNA suggests that favorable cationic interactions with the bases are key. In the case of adenine homopolymer DNA strands, the absence of fluorescent products is puzzling given the presence of endocyclic nitrogen atoms in A, as in C and G but not T, and suggests that the specific geometry of Ag^+-adenine binding is incompatible with the formation of the $Ag°$-$Ag°$ bonds that are integral to clusters.

It appears that the different binding interactions of silver cations with the different bases lead to multi-base motifs [70] that favor selection of specific "magic" neutral silver clusters sizes upon reduction [26]. The roughly 60 nm spread in emission wavelength of clusters containing six $Ag°$ shows that there are additional color-tuning mechanisms at play. Shape variations, such as curvature of rod-shaped clusters [26] and or variation in dihedral $Ag°$-$Ag°$-$Ag°$ bond angles [77], may contribute. Additionally, the specific base content in the cluster-stabilizing motif presumably also tunes color through inductive chemical interactions, in analogy to the rather weak effects on optical properties of Au_{38} clusters observed for addition of electron donating or withdrawing groups to the stabilizing ligands [78].

2.3. Ag_N-DNA Assemblies

DNA nanotechnology allows precise nano-scale arrangement of optical elements such as nanoparticles and dye molecules onto scaffolds with diverse, DNA-programmed sizes and shapes [11,79]. Such DNA constructs have already demonstrated intriguing phenomena from interactions of arrayed elements, including FRET [17], optical chirality [18], and surface-enhanced Raman scattering [20]. Achieving the desired assembly properties requires stringent control over individual elements, a challenge for nanoparticles due to the difficulty of achieving precise sizes, shapes and surface morphologies [80] with DNA-compatible chemistries.

In contrast, ligand-protected metal clusters offer *atomically precise* control of optical behavior. A great advantage of metal clusters is their unique combination of metallic and molecular attributes, which can bring together strong collective excitations of valence electrons with high fluorescence quantum yields related to the sparse density of states. Developing methods for arranging metal clusters in designed nanoscale patterns is a necessary step for harnessing their potential for applications in nanotechnology.

2.3.1. Bi-Color, Dual Cluster Ag_N-DNA Assemblies

With their DNA template-tuned colors, compact sizes, biocompatibility and high fluorescence quantum yields, Ag_N-DNA are exciting prospects for multicolor decoration of DNA assemblies. However, this requires Ag clusters of different size to retain stability when held in nanoscale proximity. DNA templates that provide high enough barriers are needed to prevent reorganization driven by the size-dependence of bare cluster free energies [47,81]. But solutions containing DNA-stabilized Ag clusters of unknown composition exhibited color and brightness changes when mixed with additional DNA strands [9]. This raises the question of whether it is possible to select templates that preserve the stability of different sized clusters when they are brought together into a single nanoscale construct.

We developed 2-color constructs as test-of-principle experiments. The templates for each cluster have appended hybridization "tails", designed to bring red ($N = 15$) and green ($N = 10$) Ag_N-DNA together in a "clamp" assembly (Figure 9a) [82]. Observation of fluorescence resonance energy transfer (FRET; Figure 9a,b) established that these assemblies hold the clusters at separations below 6.2 nm, the Förster radius of the Ag_{10}-Ag_{15} pair [82]. Thermally-modulated FRET confirmed that the dual cluster structures disassemble and reassemble under thermal cycling [82]. The absence of spectral shifts in these dual-cluster FRET pairs, relative to the individual clusters, showed that select few-atom silver clusters of different sizes are sufficiently stable to retain structural integrity when held within a single nanoscale DNA construct.

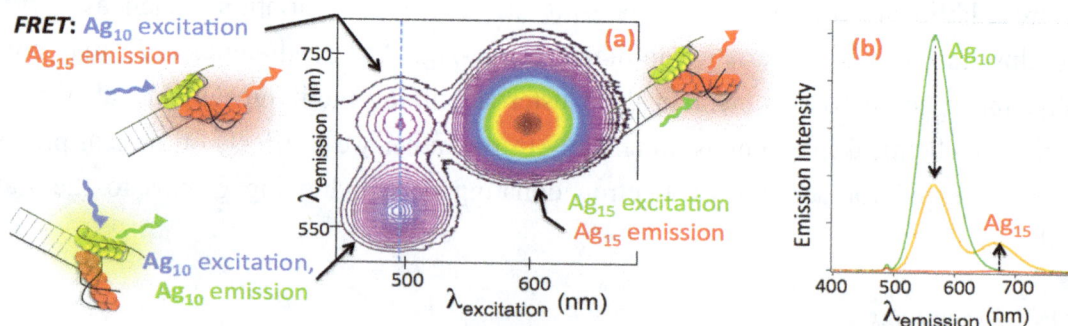

Figure 9. Development of dual cluster, bicolor silver cluster assemblies. Schematic to the left: Cartoons: paired cluster assemblies. Red: Ag_{15}; Green: Ag_{10}. (**a**) Contour map of emission intensity from the purified solution of paired clusters shows the expected peaks for direct excitation of Ag_{10} and Ag_{15}, and FRET: emission from Ag_{15} due to excitation of Ag_{10}, via radiationless energy transfer [82]; (**b**) Emission excited at the Ag_{10} absorbance peak. Green and (baseline) red curves are emission from separate solutions of the individual clusters. Orange: FRET of paired clusters produces emission from Ag_{15} and reduces Ag_{10} emission. The unshifted wavelengths relative to the individual clusters verify intact, stable assembly. (Adapted with permission from reference [82]).

2.3.2. Ag$_N$-DNA on DNA scaffolds

While dense, atomically precise arrays of Ag$_N$-DNAs have not yet been realized on larger DNA scaffolds assembled from many constituent strands, there has been progress towards this goal. The preference for Ag cluster formation onto ssDNA [21] was exploited by O'Neill and Fygenson to synthesize Ag$_N$-DNAs on DNA nanotubes with hairpin protrusions [83]. Sparse cluster decoration, with average separations of ~1 μm, enabled an estimate of ~45% initial occupation of all hairpin protrusions by unstable red fluorescent clusters (Figure 10). Shifts in emission wavelength on tubes, relative to the hairpin free in solution, may indicate formation of a slightly different mixture of red-emitting cluster sizes. Despite the lack of long-term stability, this is an encouragingly high synthesis yield of fluorescent Ag$_N$-DNA products. It was achieved by using a native hairpin that produces unusually high yields of fluorescent products [44], for reasons that are not well understood. However the progress described above in realizing more stable Ag$_N$-DNA, and in the emerging understanding of multi-base discriminative motifs, suggests that the need for high synthesis yields of stable Ag$_N$-DNA may be within reach.

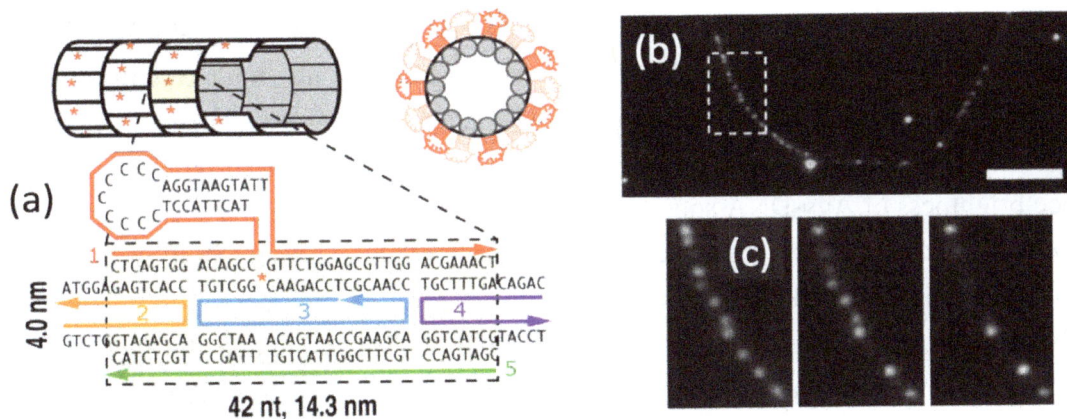

Figure 10. Decoration of a DNA nanotube with fluorescent Ag$_N$-DNA at ~1 μm separations. (Adapted with permission from reference [83]). (**a**) Nanotube design, with a native DNA hairpin extruding from every tile; (**b**) Fluorescence microscopy of nanotubes assembled with hairpins on 0.5% of tiles. Direct synthesis onto the assembled tubes produces bright emission spots from fluorescent red silver clusters; (**c**) Photobleaching of emission spots along the linear nanotube contour enables estimation of ~45% initial occupation of hairpin protrusions by the unstable red clusters.

Achieving atomically precise arrays of metal clusters at the <10 nm separations afforded by DNA scaffolds will require methods that achieve ~100 times more closely spaced clusters, with monodisperse cluster sizes. For much larger metal nanoparticles coated with DNA, attachment to DNA origami was limited to edge-to-edge particle separations of ~30 nm or greater by the repulsion of the negatively charged DNA [84]. However as shown above, silver clusters can be held by DNA at separations below 10 nm, perhaps in part due to a reduction of backbone repulsion by the Ag$^+$ that are incorporated into Ag$_N$-DNA. Therefore we anticipate that electrostatic effects will not prevent formation of dense arrays of silver clusters on DNA scaffolds.

Fluorescent Ag$_N$-DNA have also been used to decorate DNA nanowires [85] and DNA hydrogels [86], by incorporating loops of unpaired C bases into the strands that Watson-Crick pair to form these one-dimensional and three-dimensional scaffolds. Synthesis of silver clusters was performed directly onto the scaffolds, and fluorescence was observed in both of these widely differing geometries. Because synthesis typically produces abundant dark silver-DNA byproducts that occupy designed sites for fluorescent silver clusters, it is unknown what fraction of the intended scaffold sites were occupied by fluorescent Ag$_N$-DNA.

3. Conclusions

Rod-like metal clusters were initially realized using a STM to position Au atoms on atomically flat substrates in ultra-high vacuum [87]. It appears that that silver cluster rods [14] can instead be assembled at low cost in aqueous solution by using base motifs [70] within DNA templates as cluster-protecting ligands. The fluorescence and color tunability of Ag$_N$-DNA are already being exploited in novel sensing applications. Thus the atomic metal rod, a beautiful model system, is now becoming technologically relevant.

While the sensitivity of fluorescence wavelengths to the specifics of the host DNA template is an exciting aspect of Ag$_N$-DNA, a coherent picture is still lacking of just how certain templates favor fluorescent Ag$_N$-DNA of a given size and color. Towards achieving this understanding, we have used large array data sets and machine learning tools to establish that multi-base motifs govern the fluorescence brightness of Ag$_N$-DNA solutions [70]. We separately identified sets of motifs that select for brightness and sets of motifs that discriminate against bright products. Both motif types will be important for realizing designed multi-cluster constructs.

Many basic open questions remain regarding Ag$_N$-DNA. These include the factors that govern Stokes shifts and the widely-varying fluorescence quantum yields as well as the specific roles of the bases in cluster structure and color. However the work to date showing that clusters of distinct size and color retain their stability when held in nanoscale proximity [82] is promising for development of atomically precise, metal cluster nanophotonics on the versatile scaffold geometries provide by DNA nanotechnology.

Acknowledgments

The work reported from our group was supported by NSF-CHE-1213895, NSF-CHE-0848375 and NSF-DMR-1309410. SMC acknowledges support from NSF-DGE-1144085.

Conflicts of Interest

The authors declare no conflict of interest.

References

1.   Petty, J.T.; Zheng, J.; Hud, N.V.; Dickson, R.M. DNA-templated Ag nanocluster formation. *J. Am. Chem. Soc.* **2004**, *126*, 5207–5212.

2.   Yu, J.; Choi, S.; Richards, C.I.; Antoku, Y.; Dickson, R.M. Live cell surface labeling with fluorescent Ag nanocluster composites. *Photochem. Photobiol.* **2008**, *84*, 1435.

3.   Li, T.; Zhang, L.B.; Ai, J.; Dong, S.; Wang, E.K. Ion-tuned DNA/Ag fluorescent nanoclusters as versatile logic device. *ACS Nano* **2011**, *5*, 6334–6338.

4.   Guo, W.; Yuan, J.; Wang, E.K. Strand exchange reaction modulated fluorescence "on-off" switching of hybridized DNA duplex stabilized silver nanoclusters. *Chem. Comm.* **2011**, *47*, 10930–10933.

5.   Guo, W.W.; Yuan, J.P.; Dong, Q.Z.; Wang, E.K. Highly sequence dependent formation of fluorescent silver nanoclusters in hybridized DNA duplexes for single nucleotide mutation identification. *J. Am. Chem. Soc.* **2010**, *132*, 932–934.

6.   Ma, K.; Cui, Q.H.; Liu, G.Y.; Wu, F.; Xu, S.J.; Shao, Y. DNA abasic site-directed formation of fluorescent silver nanoclusters for selective nucleobase recognition. *Nanotechnology* **2011**, *22*, 305502–305507.

7.   Yeh, H.-C.; Sharma, J.; Shih, I.-M.; Vu, D.M.; Martinez, J.S.; Werner, J.H. A fluorescence light-up Ag nanocluster probe that discriminates single-nucleotide variants by emission color. *J. Am. Chem. Soc.* **2012**, *134*, 11550–11558.

8.   Yang, S.W.; Vosch, T. Rapid detection of microRNA by a silver nanocluster DNA probe. *Anal. Chem.* **2011**, *83*, 6935–6939.

9.   Yeh, H.-C.; Sharma, J.; Han, J.J.; Martinez, J.S.; Werner, J.H. A DNA-silver nanocluster probe that fluoresces upon hybridization. *Nano Lett.* **2010**, *10*, 3106–3110.

10. Yuan, Z., Chen, Y.-C., Li, H.-W.; Chang, H.-T. Fluorescent silver nanoclusters stabilized by DNA scaffolds. *Chem. Comm.* **2014**, *50*, 8900–9815.

11. Rothemund, P.W.K. Folding DNA to create nanoscale shapes and patterns. *Nature* **2006**, *440*, 297–302.

12. Winfree, E.; Liu, F.; Wenzler, L.A.; Seeman, N.C. Design and self-assembly of two-dimensional DNA crystals. *Nature* **1998**, *394*, 539–544.

13. Douglas, S.M.; Dietz, H.; Liedl, T.; Högberg, B.; Graf, F.; Shih, W.M. Self-assembly of DNA into nanoscale three-dimensional shapes. *Nature* **2009**, *459*, 414–418.

14. Schultz, D.; Gardner, K.; Oemrawsingh, S.S.R.; Markešević, N.; Olsson, K.; Debord, M.; Bouwmeester, D.; Gwinn, E. Evidence for rod-shaped DNA-stabilized silver nanocluster emitters. *Adv. Mater.* **2013**, *25*, 2797–2803.

15. Stamplecoskie, K.G.; Kamat, P.V. Size-dependent excited state behavior of glutathione-capped gold clusters and their light-harvesting capacity. *J. Am. Chem. Soc.* **2014**, *136*, 11093–11099.

16. Häkkinen, H. Atomic and electronic structure of gold clusters: Understanding flakes, cages and superatoms from simple concepts. *Chem. Soc. Rev.* **2008**, *37*, 1847–1859.

17. Stein, I.H.; Schüller, V.; Böhm, P.; Tinnefeld, P.; Liedl, T. Single-molecule FRET ruler based on rigid DNA origami blocks. *Chemphyschem* **2011**, *12*, 689–695.

18. Kuzyk, A.; Schreiber, R.; Fan, Z.; Pardatscher, G.; Roller, E.-M.; Högele, A.; Simmel, F.C.; Govorov, A.O.; Liedl, T. DNA-based self-assembly of chiral plasmonic nanostructures with tailored optical response. *Nature* **2012**, *483*, 311–314.

19. Shen, X.; Asenjo-Garcia, A.; Liu, Q.; Jiang, Q.; García de Abajo, F.J.; Liu, N.; Ding, B. Three-dimensional plasmonic chiral tetramers assembled by DNA origami. *Nano Lett.* **2013**, *13*, 2128–2133.

20. Pilo-Pais, M.; Watson, A.; Demers, S.; LaBean, T.H.; Finkelstein, G. Surface-enhanced raman scattering plasmonic enhancement using DNA origami-based complex metallic nanostructures. *Nano Lett.* **2014**, *14*, 2099–2104.

21. Gwinn, E.G.; O'Neill, P.; Guerrero, A.; Bouwmeester, D.; Fygenson, D.K. Sequence-dependent fluorescence from DNA-hosted silver nanoclusters. *Adv. Mater.* **2008**, *20*, 279–283.

22. Richards, C.I.; Choi, S.; Hsiang, J.C.; Antoku, Y.; Vosch, T.; Bongiorno, A.; Teng, Y.L.; Dickson, R.M. Oligonucleotide-stabilized Ag nanocluster fluorophores. *J. Am. Chem. Soc.* **2008**, *130*, 5038–5039.

23. Sengupta, B.; Ritchie, C.M.; Buckman, J.G.; Johnsen, K.R.; Goodwin, P.M.; Petty, J.T. Base-directed formation of fluorescent silver clusters. *J. Phys. Chem. C* **2008**, *112*, 18776–18782.

24. Sengupta, B.; Springer, K.; Buckman, J.G.; Story, S.P.; Hasan, Z.W.; Prudowsky, Z.D.; Rudisill, S.E.; Degtyareva, N.N.; Petty, J.T. DNA templates for fluorescent silver clusters and i-motif folding. *J. Phys. Chem. C* **2009**, *113*, 19518–19524.

25. Schultz, D.; Gwinn, E.G. Stabilization of fluorescent silver clusters by RNA homopolymers and their DNA analogs: C,G *versus* A,T(U) dichotomy. *Chem. Comm.* **2011**, *47*, 4715–4718.

26. Copp, S.M.; Schultz, D.; Swasey, S.; Pavlovich, J.; Debord, M.; Chiu, A.; Olsson, K.; Gwinn, E. Magic numbers in DNA-stabilized fluorescent silver clusters lead to magic colors. *J. Phys. Chem. Lett.* **2014**, *5*, 959–963.

27. Jaswinder, S.; Yeh, H.C.; Yoo, H.; Werner, J.H.; Martinez, J.S. A complementary palette of fluorescent silver nanoclusters. *Chem. Comm.* **2010**, *46*, 3280.

28. Schultz, D.; Gwinn, E.G. Silver atom and strand numbers in fluorescent and dark Ag:DNAs. *Chem. Comm.* **2012**, *48*, 5748–5750.

29. Vosch, T.; Antoku, Y.; Hsiang, J.C.; Richards, C.I.; Gonzales, J.I.; Dickson, R.M. Strongly emissive individual DNA-encapsulated Ag nanoclusters as single-molecule fluorophores. *Proc. Nat. Acad. Sci. USA* **2007**, *104*, 12616–12621.

30. Fountain, K.J.; Gilar, M.; Gebler, J.C. Analysis of native and chemically modified oligonucleotides by tandem ion-pair reversed-phase high-performance liquid chromatography/electrospray ionization mass spectrometry. *Rapid Comm. Mass Spec.* **2003**, *17*, 646–653.

31. Gilar, M.; Fountain, K.J.; Budman, Y.; Neue, U.D.; Yardley, K.R.; Rainville, P.D.; Russell, R.J.; Gebler, J.C. Ion-pair reversed-phase high-performance liquid chromatography analysis of oligonucleotides: Retention prediction. *J. Chromat. A* **2002**, *958*, 167–182.

32. Petty, J.T.; Fan, C.Y.; Story, S.P.; Sengupta, B.; Iyer, A.S.; Prudowsky, Z.; Dickson, R.M. DNA encapsulation of 10 silver atoms producing a bright, modulatable, near-infrared-emitting cluster. *J. Phys. Chem. Lett.* **2010**, *1*, 2524–2529.

33. Petty, J.T.; Nicholson, D.A.; Sergev, O.O.; Graham, S.K. Near-infrared silver cluster optically signaling oligonucleotide hybridization and assembling two DNA hosts. *Anal. Chem.* **2014**, *86*, 9220–9228.

34. Petty, J.T.; Giri, B.; Miller, I.C.; Nicholson, D.A.; Sergev, O.O.; Banks, T.M.; Story, S.P. Silver clusters as both chromophoric reporters and DNA ligands. *Anal. Chem.* **2013**, *85*, 2183–2190.

35. Huang, C.Y. Determination of the binding stoichiometry by the continuous variation method: The job plot. *Meth. Enzymol.* **1982**, *87*, 509–525.

36. Ma, K.; Shao,Y.; Ciu, Q.; Wu, F.; Xu, S.; Liu, G. Base-stacking determined fluorescence emission of DNA abasic site-templated silver clusters. *Langmuir* **2012**, *28*, 15313–15322.

37. Neidig, M.L.; Sharma, J.; Yeh, H.-C.; Martinez, J.S.; Conradson, S.D.; Shreve, A.P. Ag K-edge EXAFS analysis of DNA-templated fluorescent silver nanoclusters: Insight into the structural origins of emission tuning by DNA sequence variations. *J. Am. Chem. Soc.* **2011**, *133*, 11837–11839.

38. Greig, M.J. Detection of oligonucleotide-ligand complexes by ESI-MS (DOLCE-MS) as a component of high throughput screening. *J. Biomol. Screen.* **2000**, *5*, 441–454.

39. Hofstadler, S.; Sannes-Lowry, K. Applications of ESI-MS in drug discovery: Interrogation of non-covalent complexes. *Nat. Rev. Drug Discov.* **2006**, *5*, 585–595.

40. Rosu, F.; Gabelica, V.; Houssier, C.; de Pauw, E. Determination of affinity, stoichiometry and sequence selectivity of minor groove binder complexes with double-stranded oligodeoxy-nucleotides by electrospray ionization mass spectrometry. *Nucleic Acids Res.* **2002**, *30*, e82.

41. O'Neill, P.R.; Gwinn, E.G.; Fygenson, D.K. UV Excitation of DNA-stabilized Ag-cluster fluorescence via the DNA bases. *J. Phys. Chem. C* **2011**, *115*, 24061–24066.

42. Swasey, S.M.; Karimova, N.; Aikens, C.M.; Schultz, D.E.; Simon, A.J.; Gwinn, E.G. Chiral electronic transitions in fluorescent silver clusters stabilized by DNA. *ACS Nano* **2014**, *8*, 6883–6892.

43. Ono, A.; Cao, S.; Togashi, H.; Tashiro, M.; Fujimoto, T.; Machinami, T.; Oda, S.; Miyake, Y.; Okamoto, I.; Tanaka, Y. Specific interactions between silver (I) ions and cytosine-cytosine pairs in DNA duplexes. *Chem. Comm.* **2008**, *39*, 4825–4827.

44. O'Neill, P.R.; Velazquez, L.R.; Dunn, D.G.; Gwinn, E.G.; Fygenson, D.K. Hairpins with poly-C loops stabilize four types of fluorescent Ag_n:DNA. *J. Phys. Chem. C* **2009**, *113*, 4229–4233.

45. Driehorst, T.; O'Neill, P.R; Goodwin, P.J.; Pennathur, S.; Fygenson, D.K. Distinct conformations of DNA-stabilized fluorescent silver nanoclusters revealed by electrophoretic mobility and diffusivity measurements. *Langmuir* **2011**, *77*, 8923–8933.

46. Shah, P.; Rorvig-Lund, A.; Ben Chaabane, S.; Thulstrup, P.W.; Kjaergaard, H.G.; Fron, E.; Hofkens, J.; Yang, S.W.; Vosch, T. Design aspects of bright red emissive silver nanoclusters/DNA probes for microRNA detection. *ACS Nano* **2012**, *6*, 8803–8814.

47. De Heer, W.A. The physics of simple metal clusters: Experimental aspects and simple models. *Rev. Mod. Phys.* **1993**, *65*, 611–676.

48. Walter, M.; Akola, J.; Lopez-Acevedo, O.; Jadzinsky, P.D.; Calero, G.; Ackerson, C.J.; Whetten, R.L.; Grönbeck, H.; Häkkinen, H. A unified view of ligand-protected gold clusters as superatom complexes. *Proc. Natl. Acad. Sci. USA* **2008**, *105*, 9157–9162.

49. Link, S.; Mohamed, M.B.; El-Sayed, M.A. Simulation of the optical absorption spectra of gold nanorods as a function of their aspect ratio and the effect of the medium dielectric constant. *J. Phys. Chem. B* **1999**, *103*, 3073–3077.

50. Yan, J.; Gao, S. Plasmon resonances in linear atomic chains: Free-electron behavior and anisotropic screening of d electrons. *Phys. Rev. B* **2008**, *78*, 235413.

51. Baishya, K.; Idrobo, J.; Öğüt, S.; Yang, M.; Jackson, K.; Jellinek, J. Optical absorption spectra of intermediate-size silver clusters from first principles. *Phys. Rev. B* **2008**, *78*, 075439.

52. Desireddy, A.; Conn, B.E.; Guo, J.; Yoon, B.; Barnett, R.N.; Monahan, B.M.; Kirschbaum, K.; Griffith, W.P.; Whetten, R.L.; Landman, U.; *et al.* Ultrastable silver nanoparticles. *Nature* **2013**, *501*, 399–402.

53. Guidez, E.B.; Aikens, C.M. Theoretical analysis of the optical excitation spectra of silver and gold nanowires. *Nanoscale* **2012**, *4*, 4190–4198.

54. Yu, Y.; Luo, Z.; Chevrier, D.M.; Leong, D.T.; Zhang, P.; Jiang, D.; Xie, J. Identification of a highly luminescent $Au_{22}(SG)_{18}$ nanocluster. *J. Am. Chem. Soc.* **2014**, *22*, 5–8.

55. Oemrawsingh, S.S.R.; Markešević, N.; Gwinn, E.G.; Eliel, E.R.; Bouwmeester, D. Spectral properties of individual DNA-hosted silver nanoclusters at low temperatures. *J. Phys. Chem. C* **2012**, *116*, 25568–25575.

56. Lermé, J.; Lyon, D.; Kastler, A. Size evolution of the surface plasmon resonance damping in silver nanoparticles : Confinement and dielectric effects. *J. Phys. Chem. C* **2011**, *115*, 14098–14110.

57. Weick, G.; Molina, R.; Weinmann, D.; Jalabert, R. Lifetime of the first and second collective excitations in metallic nanoparticles. *Phys. Rev. B* **2005**, *72*, 115410.

58. Markešević, N.; Oemrawsingh, S.S.R.; Schultz, D.; Gwinn, E.G.; Bouwmeester, D. Polarization resolved measurements of individual DNA-stabilized silver clusters. *Adv. Opt. Mater.* **2014**, *2*, 765–770.

59. Walter, M.; Whetten, R.L.; Ha, H.; Gro, H. On the Structure of Thiolate-Protected Au 25. *J. Am. Chem. Soc.* **2008**, *130*, 3756–3757.

60. Sharma, J.; Rocha, R.C.; Phipps, M.L.; Yeh, H.C.; Balatsky, K.A.; Vu, D.M.; Shreve, A.P.; Werner, J.H.; Martinez, J.S. A DNA-templated fluorescent silver nanocluster with enhanced stability. *Nanoscale* **2012**, *4*, 4107–4110.

61. Lan, G.Y.; Chen, W.Y.; Chang, H.T. One-pot synthesis of fluorescent oligonucleotide Ag nanoclusters for specific and sensitive detection of DNA. *Biosens. Bioelectron.* **2011**, *26*, 2431–2435.

62. Dolamic, I.; Knoppe, S.; Dass, A.; Bürgi, T. First enantioseparation and circular dichroism spectra of Au_{38} clusters protected by achiral ligands. *Nat. Commun.* **2012**, *3*, 798.

63. Petty, J.T.; Sergev, O.O.; Nicholson, D.A.; Goodwin, P.M.; Giri, B.; McMullan, D.R. A silver cluster-DNA equilibrium. *Anal. Chem.* **2013**, *85*, 9868–9876.

64. Lan, G.Y.; Chen, W.Y.; Chang, H.T. Characterization and application to the detection of single-stranded DNA binding protein of fluorescent DNA-templated copper/silver nanoclusters. *Analyst* **2011**, *136*, 3623–3628.

65. Son, I.; Shek, Y.L.; Dubins, D.N.; Chalikian, T.V. Hydration changes accompanying helix-to-coil DNA transitions. *J. Am. Chem. Soc.* **2014**, *136*, 4040–4047.

66. Spink, C.H.; Garbett, N.; Chaires, J.B. Enthalpies of DNA melting in the presence of osmolytes. *Biophys. Chem.* **2007**, *126*, 176–185.

67. Chen, W.-Y.; Lan, G.-Y.; Chang, H.-T. Use of fluorescent DNA-templated gold/silver nanoclusters for the detection of sulfide ions. *Anal. Chem.* 2011, *83*, 9450–9455.

68. Berti, L.; Burley, G.A. Nucleic acid and nucleotide-mediated synthesis of inorganic nanoparticles. *Nat. Nanotechnol.* **2008**, *3*, 81–87.

69. Wang, S.X.; Meng, X.M.; Das, A.; Li, T.; Song, Y.B.; Cao, T.T.; Zhu, X.Y.; Zhu, M.Z.; Jin, R.C. A 200-fold quantum yield boost in the photoluminescence of silver-doped Ag(x)Au(25 − x) nanoclusters: The 13th silver atom matters. *Angew. Chem. Int. Ed.* **2014**, *53*, 2376–2380.

70. Copp, S.M.; Bogdanov, P.; Debord, M.; Singh, A.; Gwinn, E. Base motif recognition and design of DNA templates for fluorescent silver clusters by machine learning. *Adv. Mater.* **2014**, *26*, 5839–5845.

71. Cortes, C.; Vapnik, V. Support vector networks. *Mach. Learn.* **1995**, *20*, 273–297.

72. Bradford, J.R.; Westhead, D.R. Improved prediction of protein-protein binding sites using a support vector machines approach. *Bioinformatics* **2005**, *21*, 1487–1494.

73. Brown, M.P.S.; Grundy, W.N.; Lin, D.; Cristianini, N.; Sugnet, C.W.; Furey, T.S.; Ares, M.; Haussler, D. Knowledge-based analysis of microarray gene expression by using support vector machines. *Proc. Natl. Acad. Sci. USA* **2000**, *97*, 262–267.

74. Vens, C.; Rosso, M.N.; Danchin, E.G.J. Identifying discriminative classification-based motifs in biological sequences. *Bioinformatics* **2011**, *27*, 1231–1238.

75. Walczak, S.; Morishita, K.; Ahmed, M.; Liu, J. Towards understanding of poly-guanine activated fluorescent silver nanoclusters. *Nanotechnology* **2014**, *25*, 155501–155510.

76. Morishita, K.; MacLean, J.L.; Liu, B.; Jiang, H.; Liu, J. Correlation of photobleaching, oxidation and metal induced fluorescence quenching of DNA-templated silver nanoclusters. *Nanoscale* **2013**, *5*, 2840–2848.

77. Ramazanov, R.; Kononov, A. Excitation spectra argue for threadlike shape of DNA-stabilized silver fluorescent clusters. *J. Phys. Chem. C* **2013**, *117*, 18681–18687.

78. Guo, R.; Murray, R.W.; Hill, C.; Carolina, N. Substituent effects on redox potentials and optical gap energies of molecule-like $Au_{38}(SPhX)_{24}$ nanoparticles. *J. Am. Chem. Soc.* **2005**, *38*, 12140–12143.

79. Yin, P.; Hariadi, R.F.; Sahu, S.; Choi, H.M. T.; Park, S.H.; Labean, T.H.; Reif, J.H. Programming DNA tube circumferences. *Science* **2008**, *321*, 824–826.

80. Lee, Y.-J.; Schade, N.B.; Sun, L.; Fan, J.A.; Bae, D.R.; Mariscal, M.M.; Lee, G.; Capasso, F.; Sacanna, S.; Manoharan, V.N.; *et al.* Ultrasmooth, highly spherical monocrystalline gold particles for precision plasmonics. *ACS Nano* **2013**, *7*, 11064–11070.

81. Guticrez, M.; Henglein, A. Formation of colloidal silver by "push-pull" reduction of Ag^{+}. *J. Phys. Chem.* **1993**, *97*, 11368–11370.

82. Schultz, D.; Copp, S.M.; Markešević, N.; Gardner, K.; Oemrawsingh, S.S.R.; Bouwmeester, D.; Gwinn, E. Dual-color nanoscale assemblies of structurally stable, few-atom silver clusters, as reported by fluorescence resonance energy transfer. *ACS Nano* **2013**, *7*, 9798–9807.

83. O'Neill, P.R.; Young, K.; Schiffels, D.; Fygenson, D.K. Few-atom fluorescent silver clusters assemble at programmed sites on DNA nanotubes. *Nano Lett.* **2012**, *12*, 5464–5469.

84. Pilo-Pais, M.; Goldberg, S.; Samano, E.; Labean, T.H.; Finkelstein, G. Connecting the nanodots: Programmable nanofabrication of fused metal shapes on DNA templates. *Nano Lett.* **2011**, *11*, 3489–3492.

85. Orbach, R.; Guo, W.W.; Wang, F.; Lioubashevski, O.; Wilner, I. Self-assembly of luminescent Ag nanocluster-functionalized nanowires. *Langmuir* **2013**, *29*, 13066–13071.

86. Guo, W.; Orbach, R.; Mironi-Harpaz, I.; Seliktar, D.; Wilner, I. Fluorescent hydrogels composed of nucleic acid-stabilized silver nanoclusters. *Small* **2013**, *9*, 3748–3752.

87. Nilius, N.; Wallis, T.M.; Ho, W. Development of one-dimensional band structure in artificial gold chains. *Science* **2002**, *297*, 1853–1856.

Cysteine-Functionalized Chitosan Magnetic Nano-Based Particles for the Recovery of Light and Heavy Rare Earth Metals: Uptake Kinetics and Sorption Isotherms

Ahmed A. Galhoum [1,2,†,*], **Mohammad G. Mafhouz** [2,†], **Sayed T. Abdel-Rehem** [3],
Nabawia A. Gomaa [2,†], **Asem A. Atia** [4,†], **Thierry Vincent** [2] and **Eric Guibal** [2,†,*]

[1] Ecole des mines d'Alès, Centre des Matériaux des Mines d'Alès, 6 avenue de Clavières,
F-30319 Alès cedex, France

[2] Nuclear Materials Authority, P.O. Box 530, El-Maadi, Cairo, Egypt;
E-Mails: mahfouznma@yahoo.com (M.G.M.); g_nabawia@hotmail.com (N.A.G.);
thierry.vincent@mines-ales.fr (T.V.)

[3] Chemistry Department, Faculty of Science, Ain Shams University, P.O. Box 11566 Ain Shams,
Egypt; E-Mail: sayedth@sci.asu.edu.eg

[4] Chemistry Department, Faculty of Science, Menoufia University, P.O. Box 32511 Shebin El-Kom,
Egypt; E-Mail: asemali2010@yahoo.com

[†] These authors contributed equally to this work.

[*] Authors to whom correspondence should be addressed;
E-Mails: galhoum_nma@yahoo.com (A.A.G.); eric.guibal@mines-ales.fr (E.G.)

Academic Editor: Jorge Pérez-Juste

Abstract: Cysteine-functionalized chitosan magnetic nano-based particles were synthesized for the sorption of light and heavy rare earth (RE) metal ions (La(III), Nd(III) and Yb(III)). The structural, surface, and magnetic properties of nano-sized sorbent were investigated by elemental analysis, FTIR, XRD, TEM and VSM (vibrating sample magnetometry). Experimental data show that the pseudo second-order rate equation fits the kinetic profiles well, while sorption isotherms are described by the Langmuir model. Thermodynamic constants ($\Delta G°$, $\Delta H°$) demonstrate the spontaneous and endothermic nature of sorption. Yb(III) (heavy RE) was selectively sorbed while light RE metal ions La(III) and Nd(III) were concentrated/enriched in the solution. Cationic species RE(III) in aqueous solution can be adsorbed by the combination of chelating and anion-exchange mechanisms. The sorbent can be efficiently regenerated using acidified thiourea.

Keywords: cysteine-grafting; rare earth metals; magnetic chitosan nanocomposites; sorption isotherms; uptake kinetics; thermodynamics

---

1. Introduction

The recovery of heavy metals from dilute aqueous systems requires the development of new technologies for their concentration and separation [1]. Flocculation, coagulation, adsorption, ion exchange, membrane filtration, electrodeposition and chemical precipitation are the most conventional processes for the treatment of metal-bearing effluents. However, these techniques generally face economic or environmental constraints that make them ineffective for removing toxic or strategic metal ions at low or trace levels from aqueous waste streams. Adsorption is one of the most important physicochemical processes that has proved to be effective for metal recovery from dilute effluents [2].

Chitosan is a naturally abundant and biodegradable polysaccharide obtained by partial alkaline deacetylation of chitin: commercial chitosan is thus a copolymer of glucosamine and N-acetyl-D-glucosamine linked together by β $(1\rightarrow4)$ glycosidic bonds [3,4]. Most of its advantages are related to the wide availability of this renewable resource and its easy derivatization: it is readily chemically modified and can be physically conditioned under different forms [5]. In addition, this material is generally more hydrophilic than synthetic materials, such as polystyrene-divinylbenzene, polyethylene and polyurethane, which are commonly used as support for chelating and ion-exchange resins. Chitosan bearing amino groups can bind metal cations at near neutral pH by complexation/chelation on a free electronic doublet of nitrogen and metal anions by ion-exchange/electrostatic attraction on protonated amino groups in acid solutions [6]. Chitosan is a promising starting material for manufacturing new chelating/ion exchange resins [7]. The solubility of chitosan in acid media is a critical issue to be addressed for stable application; it is generally necessary to cross-link the biopolymer (by chemical modification) for extending the use of the biopolymer for a broader range of use (especially in terms of pH characteristics). It is often cross-linked to confer better microbiological and mechanical or chemical resistance [8]. On the other hand, the cross-linking of chitosan may contribute to reducing its ability to bind metal ions: in the case of Ln(III) sorption, it was attributed to the modification of chelating groups [9]. Therefore, novel chitosan resins bearing additional chelating moieties have been developed using the cross-linked chitosan resin as a starting support material [10]. A huge number of chitosan derivatives have been developed for the sorption of metal ions. The grafting of new functional groups on the backbone of chitosan increases the density of the sorption site and may change the sorption sites and the sorption mechanism, resulting in an increase of sorption capacity and a better selectivity for targeted metals [11].

Compared to conventional micron-sized supports used in separation process, nanometric sorbents possess quite good performance due to their high specific surface areas and the absence of internal diffusion resistance. However, the nano-adsorbents face serious drawbacks in terms of separation and recovery from treated solutions: they usually require sophisticated separation processes, such as very fine filtration or centrifugation. Magnetic nano-adsorbents offer interesting alternative for phase separation using an external magnetic field for the recovery of spent sorbents [12]. An additional advantage of these magnetic materials concerns their use in hazardous conditions, such as encountered in irradiated zones. Magnetic particles are usually composed of a magnetic core to ensure a strong magnetic response and a polymeric shell to provide favorable functional groups for sorption applications [13–17]. Alternatively, the magnetic core can be directly decorated with reactive groups through spacer arms [18]. In addition, iron (or other metal) oxide (or hydroxide) particles can bind metal ions [19,20], including rare earth elements (REEs) [21].

The REEs are the group of 17 chemical elements, including (a) two transition metal elements, scandium and yttrium; and (b) the 15 lanthanides (with atomic numbers 57–71). Rare earth elements are important in photo-electronic and metallurgical industries, as well as in nuclear energy programs. The demand for rare earths and their alloys as structural materials, fluxes and radiation detectors, diluents of plutonium, *etc.*, in nuclear technology is steadily increasing [22]. The designation "rare earths" refers to the elements of the periodic table known as "lanthanides", further divided as a function of their atomic number into the "cerium group "(or "light" RE elements: La, Ce, Pr, Nd, Pm, Sm, Eu, Gd), and the "yttrium group" (or "heavy" RE elements: Y, Tb, Dy, Ho, Er, Tm, Yb, Lu) [23,24]. The challenge and the specificity of these elements are associated with the fact that (a) they do not naturally occur in their metallic form; and (b) they are difficult to separate from each other due to their very similar physicochemical properties (very close electronic configurations). Their trivalent state is the more stable form of RE ions in aqueous solution. Each lanthanide element has in its electronic configuration an inner shell with electrons in the $4f^n$ orbital shielded by an outer shell composed of electrons in orbitals $5s^2$, $5p^6$, $5d^{1-10}$, and $6s^2$. The differences among lanthanides are caused by the electrostatic effect associated with the increase of the shielded nuclear charge through the electron partial supply of the 4f orbital (which results in lanthanide contraction of the atomic and ionic radius along lanthanide series) [25]. This contraction is responsible for the low differences in their chemical properties, which allow lanthanide separation by fractionating methods. Traditionally, rare earth elements with similar chemical behaviors are first separated into a group before being further separated and purified. RE trivalent ions (Pearson hard acids: high oxidation state ions, species with low electronegativity and small size) tend to readily react with oxygen, nitrogen, sulfur, and phosphorus atoms (Pearson hard bases: electron donors, with high electronegativity and low polarizability). According to the theory of hard and soft acids and bases (HSAB) defined by Pearson, metal ions (depending on their hardness) will have a preference for complexing with ligands that have more or less electronegative donor atoms [26]. It is important to establish the affinity differences among selected elements to propose a process for lanthanide separation and purification through the sorption process [27]. One of the promising methods is the use of chelating or coordinating resins with covalently bound functional groups containing one or more donor atoms that are capable of directly forming complexes with metal ions. These polymers can also be used for a specific separation of one or more metal ions from solutions with different chemical environments [28]. In chelating resins, the functional group atoms that are most frequently used are nitrogen (e.g., N present in amines, azo groups, amides, nitriles), oxygen (e.g., O present in carboxylic, hydroxyl, phenolic, ether, carbonyl, phosphoryl groups) and sulfur (e.g., S present in thiols, thiocarbamates, thioethers). Usually, the anchored molecules contain nitrogen, oxygen or sulfur atoms, or a combination of them, acting as the basic centers that complex cations and allow selective extraction [29]. Several chelating ligands such as catechol, iminodiacetic acid, iminodimethyl-phosphonic acid, phenylarsonic acid, or serine [9], Ethylenediaminetetraacetic acid (EDTA) and/or diethylene triamine pentaacetic acid (DTPA) [10,30] and amino acids moieties (glycine, valine, leucine, and serine) [5] were used to functionalize crosslinked chitosan for the sorption of lanthanide metal ions.

In the present work, nano-magnetic particles were prepared using chitosan as the encapsulating material for embedding Fe_3O_4 nanoparticles (as the core material): the magnetic particles were *in situ* synthesized in the biopolymer. The composite material was chemically modified and functionalized through successive treatments with epichlorohydrin and cysteine (bringing the reactive functional groups). The structural, surface, and magnetic characteristics were investigated by elemental analysis, FTIR spectrometry, XRD and TEM analysis. The magnetic properties were measured using a vibrating-sample magnetometer (VSM). The sorption properties were investigated in batch tests on three different lanthanide ions: (a) "light" La(III) and Nd(III); and (b) "heavy" Yb(III). The sorption efficiency was evaluated through the influence of pH, sorption isotherms, and uptake kinetics. Thermodynamic parameters were also determined before investigating the regeneration of metal-loaded sorbent.

2. Results and Discussion

2.1. Synthesis of Sorbent Particles

A simple one-pot *in situ* co-precipitation method was used to synthesize magnetic chitosan nanoparticles. Chitosan precipitates in alkaline conditions simultaneously to the synthesis of magnetic iron particles (the reaction between Fe(II) and Fe(III) under alkaline conditions and under heating), resulting in the formation of composite chitosan-magnetic nano-based particles. The dropwise addition of NaOH leads to the formation of nanometric particles of a chitosan-magnetite composite [2].

Chitosan-magnetite particles are chemically modified to prevent their dissolution in acidic media; however, aldehyde crosslinking may result in the loss of sorption capacity, because some amine groups are involved in the crosslinking reaction [8,31], so epichlorohydrin (or chloromethyloxirane) had been used as the crosslinking agent. Indeed, the crosslinking mono-functional agent is used to form covalent bonds with the carbon atoms linked to the hydroxyl groups of chitosan (associated with the rupturing of the epoxide ring and the release of a chlorine atom) [32]. Figure 1 shows the route for the synthesis of cysteine-functionalized chitosan magnetic nano-based particles.

2.2. Characterization of Sorbents

The C, H, N, and S contents in the cross-linked chitosan-magnetite material were 14.2%, 2.5%, 1.7% and 0%, respectively, while in the cysteine-grafted sorbent their values increased to 19.8%, 3.9%, 3.1% and 2.3%, respectively. The increases of carbon, hydrogen, nitrogen and sulfur contents clearly show the successful grafting of the cysteine moiety onto the cross-linked chitosan magnetite (through the epichlorohydrin spacer arm).

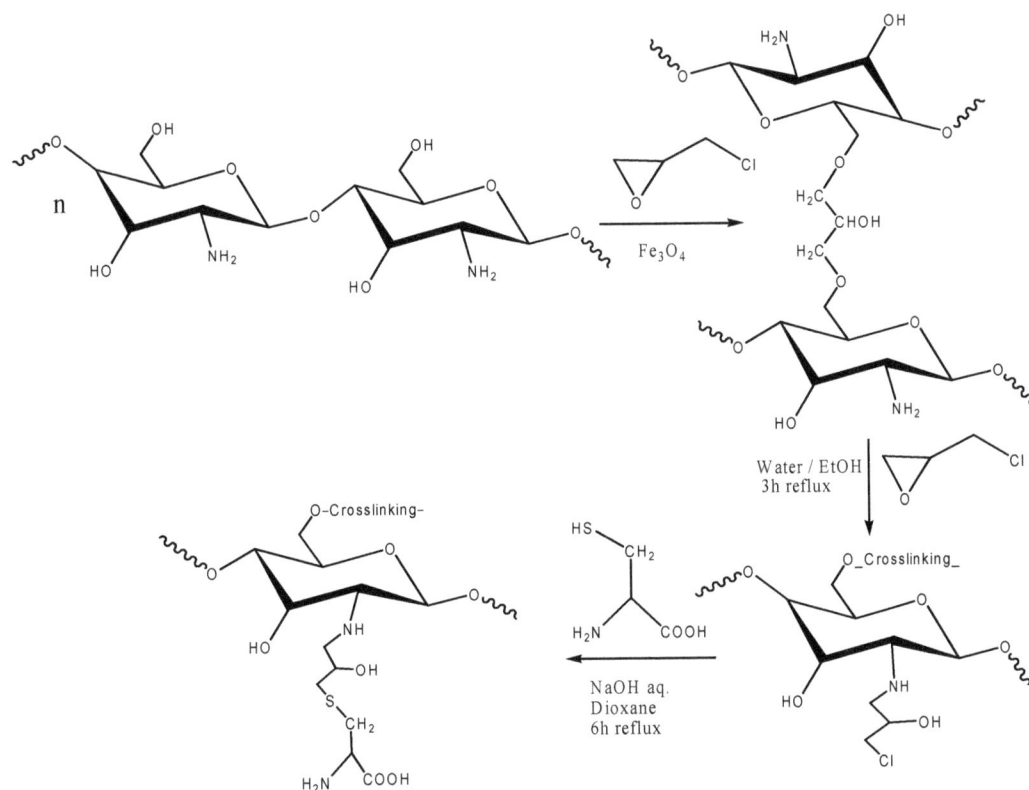

Figure 1. Scheme for the synthesis of cysteine-functionalized chitosan magnetic nano-based particles.

To confirm the existence of the surface coating, the products obtained at each reaction step were characterized by FTIR spectra, as shown in Figure 2. The band at 568 cm^{-1} is assigned to the Fe–O stretching vibration of Fe$_3$O$_4$ [1,2]. A characteristic strong and broad band appeared at around 3399 cm^{-1}, corresponding to the stretching vibration of the –OH group, the extension vibration of the N–H group and inter-hydrogen bonds of polysaccharides in the chitosan-magnetite spectra. The characteristic peaks of primary amine –NH$_2$ appear at 3399 and 1613 cm^{-1}. The bands at 1463 and 1364 cm^{-1} can be attributed to the C–O–C stretching and –OH bending vibrations, respectively. The absorption band at 893 cm^{-1} corresponds to the β-D-glucose unit [3]. The absorption bands around 1320 and 1065 cm^{-1} correspond to the stretching vibration of the primary –OH group and the secondary –OH group, respectively. However, the absorption intensities of –NH$_2$ and –OH group (for the cross-linked material) are obviously lower than those on the chitosan-magnetite spectrum: the cross-linking reaction between chitosan and epichlorohydrin involves these two functional groups [3].

The introduction of spacer arms on the cross-linked chitosan is confirmed by the appearance of a new band at 792 cm^{-1} that can be attributed to –CH$_2$–Cl stretching vibration (in comparison with the magnetic chitosan material) [33]. An additional band appears at 1631 cm^{-1}; this band is characteristic of the (–COO$^-$) carboxylate group vibration of the cysteine moiety [5]. In addition, the intensity of the band at 1387 cm^{-1} increases in the spectrum of the cysteine-type material; this is correlated with the introduction of additional amine groups. The values of amine group concentration of the cysteine-sorbent were found to be 3.53 mmol·g^{-1}; this is about 24% more than in the spacer-arm-grafted cross-linked material.

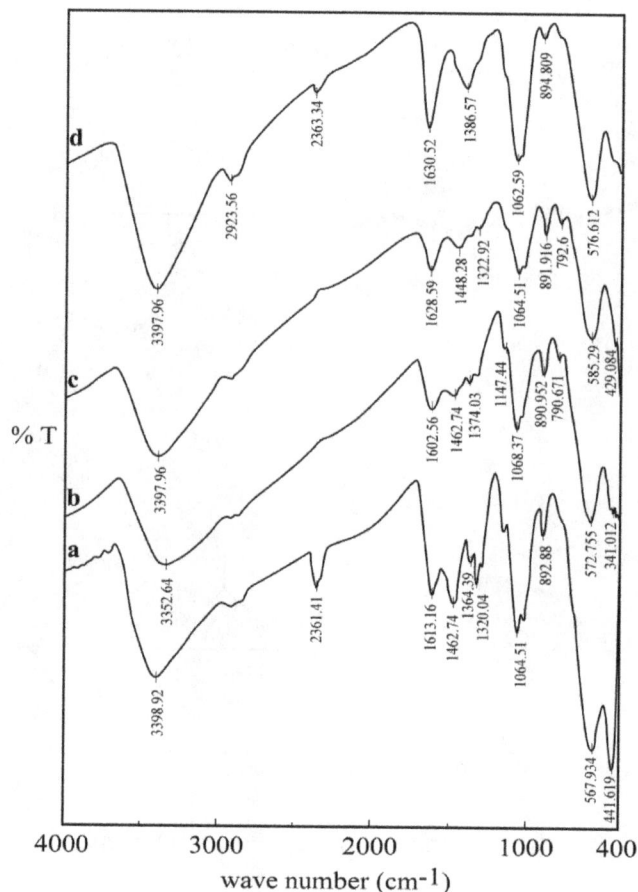

Figure 2. FTIR spectra of (**a**) chitosan-magnetite nanoparticles; (**b**) after cross-linking; (**c**) cross-linked chitosan magnetite with the spacer arm; and (**d**) cysteine-adsorbent.

The XRD pattern of cysteine-functionalized chitosan-magnetic nano-based particles is shown in Figure 3, together with the assignments of the different peaks representative of Fe_3O_4: (111), (220), (311), (400), (422), (511), (440) and (622). These peaks are consistent with the database in the JCPDS file (No. 65-3107) [1]. This confirms the existence of iron oxide particles (Fe_3O_4) with a spinel structure, which has magnetic properties and can be used for the magnetic separation and recovery of loaded particles. The half width at half maximum was used for calculating the size of nano-based particles through the Debye-Scherrer Equation [34]:

$$D = k \, \lambda/\beta \cos \theta \qquad \qquad (1)$$

where D is the average diameter of nanoparticles, λ is the wavelength of X-ray radiation (1.5418 Å), θ is the diffraction angle, $k = 0.9$ (shape parameter) and β is the full width at half maximum of the X-ray diffraction peaks. The crystal size has been found to be close to 13.5 nm (using index (311)).

The TEM image of the sorbent (Figure 4) shows that (a) the particles have a spherical morphology, and; (b) they are distributed in several classes of particles: 7–10 nm for the smallest and 20–25 nm for the largest. The sorbent particles are characterized by a partial aggregation that led to an average diameter of 150–250 nm. This aggregation may be attributed to the dipole-dipole magnetic attraction of nanoparticles. In addition, TEM also showed different contrasts on the photographs of chitosan-Fe_3O_4 composite particles: the dark areas can be attributed to the crystalline Fe_3O_4 core, while the bright or clear zones are associated with the chitosan coating. BET-analysis shows specific surface area close to

43 $m^2 \cdot g^{-1}$: this means 20–30 times the value usually reported for chitosan flakes: this is consistent with the values obtained for natural and synthetic magnetites [35]. This value is much smaller than the levels expected for nanoparticles: this means that even if some iron magnetic nanoparticles are not completely covered by the chitosan-based material this fraction of exposed iron particles is negligible compared to the coated particles. This is consistent with the TEM observation that shows the iron core coated by a thin layer of chitosan.

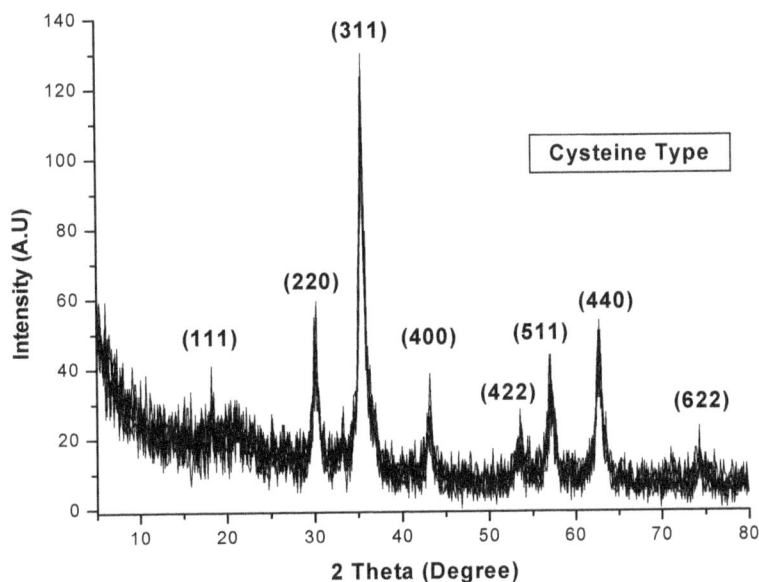

Figure 3. Powder X-ray diffraction (XRD) pattern of cysteine-sorbent nanoparticles.

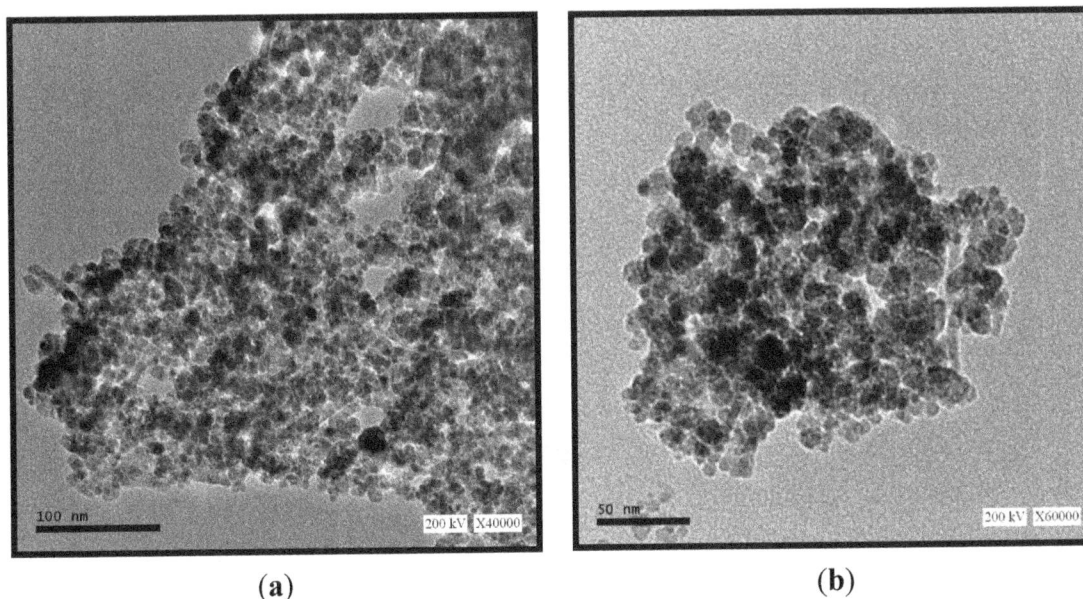

Figure 4. TEM micrographs (the scale bars are (**a**) 100 and (**b**) 50 nm, respectively).

Figure 5 shows the typical magnetization loop (hysteresis loop) for cysteine-functionalized chitosan magnetic nano-based particles. There is negligible remanence and coercivity; the chitosan-Fe_3O_4 nanoparticles can be described as super-paramagnetic nano-based particles.

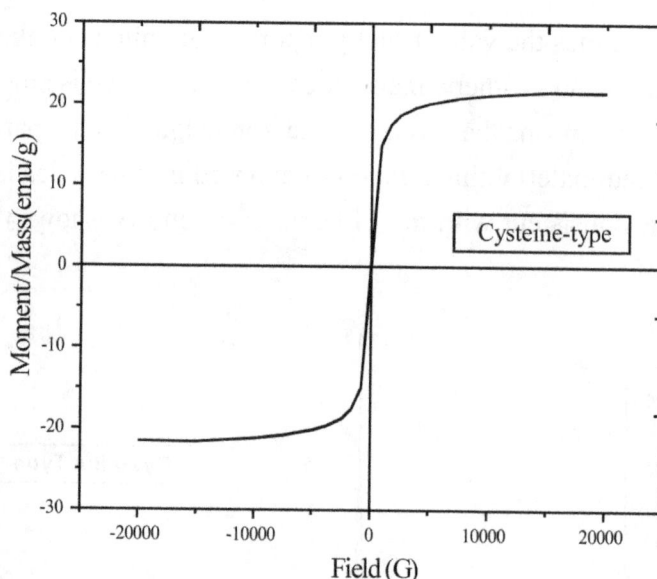

Figure 5. Magnetization curve of cysteine-functionalized chitosan magnetic nano-based particles.

This is consistent with the conclusions already reported by Kim *et al.* [36]: for nanometric materials below a critical size, the particles become individual single magnetic domains with remarkable superparamagnetic properties (*i.e.*, a large constant magnetic moment, a behavior similar to giant paramagnetic atoms with a fast response to the applied magnetic fields and negligible remanence and coercivity). The saturation magnetization was calculated to be about 21.51 emu·g^{-1} (21.51 A·m^2·kg^{-1}). This value is smaller than the levels reached by Chen and Wang [37] with a magnetic-chitosan composite (*i.e.*, around 35–40 emu·g^{-1}), but a little higher than the values found in the literature (*i.e.*, 12–20 emu·g^{-1}) for similar materials [12,13,38,39], and even more compared to systems prepared by coating of pre-formed magnetite particles with chitosan-based materials (*i.e.*, 6–13 emu·g^{-1}) [17,40]. These values are significantly lower than the values obtained with pure Fe_3O_4 magnetic particles (*i.e.*, 50–70 emu·g^{-1}); the decrease in the fraction of magnetic material, the diamagnetic contribution of the grafted copolymer layer and the crystalline disorder at the surface of the particles induced by the polymer layer may explain this reduction [13,17,40]. Weight loss on pristine chitosan and cysteine-based sorbent was determined at different temperatures (*i.e.*, 110 °C for water elimination, 400 °C, 600 °C and 800 °C) to follow the thermal degradation of organic material and evaluate the actual fraction of magnetite inorganic material (not shown): the fraction of magnetite was close to 49% in the final product; this is consistent with the expected values on the basis of the fraction of chitosan and magnetite introduced in the reactive media during sorbent synthesis. This may partially explain the substantial decrease of the saturation magnetization. In any case, the magnetic properties of synthesized hybrid materials make the sorbent easily recoverable with the help of an external magnetic field. This may be very helpful for solid/phase separation and/or handling the material in hazardous environments.

2.3. Sorption Properties

2.3.1. Sorption as a Function of pH

Hereafter, the magnetic sorbent (*i.e.*, cysteine-type sorbent) has been carried out for the sorption of several metal cations (*i.e.*, La(III), Nd(III) and Yb(III)) from dilute sulfate solutions. It is well known that the sorption efficiency of sorbents can be affected by a variety of parameters: the pH is one of the most important parameters, especially for sorbents having acid-base properties (ion exchange or proton exchange characteristics).

The initial pH of the aqueous solution was varied between 1.0 and 7.0, controlled with either 0.5 M H_2SO_4 or 0.5 M NaOH. At a pH higher than 7.0, precipitation of Nd(III) and Yb(III) ions as $M(OH)_3$ may spontaneously occur, making the interpretation of the sorption for metal concentrations higher than 100 mg·L^{-1} difficult. On the other hand, the inorganic magnetic material may also dissolve at a pH below 1.5. Figure 6 shows that sorption capacities increase with the increase in pH from 1.0 to 7.0, and there is a drastic increase at a pH of 5.0, while for higher pH values, the sorption capacity tends to stabilize. Figure 6 also shows pH variation during metal sorption. In the range pH 1–2, the final pH remained constant. When the initial pH was in the range 3–4, the equilibrium pH strongly increased up to 5–6. For initial values in the range of 6–7, the pH tended to stabilize around pH 6–6.5. Actually, the material has a kind of buffering effect around pH 5.5–6.5 when the initial value of the pH was set in the range of 3–7. This is probably due to the acid-base properties of chitosan (the pK_a of which depends on the degree of acetylation and varies between 6.3 and 6.8 for common chitosan samples [41]). For further experiments the pH was set to five to avoid any misinterpretation of the sorption performance that could be associated with metal hydrolysis or precipitation and to profit from both the optimum sorption performance and pH stabilization (pH variation of less than one pH unit).

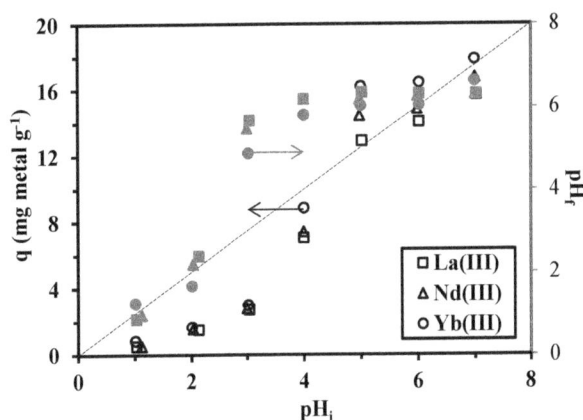

Figure 6. Effect of pH on the sorption of La(III), Nd(III) and Yb(III) ions using cysteine-functionalized chitosan magnetic nano-based particles (C_i = 100 mg·L^{-1}; T = 300 K; t = 4 h; m = 0.05 g; V = 100 mL).

In strong acidic solutions, both carboxyl groups and amino groups of the sorbent are protonated, resulting in a positively-charged surface for the sorbent. Therefore, the sorption capacities for La(III), Nd(III) and Yb(III) ions dramatically drop in low pH conditions; the sorption capacities are negligible at a pH close to 1.0 (Figure 6). At a low pH value, the coordinating atoms in the sorbent are partially protonated, as are the charged metal(III) species; this leads to repulsive electrostatic forces that limit the sorption of the metal on the cysteine-type material. However, this behavior is interesting since it allows predicting the possibility for the metal to be desorbed from the loaded-sorbent and the sorbent to be recycled using acid solutions (providing the acid-base stability of the sorbent is respected). As the pH increases, the protonated amine and carboxylic acid groups would gradually deprotonate. Therefore, the surface charge on the sorbent turns negative, which significantly enhances the electrostatic interaction between the sorbent and positively-charged metal ions [38].

Based on Figure 6, the sorption capacities (under selected experimental conditions) are found close to 16.2, 14.6 and 12.9 mg·g^{-1} for Yb(III), Nd(III) and La(III), respectively. In molar units, the sorption capacities' range is close to 0.1 mmol·RE·g^{-1}; this means that the three REs are equally sorbed by the material and that the differences in the ranking of the REs (between light REs, La(III) and Nd(III), and heavy RE, such as Yb(III)) are not sufficient for separating the metal ions.

The sorption of REs is clearly pH dependent and the cationic species can be sorbed through a chelating mechanism rather than an anion-exchange mechanism. This can probably be attributed to the presence of a free lone pair of electrons on nitrogen or sulfur that was suitable for coordination with metal ions to give the corresponding resin-metal complex. In addition, REE(III) ions can form chelates with the primary amino group (–NH$_2$) and carboxyl group (–COOH), due to the limited steric hindrance. Moreover amino-based chelating resins may have ionic interaction properties through protonated amines (in acid solutions). On the other hand, sulfur is quite efficient for coordinating with metal ions [29], in a broad range of pH values.

2.3.2. Uptake Kinetics

Sorption kinetics is another fundamental and significant issue for the evaluation of the potential of the sorbent for metal recovery. Figure 7 (displaying the plots of q_t *versus* t) shows that, regardless of the metal, the sorption equilibrium is achieved within 4 h. The kinetic profiles have been analyzed by various models, such as the pseudo-first order rate equation (PFORE), the pseudo-second order rate equation (PSORE), the Elovich equation and the intraparticle diffusion kinetic models [42].

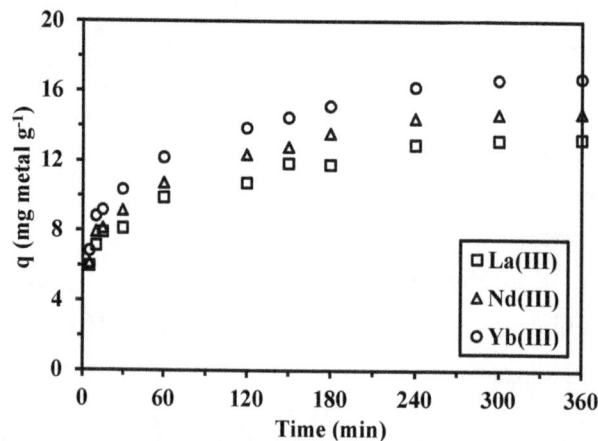

Figure 7. Effect of contact time on the adsorption of La(III), Nd(III) and Yb(III) ions (C_i = 100 mg·L^{-1}; T = 300 K; pH = 5; m = 0.05 g; V = 100 mL).

The linear forms of the PFORE and PSORE models are given in Equations (2) and (3), respectively:

$$\log (q_e - q_t) = \log q_e - \frac{k_1}{2.303}t \tag{2}$$

$$\frac{t}{q_t} = \frac{1}{k_2 q_e^2} + \frac{1}{q_e}t \tag{3}$$

where q_e and q_t (mg·g^{-1}) are the sorption capacities at equilibrium and time t (min), respectively. k_1 (min^{-1}) and k_2 (g·mg^{-1}·min^{-1}) are the rate constants of PFORE and PSORE, respectively.

The Elovich model is generally associated with chemisorption [43]. It was initially developed describing the kinetics of the chemical sorption of gases. However, recently, the Elovich equation was extensively used for modeling liquid phase sorption according to Equation (4):

$$q_t = A_E + B_E \ln t = 1/\beta \ln(\alpha\beta) + 1/\beta \ln t \qquad (4)$$

where q_t is the amount of metal ion sorbed on the sorbents (mg·g^{-1}) at time t (min), A_E (mg·g^{-1}) and B_E (mg·g^{-1}) are the Elovich constants, related to α (the initial sorption rate) and β (the function of surface coverage and activation energy).

The resistance to intraparticle diffusion is also an important step in the control of kinetics; especially for systems involving poorly porous materials, or large molecules. Several complex equations, such as the Crank equation, have been proposed for approaching diffusion models (derived from the Fick equation). In a first approximation, this equation was simplified with Equation (5) [43]:

$$q_t = k_{int} \cdot t^{0.5} + c \qquad (5)$$

where q_t (mg·g^{-1}) is the amount of metal ions adsorbed at time t (min), and k_{int} (mg·g^{-1}·min$^{-0.5}$) is the intraparticle diffusion constant.

The experimental data have been fitted by the aforementioned kinetic models: the parameters are all listed in Tables 1 and 2. Based on the analysis of the correlation coefficients for the linear forms of the different kinetics models (Table 1), PSORE best fits the kinetics profiles for the sorption of La(III), Nd(III) and Yb(III) ions onto cysteine-functionalized chitosan magnetic nano-based particles. The fitting results of the pseudo-second order model are shown in Figure S1: solid lines fits the experimental data well. This means that the rate limiting step for sorption is probably the chemical adsorption rate that involves the valence forces through the sharing or exchange of electrons (*i.e.*, complexation, coordination and chelation). The metal binding within the first 30 min of contact is associated with physical adsorption, which is supposed to occur rapidly: this step represents about 62% of the total sorption. Thereafter, strong chemical interactions take place involving chemical bonding for charge neutralization, coordination and chelation [44]. Under selected experimental conditions (taking into account the metal concentration and sorbent dosage), the equilibrium is reached within 4 h: this contact time was selected for further equilibrium studies.

Table 1. Kinetics parameters of the pseudo-first order rate equation (PFORE) and the pseudo-second order rate equation (PSORE) for the sorption of La(III), Nd(III) and Yb(III) metal ions using cysteine-based magnetic-chitosan nanoparticles (*T*: 27 °C; pH: 5).

Metal ion	$q_{eq., exp.}$ (mg·g^{-1})	PFORE			PSORE		
		$k_1 \times 10^2$ (min^{-1})	q_e (mg/g)	R^2	$k_2 \times 10^3$ (g·mg^{-1}·min^{-1})	q_e (mg/g)	R^2
La(III)	13.3	1.04	6.37	0.960	4.2	13.68	0.995
Nd(III)	14.8	1.15	7.75	0.984	3.7	15.27	0.996
Yb(III)	16.8	1.13	8.64	0.990	3.1	17.33	0.996

Table 2. Kinetics parameters of the Elovich and intraparticle diffusion models for the sorption of La(III), Nd(III) and Yb(III) metal ions using cysteine-based magnetic-chitosan nanoparticles (T: 27 °C; pH: 5).

Metal ion	Intraparticle diffusion			Elovich equation		
	c, mg·g^{-1}	$k_{int.}$, mg·g^{-1}·min$^{-0.5}$	R^2	B_E	A_T	R^2
La(III)	4.26	0.56	0.832	1.75	2.90	0.979
Nd(III)	4.49	0.64	0.850	2.07	2.64	0.988
Yb(III)	5.02	0.86	0.855	2.36	2.86	0.991

Besides, the correlation coefficients for the pseudo-first order model and for the Elovich model were also higher than 0.92, but lower than the values obtained with PSORE, as shown in Tables 1 and 2. This means that the pseudo-second-order model can be applied to predict the sorption kinetics and that the chemisorption is contributing to the kinetic control. Figure S2 shows that the relationship between q_t and $t^{0.5}$ is not linear (poor correlation coefficient): the intra-particle diffusion is not supposed to be the only rate-controlling step [45]. Actually, most sorption reactions take place through a multistep mechanism comprising [42]: (i) external film diffusion; (ii) intra-particle diffusion; and (iii) interaction between the sorbate and active site. In the present case, the resistance to film diffusion was significantly reduced by the appropriate agitation speed. The nanometric size of the sorbent limits the resistance to intraparticle diffusion: metal ions can readily diffuse to all reactive sites. Hence, the proper chemical reaction is supposed to play the major role in the control of the uptake kinetics.

2.3.3. Sorption Isotherms

Sorption isotherms are fundamental for understanding the interaction mechanisms and establishing both the maximum sorption capacities and the affinity of the sorbent for target solutes [46,47]. Different equations have been designed to model the distribution of the metal ions between liquid and solid phases (sorption isotherms), including the Langmuir, Freundlich, Temkin and Dubinin-Radushkevich (D-R) equations [46–50]. Though the fit of experimental data by a given equation does not necessarily means that the mechanisms associated with the model are verified, this may help in interpreting the sorption mechanism. Figure 8 shows the sorption isotherms for La(III), Nd(III) and Yb(III) using the cysteine-based sorbent at different temperatures, while Tables 3 and 4 report the parameters of the different models.

All of the curves, regardless of the temperature and target metal, are characterized by the progressive increase of the sorption capacity followed by the saturation of the sorbent (plateau) that occurs for residual concentrations higher than 140–150 mg·metal·L^{-1}. The asymptotic shape of the isotherm is consistent with the Langmuir equation (Equation (6)), contrary to the Freundlich equation (which supposes an exponential trend associated with the power function, Equation (7)):

$$q = \frac{q_m b C_{eq}}{1 + b C_{eq}} \tag{6}$$

where q is the amount of metal ions sorbed at equilibrium (mg·g^{-1}), C_{eq} is the equilibrium metal ion concentration in the aqueous solution (mg·L^{-1}), q_m is the maximum sorption capacity of the sorbent (mg·g^{-1}), and b is the Langmuir sorption constant (L·mg^{-1}), respectively.

$$q = k_F \, C_{eq}^{\frac{1}{n}} \tag{7}$$

where k_F ($L^{1/n} \cdot mg^{1-1/n} \cdot g^{-1}$) and n are the Freundlich constants.

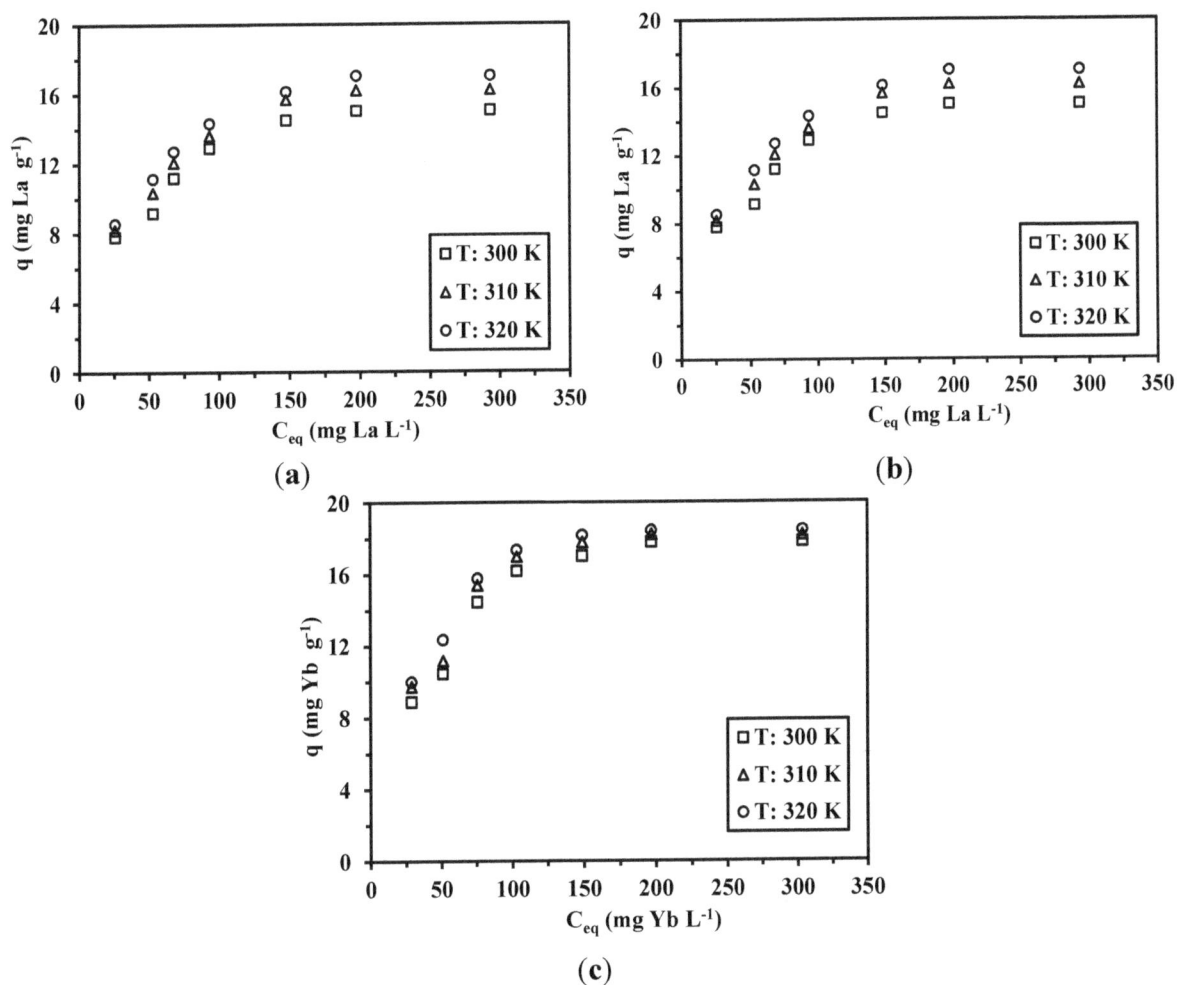

(a)

(b)

(c)

Figure 8. Adsorption isotherms for La(III), Nd(III) and Yb(III) ions at different temperatures. ($t = 4$ h; $T = 300$ K; pH = 5; $m = 0.05$ g; $V = 20$ mL).

Table 3. Parameters of the Langmuir and Freundlich equations for the sorption of La(III), Nd(III) and Yb(III) metal ions using cysteine-based magnetic-chitosan nanoparticles at different temperatures.

Metal ion	T (K)	$q_{m.,exp.}$ (mg·g^{-1})	Langmuir model			Freundlich model		
			$q_{m.,calc.}$	K_L (L·mg^{-1})	R^2	n	K_F, (mg·g^{-1})	R^2
La(III)	300	15.0	16.0	0.071	0.997	0.20	5.27	0.921
	310	16.2	17.1	0.079	0.998	0.20	5.84	0.953
	320	17.0	17.9	0.086	0.998	0.19	6.46	0.971
Nd(III)	300	15.3	16.0	0.106	0.999	0.19	5.92	0.934
	310	16.3	16.8	0.131	0.999	0.180	6.64	0.918
	320	17.1	17.6	0.144	0.999	0.17	7.17	0.920
Yb(III)	300	17.8	18.7	0.086	0.998	0.21	6.00	0.890
	310	18.1	18.9	0.120	0.998	0.18	7.40	0.871
	320	18.4	19.3	0.154	0.999	0.16	8.20	0.913

Table 4. Parameters of the Dubinin–Radushkevich (D–R) and Temkin equations for the sorption of La(III), Nd(III) and Yb(III) metal ions using cysteine-based magnetic-chitosan nanoparticles at different temperatures.

Metal ion	T (K)	q_m (mg·g^{-1})	D–R Isotherm model			Temkin model		
			$K_{ad} \times 10^4$ (mol^2·kJ^{-2})	E_{DR} (KJ·mol^{-1})	R^2	A_T (L·mg^{-1})	B_T (J·mol^{-1})	R^2
La(III)	300	15.0	0.8	0.079	0.994	3.94	2.25	0.914
	310	16.2	0.6	0.091	0.939	5.00	2.35	0.945
	320	17.0	0.5	0.100	0.944	7.18	2.34	0.962
Nd(III)	300	15.3	1.0	0.071	0.968	6.67	2.19	0.948
	310	16.3	0.7	0.085	0.973	10.45	2.20	0.950
	320	17.1	0.5	0.100	0.981	13.62	2.23	0.955
Yb(III)	300	17.8	0.6	0.091	0.997	3.36	2.79	0.902
	310	18.1	0.4	0.112	0.997	10.41	2.44	0.876
	320	18.4	0.2	0.158	0.994	20.52	2.28	0.918

The Langmuir model supposes that: (a) all of the reactive sites are energetically equivalent (the same affinity for the target solute); (b) the sorption occurs as a monolayer; and (c) there are no interactions between the sorbed molecules. On the other hand, the empiric Freundlich model is generally associated with a heterogeneous distribution (and energy) of sorption sites with possible interactions between sorbed molecules, as well as a possible multilayer accumulation. The correlation coefficients (R^2) of the linear form (obtained by plotting C_{eq}/q vs. C_{eq}, see Figure S3) for the Langmuir model were higher (and closer to one) than the values obtained for the Freundlich equation (Table 3). The values of Langmuir constants (i.e., q_m and b) are reported in Table 3. Both q_m and b increased with increasing temperature: this means that metal binding is endothermic, regardless of the REE.

The dimensionless constant, R_L, which reflects the essential characteristic of the Langmuir model, can be obtained from the constant b, according to Equation (8) [47]:

$$R_L = \frac{1}{1 + bC_0} \tag{8}$$

where C_0 is the initial concentration of the metal ion. The calculated values of the dimensionless factor R_L lie between 0.04 and 0.35 for La(III), 0.02 and 0.26 for Nd(III), and in the range of 0.02–0.29 for Yb(III), regardless of the concentration and the temperature. All of these R_L values for the sorbent are smaller than 1.0: this is the first indication that the cysteine-based chitosan magnetic nano-based particles have a "favorable" sorption profile for La(III), Nd(III) and Yb(III).

The D–R isotherm model is usually employed for discriminating the nature of the sorption processes between physical and chemical mechanisms. The D–R isotherm equation is expressed by Equation (9a,b) [47,50]:

$$\ln q_{eq} = \ln q_{DR} - K_{DR}\,\varepsilon^2 \tag{9a}$$

$$\text{with } \varepsilon = RT \ln\left(1 + \frac{1}{C_{eq}}\right) \tag{9b}$$

where q_{DR} is the theoretical saturation capacity, and ε is the Polanyi, K_{DR} is related to the mean free adsorption energy per molecule of sorbate, E_{DR} (kJ/mol). E_{DR} provides information about the chemical or physical sorption, and can be determined according to Equation (9c):

$$E_{DR} = (2K_{DR})^{-1/2} \tag{9c}$$

Meanwhile, from the D–R isotherm, the plot of $\ln q_{eq}$ *versus* ε^2 gives a straight line with the slope K_{DR} and the intercept $\ln q_{DR}$, as shown in Figure S4. The constants of the D–R isotherm (q_{DR}, and K_{DR}) are reported in Table 4.

The mean adsorption energy (E_{DR}) corresponds to the transfer of the free energy of one mole of solute from infinity (in solution) to the surface of the sorbent. It is commonly accepted that physical sorption corresponds to mean adsorption energy below 8 kJ·mol^{-1}, while chemical sorption requires mean adsorption energy greater than 8 kJ·mol^{-1} [47,50]. Regardless of the metal, the E_{DR} for the cysteine-based magnetic-chitosan nanoparticles are in the range of 1–8 kJ·mol^{-1}, the sorbent is supposed to bind La(III), Nd(III) and Yb(III) through a physisorption mechanism. In addition, the positive value of E_{DR} confirms that the sorption process is endothermic, consistent with the improvement of the sorption capacities with temperature.

The Temkin model was also used to fit the experimental data. This model assumes that the free energy of sorption is a function of the surface coverage [47,48]. The isotherm is described by Equation (10).

$$q = B_T \ln C_{eq} + B_T \ln A_T = \frac{RT}{\Delta Q} \ln C_{eq} + \frac{RT}{\Delta Q} \ln A_T \tag{10}$$

where A_T is the Temkin equilibrium constant (L·mg^{-1}), B_T is a constant related to the surface heterogeneity of the adsorbent, ΔQ ($-\Delta H$) is the variation of sorption energy (kJ·mol^{-1}), T is the temperature (K) and R is the ideal gas constant (8.314 J·mol^{-1}·K^{-1}). Thus, the constants can be obtained from the slope and intercept of a straight-line plot of q_{eq} *versus* $\ln C_e$. The constants of the Temkin model are listed in Table 4. The greater the constant A_T, the higher the affinity of the sorbent for the solute; cysteine-based chitosan magnetic nano-based particles have a decreasing affinity according to Yb(III) > Nd(III) > La(III) based on A_T values. It is noteworthy that A_T increases as temperature does. The modeling of the uptake kinetics with the pseudo-second order rate equation (PSORE) suggested that chemisorption was the rate limiting step. On the other hand, the discussion of sorption isotherms led to the conclusion that metal binding occurs through physisorption (the value of the mean adsorption energies in the range 1–8 kJ·mol^{-1}). This apparent contradiction in the interpretation of sorption mechanism can be explained by a dual mechanism that involves both ionic interaction (chemisorption) and electrostatic (physisorption) between the sorbent and RE metal ions. At a low metal concentration the solute is physically sorbed as a monolayer at the surface of the sorbent, while close to the saturation metal ions bind to the sorbent by coordination [47].

2.3.4. Effect of Temperature—Thermodynamic Studies

The effect of temperature on the sorption of La(III), Nd(III) and Yb(III) by cysteine-functionalized chitosan magnetic nano-based particles was investigated at $T = 300$, 310 and 320 K, respectively. The values of q_m are plotted *vs.* temperature in Figure S5. The amounts of metal ions sorbed gradually increased with increasing the temperature, as expected by the endothermic characteristics of the isotherms (values of K_L and E_{DR}, see Table 5). Wang *et al.* [45] attributed the improvement of affinity with temperature to the increase in the Lewis acid-base interaction between metal ions and the ligands.

Table 5. Thermodynamic parameters for the sorption of La(III), Nd(III) and Yb(III) metal ions using cysteine-based magnetic-chitosan nano-based particles at different temperatures.

Metal ion	T (K)	$\Delta H°$ (kJ·mol^{-1})	$\Delta S°$ (kJ·mol^{-1})	$\Delta G°$ (kJ·mol^{-1})	$T\Delta S°$ (kJ·mol^{-1})	R^2
La(III)	300			−22.94	30.8	
	310	7.85	0.103	−23.97	31.8	0.998
	320			−24.99	32.8	
Nd(III)	300			−24.01	36.1	
	310	12.04	0.120	−25.29	37.3	0.957
	320			−26.49	38.5	
Yb(III)	300			−23.99	47.4	
	310	23.43	0.158	−25.57	49.0	0.995
	320			−27.15	50.6	

These experimental data (obtained at different temperatures) were used for calculating the thermodynamic parameters, such as standard Gibbs free energy change ($\Delta G°$), enthalpy change ($\Delta H°$) and entropy change ($\Delta S°$). The thermodynamic parameters were then calculated from the van't Hoff equation, and derived from Equations (11) and (12):

$$\ln b = \frac{-\Delta H°}{RT} + \frac{\Delta S°}{R} \tag{11}$$

$$\Delta G° = \Delta H° - T\Delta S° \tag{12}$$

where b is the equilibrium constant, which can be obtained from Langmuir isotherm at different temperatures, and T is the absolute temperature (K). The values of enthalpy change ($\Delta H°$) and entropy change ($\Delta S°$) were obtained by plotting $\ln b$ *vs.* $1/T$ (Figure 9). The values of the thermodynamic constants ($\Delta H°$, $\Delta S°$ and $\Delta G°$) are reported on Table 5.

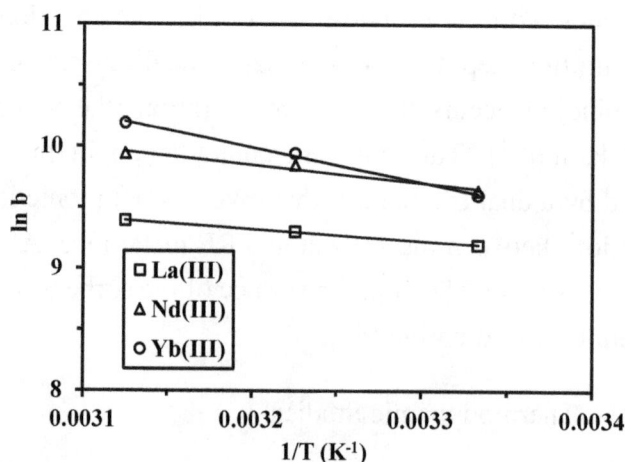

Figure 9. Van't Hoff plots of $\ln K_L$ against $1/T$, for sorption of metal (III) ions.

Table 5 confirms that the sorption reaction of La(III), Nd(III) and Yb(III) ions on cysteine-functionalized chitosan magnetic nano-based particles is endothermic (a positive value of $\Delta H°$). The sorption enthalpy increases according to the sequence: La(III) < Nd(III) < Yb(III). The free energy is systematically negative: regardless of the metal, the value of free energy ranged between −23 and −27 kJ·mol^{-1}. The decrease in the value of $\Delta G°$ with the increase in temperature shows that the reaction is enhanced at high temperature (confirmation of the endothermic nature of the sorption mechanism). The positive value of $\Delta S°$ may be related to the release of the water of hydration during the sorption process causing the increase in the randomness of the system. Metal ions in aqueous media are hydrated. When the ions get sorbed on the sorbent surface, water molecules previously bonded to metal ions are progressively released and dispersed in the solution; this results in an increase in the entropy of the system. A substantial difference is observed in entropy values between Nd(III) and Yb(III), on one side, and La(III), on the other side. Table 5 shows that $|\Delta H°|$ is systematically lower than $|T\Delta S°|$ in the studied temperature range. This means that the sorption process is dominated by entropic rather than enthalpy changes [51].

Table 6 reports the comparison of La(III), Nd(III) and Yb(III) sorption capacities for different sorbents. The direct comparison of sorption performance is difficult, since the experimental conditions may substantially differ; however, the data show that the cysteine-functionalized chitosan magnetic nano-based particles are lower than the levels reached with synthetic resins; they are comparable to the sorption capacities obtained with some biosorbents: the presence of a significant fraction of magnetite (about 50%) in the sorbent may explain the relative decrease of the sorption capacity. Indeed, the amount of reactive functional groups is halved.

Table 6. Comparison of the La(III), Nd(III) and Yb(III) sorption properties of different sorbents with cysteine-functionalized chitosan magnetic nano-based particles (CFCMNBP).

Sorbent	Metal	pH range	q_m (mg·g^{-1})	Reference
Tangerine peel	La(III)	5	155	[52]
Sargassum sp.	La(III)	5	74–100	[53]
Magnetic alginate beads	La(III)	2.8	97	[16]
Platanus orientalis leaf	La(III)	4	29	[54]
Lewatit resins	La(III)	1.5–5	100–120	[55]
Chelating ion-exchange resin	La(III)	HCl/HNO$_3$	188–240	[56]
Functionalized Amberlite XAD-4 resin	La(III)	6.1	49	[57]
CFCMNBP	La(III)	6	17	This work
EDTA:DTPA functionalized chitosan	Nd(III)	3–6	55	[10]
Phosphonic acid functionalized silica microspheres	Nd(III)	2.8	45	[58]
Ion imprinted polymer particles	Nd(III)	7.5	33	[59]
Phosphorus functionalized adsorbent	Nd(III)	6	160	[60]
Yeast cells	Nd(III)	1.5	10–12	[61]
Mordenite containing tuff	Nd(III)	5.5–6.5	13	[62]
CFCMNBP	Nd(III)	6	17	This work
Sargassum	Yb(III)	5	160	[27]
Turbinaria conoides	Yb(III)	4.9	34	[63]
Pseudomonas aeruginosa	Yb(III)	6–7	56	[64]
Imino-diacetic acid resin	Yb(III)	5.1	187	[65]
Gel-type weak acid resin	Yb(III)	5.5	266	[66]
CFCMNBP	Yb(III)	6	18	This work

2.3.5. Metal Desorption and Resin Recycling

Generally, the sorbed metal ions can be desorbed using a selective eluent with the objective of concentrating/enriching the metal concentration. However, acids such as HCl and HNO3, may react with Fe3O4, which is the core component of magnetic chitosan nanoparticles. Ethylenediamine tetraacetic acid (EDTA) and thiourea are known as very strong chelating agents for many metal ions and are supposed to displace metal ions from reactive groups (based on the greater affinity of the metal for the ligands over reactive groups on sorbent particles).

Consequently, thiourea (0.5 M) acidified with a few drops of H2SO4 (0.2 M) was chosen as the eluent for metal ions, with a contact time of 1 h. The sorbent was tested for four successive sorption/desorption cycles: regardless of the metal, the sorption efficiencies and capacities slightly decreased along the cycles, but the variations remained quite low. In the worst case, the decrease in the sorption efficiency (compared to the first cycles) was less than 7% (Table 7). Though the mechanisms of desorption were not elucidated, it is supposed that metal is removed from the loaded sorbent through electrostatic and complexation mechanisms.

Table 7. Recycling of cysteine-based magnetic-chitosan nano-based particles for the sorption of La(III), Nd(III) and Yb(III) metal ions over four cycles.

Cycle	La(III)		Nd(III)		Yb(III)	
	q_e (mg·g^{-1})	Ads. (%)	q_e (mg·g^{-1})	Ads. (%)	q_e (mg·g^{-1})	Ads. (%)
Cycle I	12.9	100.0	14.4	100.0	16.2	100.0
Cycle II	12.7	98.2	14.2	98.1	15.4	95.2
Cycle III	12.6	97.5	14.1	97.5	15.3	94.3
Cycle IV	12.6	97.1	14.0	96.9	15.2	93.8

2.3.6. Sorption Selectivity

To investigate the selective sorption of Yb(III) ions from aqueous complex solutions, the sorption properties of cysteine-functionalized chitosan magnetic nano-based particles have been investigated using binary solutions containing equivalent concentrations of Nd(III) and Yb(III) (*i.e.*, nearly equimolar concentrations: $C_0 \approx 0.23$ mmol·Yb·L^{-1} (59.5 mg·Yb·L^{-1}) and $C_0 \approx 0.25$ mmol·Nd·L^{-1}) (53.8 mg·Nd·L^{-1}). The results showed that the amount of Yb(III) sorbed (*i.e.*, 11.6 mg·Yb·g^{-1}; *i.e.*, 0.067 mmol·Yb·g^{-1}) is much higher than the amount of Nd(III) ions (*i.e.*, 5.0 mg Nd·g^{-1}; *i.e.*, 0.035 mmol·Nd·g^{-1}). At equilibrium, the fractions of Yb(III) and Nd(III) on the sorbent reached 69.7% and 30.3%, respectively. This means that the sorbent has a greater affinity for the heavy RE metal ion, Yb(III), compared to the light RE metal ion, Nd(III). The sorbent is enriched in Yb(III), while Nd(III) concentrates in the aqueous phase. This is positive, but not sufficient for achieving perfect metal separation.

Another important issue, which was not investigated in this study, concerns the effect of the presence of anions, such as bicarbonate or sulfate. It is well known that the presence of ligands (for example, lactate) may influence the selective binding of REEs (this may contribute to increase the differences in the chemical behavior of REEs) [67]. The key parameter is the speciation of metal ions in the presence of ligands, especially when ion-exchange mechanisms are involved: the formation of anionic complexes may increase the sorption properties of protonated reactive groups for anionic complexes. For example the presence of high concentrations of carbonate or phosphate leads to the formation of anionic complexes [68]. In some cases the presence of carbonate is required for producing ternary REE-carbonate-surface complexes to improve REE recovery [69]. On the other hand, the presence of nitrate, chloride or sulfate anions (in three-times excess of anion compared to REE) did not show a significant impact of REE sorption when using *Pseudomonas aeruginosa* biomass [64]. Schijf and Marshall [70] reported the decrease of sorption for yttrium and REEs on hydrous ferric oxide when increasing the ionic strength of the solution: they observe that at high ionic strength the sorption is less influenced by the pH and that the sorption decreases due to enhanced deprotonation of the sorbent surface. Tang and Johannesson [71] commented on the effect of carbonate/bicarbonate (in relation to pH) on the sorption of lanthanide onto sand: at a pH lower than 7.3, inorganic ligands did not significantly compete with surface sites for metal binding, while at a higher pH, lanthanide complexation with carbonate (to form anionic species) leads to simultaneous sorption of free and complexed REE forms. From the literature, it appears that both the pH and the presence of anions (at a high relative concentration) may influence the speciation of REEs and, consequently, their affinity for sorption on the reactive groups at the surface of composite nanoparticles. This requires a specific study for the application in complex effluents.

3. Experimental Section

3.1. Materials

Chitosan (90.5% deacetylation) was supplied by Sigma-Aldrich (France). Cysteine was obtained from Sigma-Aldrich, and epichlorohydrin (>98%), 1,4-dioxane (99.9%) and ethanol were purchased from Fluka Chemika AG (Germany). Sodium hydroxide solution (30%) was supplied by Chem-Lab. NV and all other chemicals were Prolabo products and were used as received.

3.2. Rare Earth Solutions and Analytical Procedures

La_2O_3, $NdCl_3$ and $YbCl_3.6 H_2O$ were purchased from Sigma-Aldrich and were burned off at 900 °C for 3 h. Stock solutions of rare earth ions La(III), Nd(III) and Yb(III) were prepared by mineralizing the corresponding salts in concentrated sulfuric acid under heating, before diluting with demineralized water until a final concentration of 1000 $mg \cdot L^{-1}$. The working solutions were prepared by appropriate dilution of the stock solutions immediately prior to use. The metal concentrations in both initial and withdrawn samples were determined by an Inductively Coupled Plasma Atomic Emission Spectrometer (ICP-AES JY Activa M, Jobin-Yvon, Longjumeau, France).

3.3. Sorbent Synthesis and Characterization

3.3.1. Preparation of Cross-Linked Chitosan-Magnetite Nanocomposites

Chitosan-magnetite nanocomposites were prepared by chemical co-precipitation of Fe(II) and Fe(III) ions by NaOH in the presence of chitosan followed by treatment under hydrothermal conditions using a method derived from Namdeo and Bajpai [2]. Briefly, Chitosan (4 g) was dissolved in 200 mL (20%) acetic acid and mixed with $FeSO_4$ and $FeCl_3$ salts (added in a 1:2 molar ratio; *i.e.*, 6.62 g of $FeSO_4.7H_2O$ and 8.68 g of $FeCl_3$). The resulting solution was chemically precipitated at 40 °C by adding dropwise a

1 M NaOH solution under continuous stirring, at controlled pH (10–10.4). The suspension was heated at 90 °C for 1 h under stirring and finally recovered by decantation and magnetic separation. Then, a solution of 0.01 M epichlorohydrin containing 0.067 M sodium hydroxide was prepared (pH close to 10) and added to freshly prepared wet magnetite-chitosan nano-based particles in a ratio of 1:1. The mixture of chitosan-magnetite and epichlorohydrin was heated at 40–50 °C for 2 h under stirring [72]. Finally, the product (ii) was filtered and intensively washed with distilled water to remove any unreacted epichlorohydrin.

The amino acid moiety (cysteine) was introduced to the cross-linked chitosan magnetic material in two steps [5]. First, the cross-linked chitosan (ii) was suspended in a 150 mL ethanol/water mixture (1:1 *v/v*); then, epichlorohydrin (15 mL) was added to the suspension, and the mixture was refluxed for 4 h. After the reaction, the product (iii) was filtered and washed 3 times with ethanol and with ultrapure water to remove any residual reagent. Secondly, the washed product (iii) and cysteine (16 g) were suspended in dioxane (200 mL). The mixture was alkalinized to pH 9.5–10 using a 1 M NaOH solution; the mixture was refluxed for 6 h. After the reaction, the final product was filtered and washed 3 times with ethanol and with ultrapure water. Finally, the product was freeze-dried.

The amine content in the adsorbent was estimated using a volumetric method [73]: 30 mL of 0.05 M HCl solution were added to 0.1 g of the material and conditioned for 15 h on a shaker. The residual concentration of HCl was estimated through titration against 0.05 M NaOH solution using phenolphthalein as the indicator. The number of moles of HCl having interacted with amino groups and consequently the concentration of amino groups ($mmol \cdot g^{-1}$) was calculated from Equation (13):

$$\text{Concentration of amino group} = (M_1 - M_2) \times 30/0.1 \tag{13}$$

where M_1 and M_2 are the initial and final concentrations of HCl, respectively.

3.3.2. Characterization Methods

Elemental composition of the resin (*i.e.*, C, H, S and N contents) was determined using a Heraeus CHN-O-Rapid elemental analyzer (Hanau, Germany). Powder X-ray diffraction (XRD) patterns were obtained at room temperature using an X-ray diffractometer Philips model PW 3710/31 (Eindhoven, The Netherlands), using the Cu K_α radiation in the range of $2\theta = 10°–90°$. The size and morphology of sorbent particles were obtained using a Hitachi H-800 transmission electron microscope (Hitachi, Japan). The magnetic property was measured on a vibrating-sample magnetometer (VSM) (Lake Shore 730T, Westerville, OH, USA) at room temperature. Functional groups of sorbent were analyzed by Fourier transform infrared spectrometry using a Nicolet Nexus 870 FTIR spectrometer (Thermo Electron Scientific Instruments Corporation, Madison, WI, USA) in the wavelength range 400–4000 cm^{-1} (using the KBr pellet technique). The BET specific surface area was determined through nitrogen adsorption isotherms using a Quantachrome NOVA 3200 (Boynton Beach, FL, USA) analyzer. The magnetite content was determined by measurement of the weight losses at different successive ignition temperatures (*i.e.*, 110, 400, 600 and 800 °C, exposure for 1 h).

3.3.3. Sorption and Desorption Methods

Standard batch experiments were carried out by contact of 50 mg of sorbent (m) with 20 mL (V) aqueous metal ion solution (C_0: 100 mg·metal·L^{-1}; initial pH: 5) in a polypropylene centrifuge tube. The samples were agitated at 300 rpm (stirring speed) for 4 h (the temperature being set at 27 ± 1 °C). After magnetic separation, the residual metal ion concentration (C_{eq}, mg·meta·L^{-1}) in the aqueous phase was determined by ICP-AES, whilst the concentration of metal ions sorbed onto the sorbent (q_{eq}, mg·metal·g^{-1}) was obtained by the mass balance equation:

$$q_{eq} = (C_0 - C_{eq}) \times V/m \tag{14}$$

Other experiments were based on the same procedures (varying the relevant parameter) for investigating the effect of pH, the influence of equilibration time (uptake kinetics), and the impact of metal concentration (sorption isotherms). Isotherms were obtained by contact of 50 mg of sorbent with 20 mL of a pH 5 solution containing the metal ions at different initial concentrations (25, 50, 75, 100, 150, 200 and 300 mg·L^{-1}) under shaking (at a speed of 300 rpm) for 4 h. The experiments were performed at different temperatures (*i.e.*, 300, 310 and 320 ± 1 K).

The regeneration of the sorbent was tested by mixing the metal-loaded sorbent (sorbent dosage: 2.5 g·L^{-1}; C_0: 100 mg·metal·L^{-1}, pH: 5; agitation speed: 300 rpm; contact time: 4 h) with the eluent (0.5 M thiourea solution slightly acidified with a few drops of 0.2 M sulfuric acid solution; pH close to 3) for 1 h under agitation. The sorbent was recovered by magnetic separation, and the residual concentration was analyzed by ICP-AES while the sorbent was recycled (after intensive washing). Four sorption/desorption cycles were successively carried out using standard procedures: the sorption capacity and sorption yield at each stage were compared (the desorption yield was not determined).

4. Conclusions

Cysteine-functionalized chitosan magnetic nano-based particles have been synthesized, characterized and used as a sorbent with a superparamagnetic property. The sorption properties have been efficiently tested for the sorption (and possible separation) of light (La(III) and Nd(III)) and heavy (Yb(III)) RE metal ions from aqueous solutions. The Langmuir isotherm model provided the best fit for the sorption isotherms of these three metal ions. The maximum sorption capacities at pH 5 were found to be 17.0, 17.1 and 18.4 mg·g^{-1} for La(III), Nd(III) and Yb(III) ions at 320 K, respectively. Thermodynamic parameters ($\Delta G°$ and $\Delta H°$) indicate the spontaneous and endothermic nature of the sorption process, while the positive values of $\Delta S°$ indicate increased randomness due to metal sorption: the entropy of the system increases, probably due to the release of the water of hydration of metal ions after metal sorption. Finally, the sorbent can be regenerated with high efficiency by acidified thiourea as the eluent, and after three cycles, the adsorption capacities were not significantly reduced.

This sorbent showed promising properties; however, some parameters should be optimized. For example, the relative fraction of the magnetic core and chitosan layer should be varied in order to measure the impact of this parameter on the sorption and magnetization properties. Hence, increasing the modified chitosan content is expected to increase the sorption yield, but reduce the magnetization efficiency. The optimal formulation will be a compromise between the sorption and magnetization properties.

At this stage of the development of the process the feasibility of the synthesis of nano-based particles of chitosan has been demonstrated. The material can be readily modified by chemical grafting. The sorbent properties are not high enough, at this stage, for being competitive against more conventional systems. The magnetic nature of the sorbent particles is expected to make the handling and operating of the material in hazardous environments (such as radioactive environments) possible with minimized resistance to intraparticle diffusion (and enhanced kinetics; this means also the possibility of reducing the scaling up of the treatment unit by reducing required contact times).

Supplementary Materials

Supplementary materials can be accessed at: http://www.mdpi.com/2079-4991/5/1/154/s1.

Acknowledgments

This research was supported by the French Government through a fellowship granted by the French Embassy in Egypt (Institut Français d'Égypte). A special dedication is given in memory of Ahmed Donia.

Author Contributions

Ahmed A. Galhoum, as former PhD Student (degree obtained in January 2015) performed most of the sorption experiments under the supervision of Eric Guibal (who also prepared the manuscript) with the contribution of Thierry Vincent for sorption experiments and analysis. Nabawia A Gomaa and Mohammad G. Mafhouz were in charge of XRD and magnetism analysis and discussion. Sayed T. Abdel-Rehem and Asem A. Atia performed and discussed elemental and FTIR spectroscopy analysis. M.G. Mafhouz, S.T. Abdel-Rehem, N.A. Gomaa and A.A. Atia supervised the research work in Egypt (PhD project of A.A. Galhoum).

Conflicts of Interest

The authors declare no conflicts of interest.

References

1. Zhang, X.; Jiao, C.; Wang, J.; Liu, Q.; Li, R.; Yang, P.; Zhang, M. Removal of uranium(VI) from aqueous solutions by magnetic schiff base: Kinetic and thermodynamic investigation. *Chem. Eng. J.* **2012**, *198*, 412–419.

2. Namdeo, M.; Bajpai, S.K. Chitosan-magnetite nanocomposites (CMNs) as magnetic carrier particles for removal of Fe(III) from aqueous solutions. *Colloids Surf. A* **2008**, *320*, 161–168.

3. Wang, G.H.; Liu, J.S.; Wang, X.G.; Xie, Z.Y.; Deng, N.S. Adsorption of uranium (VI) from aqueous solution onto cross-linked chitosan. *J. Hazard. Mater.* **2009**, *168*, 1053–1058.

4. Hosoba, M.; Oshita, K.; Katarina, R.K.; Takayanagi, T.; Oshima, M.; Motomizu, S. Synthesis of novel chitosan resin possessing histidine moiety and its application to the determination of trace silver by ICP-AES coupled with triplet automated-pretreatment system. *Anal. Chim. Acta* **2009**, *639*, 51–56.

5. Oshita, K.; Takayanagi, T.; Oshima, M.; Motomizu, S. Adsorption behavior of cationic and anionic species on chitosan resins possessing amino acid moieties. *Anal. Sci.* **2007**, *23*, 1431–1434.

6. Guibal, E. Interactions of metal ions with chitosan-based sorbents: A review. *Sep. Purif. Technol.* **2004**, *38*, 43–74.

7. Gao, Y.H.; Oshita, K.; Lee, K.H.; Oshima, M.; Motomizu, S. Development of column-pretreatment chelating resins for matrix elimination/multi-element determination by inductively coupled plasma-mass spectrometry. *Analyst* **2002**, *127*, 1713–1719.

8. Wang, J.-S.; Peng, R.-T.; Yang, J.-H.; Liu, Y.-C.; Hu, X.-J. Preparation of ethylenediamine-modified magnetic chitosan complex for adsorption of uranyl ions. *Carbohydr. Polym.* **2011**, *84*, 1169–1175.

9. Oshita, K.; Sabarudin, A.; Takayanagi, T.; Oshima, M.; Motomizu, S. Adsorption behavior of uranium(VI) and other ionic species on cross-linked chitosan resins modified with chelating moieties. *Talanta* **2009**, *79*, 1031–1035.

10. Roosen, J.; Binnemans, K. Adsorption and chromatographic separation of rare earths with edta- and dtpa-functionalized chitosan biopolymers. *J. Mater. Chem. A* **2014**, *2*, 1530–1540.

11. Jayakumar, R.; Prabaharan, M.; Reis, R.L.; Mano, J.F. Graft copolymerized chitosan—Present status and applications. *Carbohydr. Polym.* **2005**, *62*, 142–158.

12. Zhou, L.; Xu, J.; Liang, X.; Liu, Z. Adsorption of platinum(IV) and palladium(II) from aqueous solution by magnetic cross-linking chitosan nanoparticles modified with ethylenediamine. *J. Hazard. Mater.* **2010**, *182*, 518–524.

13. Xue, X.; Wang, J.; Mei, L.; Wang, Z.; Qi, K.; Yang, B. Recognition and enrichment specificity of Fe_3O_4 magnetic nanoparticles surface modified by chitosan and *Staphylococcus aureus* enterotoxins a antiserum. *Colloids Surf. B* **2013**, *103*, 107–113.

14. Karaca, E.; Şatır, M.; Kazan, S.; Açıkgöz, M.; Öztürk, E.; Gürdağ, G.; Ulutaş, D. Synthesis, characterization and magnetic properties of Fe_3O_4 doped chitosan polymer. *J. Magn. Magn. Mater.* **2015**, *373*, 53–59.

15. Zhou, Z.; Lin, S.; Yue, T.; Lee, T.-C. Adsorption of food dyes from aqueous solution by glutaraldehyde cross-linked magnetic chitosan nanoparticles. *J. Food Eng.* **2014**, *126*, 133–141.

16. Wu, D.; Zhang, L.; Wang, L.; Zhu, B.; Fan, L. Adsorption of lanthanum by magnetic alginate-chitosan gel beads. *J. Chem. Technol. Biotechnol.* **2011**, *86*, 345–352.

17. Dodi, G.; Hritcu, D.; Lisa, G.; Popa, M.I. Core-shell magnetic chitosan particles functionalized by grafting: Synthesis and characterization. *Chem. Eng. J.* **2012**, *203*, 130–141.

18. Dupont, D.; Brullot, W.; Bloemen, M.; Verbiest, T.; Binnemans, K. Selective uptake of rare earths from aqueous solutions by edta-functionalized magnetic and nonmagnetic nanoparticles. *ACS Appl. Mater. Interface* **2014**, *6*, 4980–4988.

19. Cowan, C.E.; Zachara, J.M.; Resch, C.T. Cadmium adsorption on iron-oxides in the presence of alkaline-earth elements. *Environ. Sci. Technol.* **1991**, *25*, 437–446.

20. Trivedi, P.; Axe, L. Modeling Cd and Zn sorption to hydrous metal oxides. *Environ. Sci. Technol.* **2000**, *34*, 2215–2223.

21. Koeppenkastrop, D.; Decarlo, E.H. Uptake of rare-earth elements from solution by metal-oxides. *Environ. Sci. Technol.* **1993**, *27*, 1796–1802.

22. Zhou, J.; Duan, W.; Zhou, X.; Zhang, C. Application of annular centrifugal contactors in the extraction flowsheet for producing high purity yttrium. *Hydrometallurgy* **2007**, *85*, 154–162.

23. Greenwood, N.N.; Earnshaw, A. *Chemistry of the Elements*, 2nd ed.; Elsevier Butterworth-Heinemann: Oxford, UK, 1997; p. 1305.

24. Cotton, S. *Lanthanide and Actinide Chemistry*; John Wiley & Sons, Ltd.: Chichester, UK, 2006; p. 263.

25. Martins, T.S.; Isolani, P.C. Rare earths: Industrial and biological applications. *Quim. Nova* **2005**, *28*, 111–117.

26. Pearson, R.G. Acids and bases. *Science* **1966**, *151*, 172–177.

27. Diniz, V.; Volesky, B. Biosorption of La, Eu and Yb using Sargassum biomass. *Water Res.* **2005**, *39*, 239–247.

28. Donia, A.M.; Atia, A.A.; Daher, A.M.; Desouky, O.A.; Elshehy, E.A. Synthesis of amine/thiol magnetic resin and study of its interaction with Zr(IV) and Hf(IV) ions in their aqueous solutions. *J. Dispers. Sci. Technol.* **2011**, *32*, 634–641.

29. Filha, V.; Wanderley, A.F.; de Sousa, K.S.; Espinola, J.G.P.; da Fonseca, M.G.; Arakaki, T.; Arakaki, L.N.H. Thermodynamic properties of divalent cations complexed by ethylenesulfide immobilized on silica gel. *Colloids Surf. A* **2006**, *279*, 64–68.

30. Chethan, P.D.; Vishalakshi, B. Synthesis of ethylenediamine modified chitosan and evaluation for removal of divalent metal ions. *Carbohydr. Polym.* **2013**, *97*, 530–536.

31. Martinez, L.; Agnely, F.; Leclerc, B.; Siepmann, J.; Cotte, M.; Geiger, S.; Couarraze, G. Cross-linking of chitosan and chitosan/poly(ethylene oxide) beads: A theoretical treatment. *Eur. J. Pharm. Biopharm.* **2007**, *67*, 339–348.

32. Gonçalves, V.L.; Laranjeira, M.C.M.; Fávere, V.T.; Pedrosa, R.C. Effect of crosslinking agents on chitosan microspheres in controlled release of diclofenac sodium. *Polim. Cienc. Tecnol.* **2005**, *15*, 6–12.

33. Coates, J. Interpretation of Infrared Spectra, a Practical Approach. In *Encyclopedia of Analytical Chemistry*; Meyers, R.A., Ed.; John Wiley & Sons Ltd.: Chichester, UK, 2000; pp. 10815–10837.

34. Guinier, A.; Lorrain, P.; Lorrain, D.S.-M. *X-Ray Diffraction: In Crystals, Imperfect Crystals and Amorphous Bodies*; W.H. Freeman & Co.: San Francisco, CA, USA, 1963; p. 356.

35. Salazar-Camacho, C.; Villalobos, M.; Rivas-Sánchez, M.D.L.L.; Arenas-Alatorre, J.; Alcaraz-Cienfuegos, J.; Gutiérrez-Ruiz, M.E. Characterization and surface reactivity of natural and synthetic magnetites. *Chem. Geol.* **2013**, *347*, 233–245.

36. Kim, D.-H.; Nikles, D.E.; Brazel, C.S. Synthesis and characterization of multifunctional chitosan-MnFe$_2$O$_4$ nanoparticles for magnetic hyperthermia and drug delivery. *Materials* **2010**, *3*, 4051–4065.

37. Chen, Y.; Wang, J. Preparation and characterization of magnetic chitosan nanoparticles and its application for Cu(II) removal. *Chem. Eng. J.* **2011**, *168*, 286–292.

38. Yan, H.; Li, H.; Yang, H.; Li, A.; Cheng, R. Removal of various cationic dyes from aqueous solutions using a kind of fully biodegradable magnetic composite microsphere. *Chem. Eng. J.* **2013**, *223*, 402–411.

39. Reddy, D.H.K.; Lee, S.-M. Application of magnetic chitosan composites for the removal of toxic metal and dyes from aqueous solutions. *Adv. Colloid Interface Sci.* **2013**, *201*, 68–93.

40. Kyzas, G.Z.; Deliyanni, E.A. Mercury(II) removal with modified magnetic chitosan adsorbents. *Molecules* **2013**, *18*, 6193–6214.

41. Sorlier, P.; Denuzière, A.; Viton, C.; Domard, A. Relation between the degree of acetylation and the electrostatic properties of chitin and chitosan. *Biomacromolecules* **2001**, *2*, 765–772.

42. Qiu, H.; Lv, L.; Pan, B.; Zhang, Q.; Zhang, W.; Zhang, Q. Review: Critical review in adsorption kinetic models. *J. Zhejiang Univ. Sci. A* **2009**, *10*, 716–724.

43. Anagnostopoulos, V.A.; Symeopoulos, B.D. Sorption of europium by malt spent rootlets, a low cost biosorbent: Effect of pH, kinetics and equilibrium studies. *J. Radioanal. Nucl. Chem.* **2013**, *295*, 7–13.

44. Hu, X.-J.; Wang, J.-S.; Liu, Y.-G.; Li, X.; Zeng, G.-M.; Bao, Z.-L.; Zeng, X.-X.; Chen, A.-W.; Long, F. Adsorption of chromium (VI) by ethylenediamine-modified cross-linked magnetic chitosan resin: Isotherms, kinetics and thermodynamics. *J. Hazard. Mater.* **2011**, *185*, 306–314.

45. Wang, H.; Ma, L.; Cao, K.; Geng, J.; Liu, J.; Song, Q.; Yang, X.; Li, S. Selective solid-phase extraction of uranium by salicylideneimine-functionalized hydrothermal carbon. *J. Hazard. Mater.* **2012**, *229*, 321–330.

46. Langmuir, I. The adsorption of gases on plane surfaces of glass, mica and platinum. *J. Amer. Chem. Soc.* **1918**, *40*, 1361–1402.

47. Foo, K.Y.; Hameed, B.H. Insights into the modeling of adsorption isotherm systems. *Chem. Eng. J.* **2010**, *156*, 2–10.

48. Temkin, V.P. Kinetics of ammonia synthesis on promoted iron catalysts. *Acta Physiochim.* **1940**, *12*, 217–222.

49. Freundlich, H.M.F. Uber die adsorption in lasungen. *Z. Phys. Chem.* **1906**, *57*, 385–470. (In German)

50. Dubinin, M.M.; Zaverina, E.D.; Radushkevich, L.V. Sorption and structure of active carbons. I. Adsorption of organic vapors. *Zh. Fiz. Khim.* **1947**, *21*, 1351–1362.

51. Rahmati, A.; Ghaemi, A.; Samadfam, M. Kinetic and thermodynamic studies of uranium(VI) adsorption using amberlite ira-910 resin. *Ann. Nucl. Energy* **2012**, *39*, 42–48.

52. Torab-Mostaedi, M. Biosorption of lanthanum and cerium from aqueous solutions using tangerine (citrus reticulate) peel: Equilibrium, kinetic and thermodynamic studies. *Chem. Ind. Chem. Eng. Q.* **2013**, *19*, 79–88.

53. Palmieri, M.C.; Volesky, B.; Garcia, O. Biosorption of lanthanum using sargassum fluitans in batch system. *Hydrometallurgy* **2002**, *67*, 31–36.

54. Sert, Ş.; Kütahyali, C.; İnan, S.; Talip, Z.; Çetinkaya, B.; Eral, M. Biosorption of lanthanum and cerium from aqueous solutions by platanus orientalis leaf powder. *Hydrometallurgy* **2008**, *90*, 13–18.

55. Esma, B.; Omar, A.; Amine, D.M. Comparative study on lanthanum(III) sorption onto lewatit TP 207 and lewatit TP 260. *J. Radioanal. Nucl. Chem.* **2014**, *299*, 439–446.

56. Maheswari, M.A.; Subramanian, M.S. Selective enrichment of U(VI), Th(IV) and La(III) from high acidic streams using a new chelating ion-exchange polymeric matrix. *Talanta* **2004**, *64*, 202–209.

57. Dev, K.; Pathak, R.; Rao, G.N. Sorption behaviour of lanthanum(III), neodymium(III), terbium(III), thorium(IV) and uranium(VI) on Amberlite XAD-4 resin functionalized with bicine ligands. *Talanta* **1999**, *48*, 579–584.

58. Melnyk, I.V.; Goncharyk, V.P.; Kozhara, L.I.; Yurchenko, G.R.; Matkovsky, A.K.; Zub, Y.L.; Alonso, B. Sorption properties of porous spray-dried microspheres functionalized by phosphonic acid groups. *Microporous Mesoporous Mater.* **2012**, *153*, 171–177.

59. Krishna, P.G.; Gladis, J.M.; Rao, T.P.; Naidu, G.R. Selective recognition of neodymium(III) using ion imprinted polymer particles. *J. Mol. Recognit.* **2005**, *18*, 109–116.

60. Park, H.-J.; Tavlarides, L.L. Adsorption of neodymium(III) from aqueous solutions using a phosphorus functionalized adsorbent. *Ind. Eng. Chem. Res.* **2010**, *49*, 12567–12575.

61. Vlachou, A.; Symeopoulos, B.D.; Koutinas, A.A. A comparative study of neodymium sorption by yeast cells. *Radiochim. Acta* **2009**, *97*, 437–441.

62. Kozhevnikova, N.M.; Tsybikova, N.L. Sorption of neodymium(III) ions by natural mordenite-containing tuff. *Russ. J. Appl. Chem.* **2008**, *81*, 42–45.

63. Vijayaraghavan, K.; Sathishkumar, M.; Balasubramanian, R. Interaction of rare earth elements with a brown marine alga in multi-component solutions. *Desalination* **2011**, *265*, 54–59.

64. Texier, A.C.; Andres, Y.; le Cloirec, P. Selective biosorption of lanthanide (La, Eu, Yb) ions by pseudomonas aeruginosa. *Environ. Sci. Technol.* **1999**, *33*, 489–495.

65. Xiong, C.; Yao, C.; Wang, Y. Sorption behaviour and mechanism of ytterbium(III) on imino-diacetic acid resin. *Hydrometallurgy* **2006**, *82*, 190–194.

66. Zheng, Z.; Xiong, C. Adsorption behavior of ytterbium (III) on gel-type weak acid resin. *J. Rare Earths* **2011**, *29*, 407–412.

67. Konishi, Y.; Shimaoka, J.-I.; Asai, S. Sorption of rare-earth ions on biopolymer gel beads of alginic acid. *React. Funct. Polym.* **1998**, *36*, 197–206.

68. Johannesson, K.H.; Stetzenbach, K.J.; Hodge, V.F.; Lyons, W.B. Rare earth element complexation behavior in circumneutral pH groundwaters: Assessing the role of carbonate and phosphate ions. *Earth Planet. Sci. Lett.* **1996**, *139*, 305–319.

69. Piasecki, W.; Sverjensky, D.A. Speciation of adsorbed yttrium and rare earth elements on oxide surfaces. *Geochim. Cosmochim. Acta* **2008**, *72*, 3964–3979.

70. Schijf, J.; Marshall, K.S. Yree sorption on hydrous ferric oxide in 0.5 M NaCl solutions: A model extension. *Mar. Chem.* **2011**, *123*, 32–43.

71. Tang, J.W.; Johannesson, K.H. Adsorption of rare earth elements onto carrizo sand: Experimental investigations and modeling with surface complexation. *Geochim. Cosmochim. Acta* **2005**, *69*, 5247–5261.

72. Wan Ngah, W.S.; Endud, C.S.; Mayanar, R. Removal of copper(II) ions from aqueous solution onto chitosan and cross-linked chitosan beads. *React. Funct. Polym.* **2002**, *50*, 181–190.

73. Donia, A.M.; Atia, A.A.; Abouzayed, F.I. Preparation and characterization of nano-magnetic cellulose with fast kinetic properties towards the adsorption of some metal ions. *Chem. Eng. J.* **2012**, *191*, 22–30.

Permissions

All chapters in this book were first published in Nanomaterials, by MDPI; hereby published with permission under the Creative Commons Attribution License or equivalent. Every chapter published in this book has been scrutinized by our experts. Their significance has been extensively debated. The topics covered herein carry significant findings which will fuel the growth of the discipline. They may even be implemented as practical applications or may be referred to as a beginning point for another development.

The contributors of this book come from diverse backgrounds, making this book a truly international effort. This book will bring forth new frontiers with its revolutionizing research information and detailed analysis of the nascent developments around the world.

We would like to thank all the contributing authors for lending their expertise to make the book truly unique. They have played a crucial role in the development of this book. Without their invaluable contributions this book wouldn't have been possible. They have made vital efforts to compile up to date information on the varied aspects of this subject to make this book a valuable addition to the collection of many professionals and students.

This book was conceptualized with the vision of imparting up-to-date information and advanced data in this field. To ensure the same, a matchless editorial board was set up. Every individual on the board went through rigorous rounds of assessment to prove their worth. After which they invested a large part of their time researching and compiling the most relevant data for our readers.

The editorial board has been involved in producing this book since its inception. They have spent rigorous hours researching and exploring the diverse topics which have resulted in the successful publishing of this book. They have passed on their knowledge of decades through this book. To expedite this challenging task, the publisher supported the team at every step. A small team of assistant editors was also appointed to further simplify the editing procedure and attain best results for the readers.

Apart from the editorial board, the designing team has also invested a significant amount of their time in understanding the subject and creating the most relevant covers. They scrutinized every image to scout for the most suitable representation of the subject and create an appropriate cover for the book.

The publishing team has been an ardent support to the editorial, designing and production team. Their endless efforts to recruit the best for this project, has resulted in the accomplishment of this book. They are a veteran in the field of academics and their pool of knowledge is as vast as their experience in printing. Their expertise and guidance has proved useful at every step. Their uncompromising quality standards have made this book an exceptional effort. Their encouragement from time to time has been an inspiration for everyone.

The publisher and the editorial board hope that this book will prove to be a valuable piece of knowledge for researchers, students, practitioners and scholars across the globe.

List of Contributors

Ihab M. Obaidat
Department of Physics, United Arab Emirates University, Al-Ain 15551, United Arab Emirates

Bashar Issa
Department of Physics, United Arab Emirates University, Al-Ain 15551, United Arab Emirates

Yousef Haik
Department of Mechanical Engineering, United Arab Emirates University, Al-Ain 15555, United Arab Emirates
Center for Research Excellence in Nanobiosciences, University of North Carolina at Greensboro, Greensboro, NC 27412, USA

Eisa Kohan-Baghkheirati
Department of Plant Biology, Southern Illinois University Carbondale, Carbondale, IL 62901, USA
Department of Biology, Golestan University, Gorgan 49138-15739, Iran

Jane Geisler-Lee
Department of Plant Biology, Southern Illinois University Carbondale, Carbondale, IL 62901, USA

Taissia G. Popova
Center for Applied Proteomics and Molecular Medicine, Department of Molecular Microbiology, School of Systems Biology, George Mason University, Manassas, VA 20110, USA

Allison Teunis
Center for Applied Proteomics and Molecular Medicine, Department of Molecular Microbiology, School of Systems Biology, George Mason University, Manassas, VA 20110, USA

Ruben Magni
Center for Applied Proteomics and Molecular Medicine, Department of Molecular Microbiology, School of Systems Biology, George Mason University, Manassas, VA 20110, USA

Alessandra Luchini
Center for Applied Proteomics and Molecular Medicine, Department of Molecular Microbiology, School of Systems Biology, George Mason University, Manassas, VA 20110, USA

Virginia Espina
Center for Applied Proteomics and Molecular Medicine, Department of Molecular Microbiology, School of Systems Biology, George Mason University, Manassas, VA 20110, USA

Lance A. Liotta
Center for Applied Proteomics and Molecular Medicine, Department of Molecular Microbiology, School of Systems Biology, George Mason University, Manassas, VA 20110, USA

Serguei G. Popov
National Center for Biodefense and Infectious Diseases, Department of Molecular Microbiology, School of Systems Biology, George Mason University, Manassas, VA 20110, USA

Arben Kojtari and Hai-Feng Ji
Department of Chemistry, Drexel University, Philadelphia, PA 19104, USA

Ivan S. Maksymov
School of Physics, University of Western Australia, Crawley, WA 6009, Australia

Zhifeng Wang
School of Materials Science and Engineering, Hebei University of Technology, Tianjin 300130, China
Key Laboratory for New Type of Functional Materials in Hebei Province, Hebei University of Technology, Tianjin 300130, China
CITIC Dicastal Co. Ltd., Qinhuangdao 066011, China

Jiangyun Liu
School of Materials Science and Engineering, Hebei University of Technology, Tianjin 300130, China

Chunling Qin
School of Materials Science and Engineering, Hebei University of Technology, Tianjin 300130, China

Hui Yu
School of Materials Science and Engineering, Hebei University of Technology, Tianjin 300130, China

Xingchuan Xia
School of Materials Science and Engineering, Hebei University of Technology, Tianjin 300130, China

Chaoyang Wang
School of Materials Science and Engineering, Hebei University of Technology, Tianjin 300130, China

Yanshan Zhang
School of Materials Science and Engineering, Hebei University of Technology, Tianjin 300130, China

Qingfeng Hu
School of Materials Science and Engineering, Hebei University of Technology, Tianjin 300130, China

Weimin Zhao
School of Materials Science and Engineering, Hebei University of Technology, Tianjin 300130, China
CITIC Dicastal Co. Ltd., Qinhuangdao 066011, China

Elisabeth Gwinn
Department of Physics, The University of California, Santa Barbara, Santa Barbara, CA 93106, USA

Danielle Schultz
Department of Chemistry and Biochemistry, The University of California, Santa Barbara, Santa Barbara, CA 93106, USA

Stacy M. Copp
Department of Physics, The University of California, Santa Barbara, Santa Barbara, CA 93106, USA

Steven Swasey
Department of Chemistry and Biochemistry, The University of California, Santa Barbara, Santa Barbara, CA 93106, USA

Ahmed A. Galhoum
Ecole des mines d'Alès, Centre des Matériaux des Mines d'Alès, 6 avenue de Clavières, F-30319 Alès cedex, France

Mohammad G. Mafhouz
Nuclear Materials Authority, P.O. Box 530, El-Maadi, Cairo, Egypt

Sayed T. Abdel-Rehem
Chemistry Department, Faculty of Science, Ain Shams University, P.O. Box 11566 Ain Shams, Egypt

Nabawia A. Gomaa
Nuclear Materials Authority, P.O. Box 530, El-Maadi, Cairo, Egypt

Asem A. Atia
Chemistry Department, Faculty of Science, Menoufia University, P.O. Box 32511 Shebin El-Kom, Egypt

Thierry Vincent
Nuclear Materials Authority, P.O. Box 530, El-Maadi, Cairo, Egypt

Eric Guibal
Nuclear Materials Authority, P.O. Box 530, El-Maadi, Cairo, Egypt